Equids and Wheeled Vehicles in the Ancient World

Essays in Memory of Mary A. Littauer

Edited by
Peter Raulwing,
Katheryn M. Linduff
and Joost H. Crouwel

BAR INTERNATIONAL SERIES 2923 | 2019

BAR
PUBLISHING

Published in 2019 by
BAR Publishing, Oxford

BAR International Series 2923

Equids and Wheeled Vehicles in the Ancient World

ISBN 978 1 4073 1643 7

COVER IMAGE *Assyrian chariot. Detail of stone relief of Ashurnasirpal II, Nimrud (after Littauer and Crouwel 1979: fig. 53).*

Printed in England

BAR
PUBLISHING

BAR titles are available from:

BAR Publishing
122 Banbury Rd, Oxford, OX2 7BP, UK
EMAIL info@barpublishing.com
PHONE +44 (0)1865 310431
FAX +44 (0)1865 316916
www.barpublishing.com

Dedicated to the memory of our mentor and friend,
whose sage contributions continue to guide scholars

Mary Aiken Littauer

1912–2005

Mary Aiken Littauer in her library. © Anya Littauer, Princeton.

Contents

Preface

The essays presented in this volume are the outcome of a symposium in memory of Mary Aiken Littauer, *la grande dame de l'hippologie ancienne*. The symposium was held in June 1–3, 2010 at the International Museum of the Horse (IMH) in Lexington, Kentucky, in the heart of American horse country. It was Sandra Olsen (KU Biodiversity Institute, The University of Kansas), who was the driving force, organising the symposium, as well as doing initial editing of the papers for publication. She deserves the fullest credit for her efforts.

The symposium coincided with an exhibition at the IMH called *A Gift from the Desert: the Art, Culture and History of the Arabian Horse,* organised by the museum's director, Bill Cooke, with Cynthia Culbertson and Sandra Olsen acting as the exhibition curators. The IMH is the largest and most comprehensive museum in the world dedicated to exploring the history of horses and their impact on human civilisation. It is no coincidence that Mary Littauer's extensive library on the subject of horses, harnessing, chariots and other wheeled vehicles was donated by her family to this museum, where it is open to all. We thank Bill Cooke and his staff for their hospitality when we were in Lexington and also for helping to make the symposium a success.

The wide range of topics, which includes discussions of horses and wheeled vehicles from Near East, Egypt and Nubia, Arabia, and China, allows for comparisons that have not previously been possible in a single volume. The volume also introduces the reader to a broad range of approaches, grounded in disciplines such as philology, archaeology, archaeozoology and art history, DNA research and experimental archaeology.

Summarising past research as well as presenting much new information, together the essays are of interest to specialists and laypersons alike. The breadth and diversity of topics and areas represented honours the formative role that Mary Littauer has played for the research of the contributors.

One essay on the well-known tablets from Uruk showing ideograms of (wheeled) sledges has been added to the set of papers from the symposium (see Krispijn, Burmeister and Raulwing) to shed light on a topic of special interest to Mary Littauer.

We owe a great debt of gratitude to Adriana Maguiña-Ugarte for masterfully creating the print-ready version of the chapters in this volume, the peer-reviewers for their constructive feedback, the editors of the BAR series for accepting the articles for publication, Birgit Thaller, Dr Jane Burkowski and Chris Myers for their thoughtful guidance through the final stages of this edition, and the BAR production team for the excellent final layout.

Last but not least, we would like to thank Anya and Andrew Littauer (Princeton) for kindly making the picture of Mary Littauer available to us for this edition.

Peter Raulwing, Katheryn M. Linduff and
Joost H. Crouwel

Part I

Introduction

1

Mary Aiken Littauer and Our Study of Ancient Horse–Drawn Vehicles

Joost H. Crouwel

Mary Aiken was born in 1912 in Pittsburgh, Pennsylvania, USA, to a wealthy family. They later moved to New York. An only child, she had a sheltered upbringing. Her only formal schooling was four years of high school, but she received private tutoring in French and German and later also learned Russian, Latin and ancient Greek. From early on Mary loved reading but also enjoyed horse riding and the outdoor life. As a young woman she joined the Frontier Nursing Service, providing care in rural Kentucky—on horseback.

Mary was 23 years old when she married Vladimir Littauer whom she had met while taking lessons at the prestigious *Boots and Saddles Riding School* in Manhattan. After fighting in the First World War and the Russian Civil War, he came to the United States, eventually founding this riding school together with two other emigré cavalry officers. When I later met him, he was still being addressed as Captain Littauer.

After renting several properties in New Jersey horse country and Jericho, Long Island, they eventually bought and settled down at "Hillside Farm" where they lived long lives. He died in 1996, aged 97, she in 2005, aged 93.

Horses and books were important to both of them. Both rode, he taught riding and wrote books and articles on riding techniques. Mary became fascinated by horse-drawn vehicles of antiquity and became a self-educated expert in this field.

In 1968, at the mature age of 56, Mary published her first scholarly paper "The function of the yoke saddle in ancient harnessing", which appeared in the widely read British journal *Antiquity*. I read this article and was struck by its unusual, common–sense approach and her ideas on "how things actually worked". At that time I was in Oxford, where I was writing my doctoral dissertation on early chariots in Greece. I had gone there after finishing my university studies in Classical Archaeology at the University of Amsterdam, The Netherlands. It was in Oxford in the spring of 1969 that she and I were introduced to each other. This meeting marked the beginning of our close collaboration on ancient chariots, other vehicles, riding and horse equipment, which lasted for over 30 years. Together we wrote two books: *Wheeled Vehicles and Ridden Animals in the Ancient Near East* (1979) and *Chariots and Related Equipment from the Tomb of Tut'ankhamūn* (1985). We also published many articles, a selection of which was prepared by Peter Raulwing for another book: *Selected Writings on Chariots and Other Early Vehicles, Riding and Harness* (2002). Our last joint paper was published in 2001—Mary was then 89.

How did two people of such different ages and backgrounds, and living so widely apart, collaborate in the days before the computer? By means of phone calls, airmail letters, texts composed on typewriters, and with copies made with the help of carbon paper. I also stayed each year for some weeks at Hillside Farm, where we walked the lovely grounds and made the utmost of limited time. As to the writing, one of us usually submitted an idea for a subject and drafted a text which was then exposed to the critical eyes of the other. We would send the text back and forth until we were both sufficiently satisfied. Mary always insisted on accurate, clear English, having *Roget's Thesaurus* and other tools at the ready.

Let me now quote from one of Mary's numerous letters to me, dated April 4 1992, after I sent her a draft paper. The text speaks for itself:

> "I feel we must be absolutely frank with each other, hence I shall tell you the reasons why I am reluctant to go into this paper. I am a slow worker, we have other work to finish, there are the Tut'ankhamūn chariots to get into, and there are three or four other subjects (one of them being the evolution of the wheel) that I am eager to tackle. It probably sounds like a lazy man's excuse, but my best work is not done when I am hard pressed. The ideas and phrases do not come unless I can brood on things, and have a little feeling of space around me. This being the case, I feel I have to be selective and work on those subjects that seem most worthwhile to me. These fall primarily in two categories: correctives to misconceptions, and breaking new ground. I am not young, I have just so much energy, I am not a professional scholar, and I see my function as interpretive—that of a person with some practical experience of the objects discussed. I would like to concentrate on subjects that have either been misinterpreted or the significance of which has been overlooked."

Mary was always generous in sharing information, and she encouraged not just me but other aspiring young scholars. Several of them—not so young anymore—attended the Kentucky Horse Park symposium: Gail Brownrigg, Trudy Kawami, Katheryn Linduff, Sandra Olsen and Peter Raulwing. Each of us, in our various ways, have continued to work in Mary's field.

During her life, Mary steadfastly refused all honours. I wonder what she would have thought of this symposium dedicated to her memory.

Acknowledgements

With many thanks to Andrew Littauer, Mary's son.

2

Hommage to Mary Aiken Littauer (1912–2005)

Peter Raulwing

This article is a tribute to Mary Littauer whom I had the privilige to know and to work with over a decade. Having visited Mary at Hillside near Syosset on Long Island many times between 1995 and 2005 I have learned more about horses and chariots than anytime before and after. The following reflections on Mary Littauer's and Joost Crouwel's studies on Egyptian chariots cannot be more than humble token for her generosity and sharing of her vast knowledge.

Introduction

The symposium held at the *International Museum of the Horse* in the *Kentucky Horse Park* in June 2010, dedicated to the memory of Mary Aiken Littauer, offered an opportunity for scholars and scientists from various disciplines to commemorate her academic achievements and wide range of studies on early vehicles, chariots, horses, riding, and harnesses; many of them published with her longtime colleague and friend Joost H. Crouwel.[1]

Fast Rewind

New York, summer 1980. In its August 18 issue, *The New Yorker* published an article on Mary Littauer entitled "The Early Horse," resulting from a visit by Eugene Kinkead (1906–1992) at her estate at Hillside Farm in Oyster Bay, near Syosset, on Long Island, New York. Kinkead, one of the most prominent journalists in the US, with a 58–year career as columnist and editor with the *New Yorker* magazine (Minichiello and White 1997: 361), provides a rare glimpse into the scholarly world of research on horse domestication, and equestrian and chariot–related studies at the time; a world that might have appeared for many readers as somewhat exotic, for some maybe even eccentric, but undoubtedly passionate and certainly authoritative. The article in *The New Yorker* vividly describes Mary Littauer's study on the second floor of her home, as well as various canines belonging her household—a reminiscence that brings back memories to all of us who had the privilege to be her friend or colleague, visiting or doing research at Hillside Farm. The article also mentions J.H. Crouwel's and her examination of Egyptian chariots; later published as *Chariots and Related Equipment from the Tomb of Tut'ankhamūn* (1985), the second epoch–making monograph after their handbook *Wheeled Vehicles and Ridden Animals in the Ancient Near East* (1979). These are, without exaggeration, two masterly publications.[2]

"The Talk of the Town"

It is not the purpose of the following pages to re–examine the reprint of the article by Eugene Kinkead in *The New Yorker,* with annotations to correct each statement and comment of Mary Littauer to accommodate the current state of research in each academic field. "The Talk of the Town" section is rather chosen for this volume to provide some of Mary Littauer's own words, to enjoy, and to bring back memories to those who remember her as one of the most renowned scholars in the field, who generously shared her wealth of knowledge with all who were interested and serious about the topic.[3]

The Early Horse[4]

> "We would like to offer a few facts about man's early association with the horse. They come from Mrs. Mary Aiken Littauer, of Syosset, Long Island. "She knows more about ancient horses and their trappings than anyone else I know," Dr. Mary Moore, a Hunter College professor of ancient art, told us a while back. At the time, we were engaged in conversation with Dr. Moore in the Metropolitan Museum's Department of Greek and Roman Art over some abstruse point connected with hippic figures on classical Attic vases.[5] The information about Mrs. Littauer we filed away for use at a later moment, which is this one.

---

[1] See the bibliography in Littauer and Crouwel 2002: xxvii–xlvi (to which Littauer and Crouwel 2001 must be added).

[2] In selection: Littauer and Crouwel 1979: Ap–Thomas 1981; Decker 1981; Häusler 1985; Huot 1981; Kawami 1981; Klengel–Brandt 1981; Moorey 1980; North 194; Röllig 1981; Zaccagnini 1982. Littauer and Crouwel 1985: Decker 1988; Eaton–Krauss 1990; Nibbi 1990; Reeves 1990b.

[3] The following footnotes aim to provide some basic information to the reader not completely familiar with the material discussed by Mrs. Littauer.

[4] *The New Yorker*, 18 August 1980: 24–27 = Kinkead 1980.

[5] See Moore 1971.

The other day, then, we called on Mrs. Littauer at her estate on the North Shore, where she lives with her husband, Vladimir,[6] who was a professional cavalry officer in the army of Czar Nicholas II,[7] and with four large, lively, and engaging dogs. Mrs. Littauer's interest in horses dates from childhood ownership of a cow pony in Nevada,[8] where she went with her family each summer during the twenties. The interest was understandably strengthened by the background of her husband—he is the author of eight books on techniques of riding and training horses—who, moreover, encouraged her study in the field of her choice: the horse in ancient times. French and German, supplied by her education under a tutor[9] and at a private secondary school, have been augmented, with her husband's aid, by some Russian, and this linguistic facility enables her to cope with most of the contents of a number of learned periodicals to which she subscribes. She is also in contact with a network of archeozoologists and allied scholars, here and abroad (including some in Eastern Europe), who keep one another informed of relevant discoveries and research. All these sources of information cause Mrs. Littauer to feel that few developments in her field have escaped her since she started her studies, some twenty years ago.

Our talk took place in Mrs. Littauer's study, whose furniture, including a four–poster bed, was in large part covered with illustrations and printed material about the early horse, the periodicals bearing such titles as *Antiquity* and *The Proceedings of the Prehistoric Society*. "In marked contrast to the situation that prevailed when I began," said Mrs. Littauer, an amiable, urbane woman with a crisp delivery, "we feel we can now pinpoint the place where the domestication of the horse first occurred. This was in the Ukraine, south of Kiev, near the lower Dnieper River, in a small settlement of pastoralists named Dereivka. It belongs to the Sredni Stog culture, a term that translates—charmingly, I think—into Middle Haystack. The name comes from another settlement, where the culture was first discovered. Over sixty per cent of the faunal remains at Dereivka are of the horse. The evidence is that the herds were kept as meat animals. I might add in passing that the horse was the last animal to be domesticated, the dog, the sheep or goat, cattle, and the pig having preceded it by millennia. The date of the finds at Dereivka is the second half of the fourth millennium B.C. That is fifty–five hundred years ago."[10]

Mrs. Littauer continued, "Ancestors of these horses evolved from prototypes that originated long before, in the western regions of our own country. In time, they spread over most of the Northern Hemisphere—eventually, for some as yet unknown reason, disappearing completely from North America. Skeletal evidence shows that the Dereivka animals were domesticated. Measurements of the face, jaw, and braincase of a seven– or eight–year–old stallion found there are much closer to those of the horse of today than to those of a wild horse of that time. Also, remains of primitive bits have been found—cheekpieces of the toggle variety, typical of this early gear. Rope or gut passed through holes in the cheekpieces and lay on what is called the bars—the convenient space between the incisors and molars on the horse's lower jaw. This arrangement would provide guidance and enable a man to ride and control the horse more easily. We believe that such riding occurred at Dereivka. Sometimes an even simpler system than this was employed by the American Indians when the horse was re–introduced here. Riding bareback, they controlled their mounts by using merely a rawhide thong, leg pressure, and body weight. Imitation of the horse's domestication at Dereivka spread, of course, to surrounding areas, in the well–known ripple effect. With this development, the horse became the partner of man."[11]

Writing started just before 3000 B.C., in Mesopotamia. A few centuries later, Mrs. Littauer told us, the horse was mentioned in the cuneiform characters.[12] The horse, she believes, was brought there in trade from the north, following several hundred years of domestication. All horse traction up to Roman times, Mrs. Littauer said, was by paired teams under yoke, standing on either side of the draft pole, in imitation of the system used for oxen pulling a plow—an agrarian advance that began in Mesopotamia in the fifth millennium B.C., the rigid pole allowing the plowshare to run straight.

At this point, we were interrupted by four noisy and jubilant dogs—an Airedale, a Scottish deerhound, a German shepherd, and a mutt—which bounded into the study, Mrs. Littauer immediately scolded them and ejected them, and went on to say that horseback riding in Mesopotamia was first recorded on a small cylinder seal from the Akkadian dynasty of Babylonia, toward the end of the third millennium B.C.[13] It showed a man perched on the rear of the horse's back—in distinct contrast to the seat of today's jockeys, at the base of the neck. "The ignorance existing only a few decades ago concerning the range of the early horse at that period is well illustrated by this story," Mrs. Littauer said. "A mold for a plaque which was made in Babylonia about

[6] Captain Vladimir Stanislavovitch Littauer (1892–1989). See http://www.nytimes.com/1989/09/02/obituaries/vladimir-s-littauer-riding-instructor-96.html?pagewanted=1.
[7] See Littauer 1965.
[8] Martin 2005; Brownrigg 2006a; 2006b.
[9] See below.
[10] Later published by Telegin 1986.
[11] See now Olsen 2006. Radio–carbon dating later showed the stallion's skull to be of Iron Age date, and thus intrusive in the Dereivka complex (see also Häusler 1994: 232: "Das Ganze macht eher den Eindruck einer Abfallgrube bzw. einer gestörten Fundsituation"). The other horses at this site are no longer considered to have been ridden.
[12] On equids mentioned in cuneiform tablets Zarins 1976; 1978; 1986; 2014. On the horse in the ancient Near East furthermore Weszeli 2004; van den Hout 2004; von den Driesch and Raulwing 2004.
[13] Drews 2004: 28 fig. 3.2.

1900 B.C. and is now in the British Museum portrays a small horse with flowing mane, full tail, and small ears, carrying a rider.[14] The authority who first described it, in 1930, said it showed a boy riding a large greyhound. It was thought at the time that there were no horses then in Babylonia."

A two–wheeled chariot inscribed on a small scarab of the sixteenth century B.C. provides the first representation of the horse in Egypt.[15] The engraving shows a pharaoh of the post–Hyksos period standing in the horse–drawn vehicle. Evidence of riders in this area comes approximately two centuries later, and they are shown with bow and arrows—probably acting as battlefield messengers during the conflict at Kadesh in the early thirteenth century B.C.[16] The likenesses are engraved in low relief on Egyptian temple walls. Dating from about the same time are other low reliefs of riders and also some of fugitive charioteers, who are apparently escaping from battle. Because charioteers wore tight, ankle–length tunics of mail, they could not sit astride a horse; they are shown riding with both legs on one side of the animal, again far to the rear. The first depiction of riders using the normal mid–back seat of the equestrian of today shows up much later–in the eighth century B.C., in Assyrian palace reliefs representing the mounted troops of Sargon II, the Assyrian king. The troops are seated on pads, the forerunners of saddles.[17]

In ancient Greece, which, according to Mrs. Littauer, was not good horse country, because it was arid, steep, and rocky, the earliest evidence of the animal is once again associated with chariots, the proof being crudely cut into stone stelae of Grave Circle A[18] at Mycenae, from the middle of the sixteenth century B.C.[19] A badly damaged little terra–cotta figurine of a mounted man, also from Mycenae, and made in the thirteenth century B.C.,[20] is, Mrs. Littauer thinks, the first known depiction of a Greek equestrian.[21] He was riding, as was normal for the period, well to the back of his mount.

More than twenty articles on the early horse and its accoutrements bear Mrs. Littauer's name.[22] Last year, she collaborated on a book, "Wheeled Vehicles and Ridden Animals in the Ancient Near East,"[23] with J. H. Crouwel, a faculty member of the University of Amsterdam, in the Netherlands. The two are at present doing research for another volume, tentatively scheduled for publication next year and to be entitled "The Chariots," for the Oxford University Press Tut'ankhamūn Tomb Series.[24] The six chariots found in Tutankhamen's tomb represent almost all the extant examples from ancient Egypt. (Only two other complete Egyptian chariots[25] and one Egyptian chariot body are known.)[26] The vehicles are in relatively excellent condition. Mrs. Littauer has examined them in the Egyptian Museum, in Cairo,[27] and has read the material on them left by Howard Carter, the professional archeologist with Lord Carnarvon's expedition, which uncovered the tomb. Carter's copious notes, scale drawings, and photographs of the tomb contents fill an entire room of the Ashmolean Museum, at Oxford.[28] An interesting point, Mrs. Littauer observed, is the unique and very complex construction of the chariot wheels.[29] Wood was heat–bent into chevron form, and each of the spokes—there were six—was composed of two half spokes glued back–to–back lengthwise. The hub was formed by the heads of the chevrons, which had been cut out in slightly concave shapes. The whole was held together by glue and by rawhide, which was put on wet and shrank in drying to form a very tight binding material. The wheel received added strength from a rawhide tire, also put on when wet. "Mechanical engineers have told me that this construction allows formation of a light, safe, durable wheel with a minimum number of spokes,"[30] Mrs. Littauer said. "Mr. Crouwel and I hope to find out if this is so."

Fast Forward

New York, fall 2010. The New Yorkers and visitors of the city are in no way unfamiliar with chariots. In addition to the original Egyptian wheel in the Brooklyn Museum published by Mary Littauer and Joost Crouwel,[31] Amarna period wall reliefs in the Metropolitan Museum[32] and Brooklyn Museum,[33] as well as many representations

[14] Littauer and Crouwel 1979a: fig. 37; Drews 2004: 33 fig. 3.6 (with further examples).
[15] Crouwel 1981: fig. 134; Hofmann 1989: fig. 001.
[16] Crouwel 1981: fig. 111; Guidotti 2002a: 107 f.: fig.1 (top, center). See also Raulwing and Clutton–Brock 2008: 10, fig. 3.2.
[17] Littauer and Crouwel 1979: figs. 76–78.
[18] The information regarding the stone stelae of Grave Circle A has been corrected from the original text.
[19] Crouwel 1981: pls. 35–39.
[20] The date has been corrected from the original text.
[21] Crouwel 1981: pl. 42.
[22] See the bibliography in Littauer and Crouwel 2002: xxxvii–xlvi.
[23] Littauer and Crouwel 1979a.
[24] Littauer and Crouwel 1985.
[25] See section 3.
[26] The chariot body of Thutmose IV (Littauer and Crouwel 1985: pl. LXVII).
[27] In spring 1973.
[28] Filed in the Archive of the Griffith Institute which is part of the Faculty of Oriental Studies, University of Oxford (http://www.griffith.ox.ac.uk/gri/carter/HomePage.html); see Littauer and Crouwel 1985: 1.
[29] Littauer and Crouwel 1979b.
[30] Spruytte 1977; 1983; Hofmann 1989: 333–39; Rovetta 2002; Rovetta et al. 2000a; 2000b; 2000c; Sandor 2004a; 2004b; 2013; Herold 2004; 2006a; 2006b.
[31] Littauer and Crouwel 1979b.
[32] See e.g. the gifts from Norbert Schimmel to the Metropolitan Museum of Fine Art: "Horses Harnessed to a Chariot", ca. 1353–1336 BCE, Limestone, paint, 22.9 cm x 52.1 cm (acc. no. 1985.328.18); "Chariots with Court Ladies", same period and material, 23.5 cm x 54.7 cm (acc. no. 1985.328.18); "Heads of Two Horses and a Charioteer", same period and material, 22.2 cm x 26.7 cm (acc. no. 1985.328.20); "Galloping Horses", same period and material, 23.2 cm x 49.5 cm (acc. no. 1985.328.17). See Cooney 1965.
[33] See, e.g., "Soldiers in Chariot", c. 1352–1336 BCE: Limestone with modern paint, 53.5 x 22.8 x 3.2 cm). Brooklyn Museum, Gift of New Hermes Foundation (acc. no. 60.28), "A Span of Two Horses Pulling a Chariot", same period: Gypsum plaster, painted, 22 x 35.1 cm. Brooklyn Museum, Charles Edwin Wilbour Fund (acc. no. 54.186), currently not on display, and the famous six–spoked chariot wheel dated into the first

Figure 1a. Detail of Neo–Hittite (or Syro–Hittite) style relief framed by two lions over the left entrance to The Cass Gilbert on 130 West 30th Street in Manhattan, a landmark building designed by the eminent American architect, completed in the late 1920s. Cass Gilbert (1859–1934) is probably best known as the architect of United States Supreme Court building in Washington. Copyright Chang W. Lee/New York Times.

Figure 1b. The original scene of The Cass Gilbert goes back to an orthostat acquired by the Louvre in 1891 in Harput (Elâzığ) in eastern Anatolia. Limestone, 43 x 78 cm, AM 255 Louvre, Département des Antiquités Orientales, Paris (http://www.louvre.fr/en/oeuvre-notices/hunting-scene-chariot). The drawing shown above represents a detail of the original Neo–Hittite orthostat published by Littauer/Crouwel 1979a, Fig. 57 (drawing by Jaap Morel). The orthostat shows king Maritis, likely the ruler of Malatya, in his chariot hunting a stag with his chariot driver, accompanied by his hunting dog. A similar orthostat (now in the Anatolian Civilizations Museum, Ankara (acc. no. 12245); formerly in the Ancient Oriental Museum, Istanbul (acc. no. 7704) = Hawkins 1999 I/1, p. 318–320; Hawkins 1999, I/3, Pl. 155), displaying a lion hunt, found in Arslantepe in 1894, seems to illustrate an original provenance of the orthostat illustrated here from Arslantepe. Both illustrations shown here omit a Hieroglyphic–Luwian inscription above the hunting scene which reads: "These *shootings* (are) of Maritis, Suwarimi's son ... (?)". The inscriptions date the orthostat most probably into the early 11th or 10th century BCE (= Hawkins 1999, I/1, p. 321 f.; Hawkins 1999, I/3, Pl. 156 (museum photograph and drawing).

of chariots on Mycenaean and Greek vases,[34] we might also mention the splendid Etruscan chariot from the 2nd quarter of the 6th century BCE in the new Greek and Roman Galleries of the Metropolitan Museum, restored by Adriana Emiliozzi.[35] Not commonly known are two stylized replicas of Neo–Hittite reliefs depicting the king in his chariot hunting a stag, framed by two lions, above the two entrances of the building at 130 West 30th Street, in Manhattan[36] (Figures 1a–1b). The relief at the right front of the building shows the same hunting scene facing to the left.

However, it was, by all means, a small sensation that Chariot A4 from the tomb of Tutʿankhamūn (KV 62)[37] was part of the exhibition "Tutankhamun and the Golden Age of the Pharaohs" in the Discovery Times Square Exposition in the fall of 2010;[38] one of the very rare occasions that one of the six chariots from KV 62 has ever left Egypt.[39] Having traveled to Cairo with Joost Crouwel to examine Tutʿankhamūn's chariots and equipment in the Egyptian Museum for their book *Chariots and Related Equipment from the Tomb of Tutʿankhamūn*, published in 1985, Mary Littauer would have certainly been one of the first to revisit and simply enjoy the 'pit stop' the chariot they designated A4 made, virtually in her front yard; as she would have enjoyed the Cairo Conference on chariots

millennium, Charles Edwin Wilbour Fund (acc. no. 37.1700E), currently not on display. See also endnote 32 above.

[34] Moore 1971; Crouwel 1981; Crouwel 1992.

[35] Vogel 2007; Emiliozzi 2011; Crouwel 2012: pl. 8.

[36] Dunlap 2011.

[37] See next section.

[38] Many media reported about this exhibition. See, e.g., Kennedy 2010 and Lorenzi 2010.

[39] Chariot **A1** was part of the exhibition *Ramses Le Grand* in Paris in 1976. I owe this information to the kindness of Dr. Nadia Lokma. Jean Spruytte examined Chariot **A1** in Paris (Spruytte 1999: fn. 7) as Gail Brownrigg generously reminded me. Interestingly enough, replicas of objects from the tomb of Tutʿankhamūn were manufactured already for the *British Empire Exhibition* opened by King George V and Queen Mary at Wembley on St George's Day, 23 April 1924. Among those objects exhibited in the "Tomb of Tut–Ankh–Amun" at the *British Empire Exhibition* was a replica of one of the chariots made by the Aumonier Studios in London based on published photographs and drawings of the objects. The Egyptologist Arthur Weigall (1880–1934), who replaced Howard Carter as Chief Inspector at Luxor in 1905, functioned as an advisor for the "Tomb of Tut–Ankh–Amun" replica exhibited at Wembley. Carter, on the other hand, was convinced that the photographic material used by Weigall and "Messrs William Aumonier and Sons" belonged to the property Lord Carnarvon's excavation, and therefore "issued a writ against the [British Empire] Exhibition organizers on the ground of breach of copyrights". However, Carter later withdrew the writ after Aumonier and Weigall declared in the *Daily Express* on 22 April 1924 that they had not used photographs published by the *The Times* in London which had an exclusive contract with Lord Carnarvon (James 1992: 307–09 and fn. 3; Hankey 2001: 287 f.). Despite replicas of Egyptian chariots from the New Kingdom in Hollywood movies, Egyptian chariots were built for scholarly purposes, among others, by Jean Spruytte (1987: 37 pl. 9; 43: pl. 2) and Robert Hurford for a documentary (see also http://www.chariotmaker.com). The "Projekt Streitwagennachbau" of the Roemer– und Pelizaeus Museum Hildesheim was initiated by Arne Eggebrecht (1935–2004), Edgar B. Pusch, Rolf Schulte, and others in the early 1980s. Anja Herold, who was part of the team, published the results in 2006 ("Projekt Streitwagennachbau": See Herold in: Petschel and von Falck 2004: fig. 89; Herold 2006a: 375–86 including plates 14–23 and 2006b with illustrations 6–15b). This chariot is currently not on display in Hildesheim.

so splendidly edited by André J. Veldmeijer and Salima Ikram (2013).[40]

Tutʿankhamūn's Chariots

Since the 19th century[41], different aspects of New Kingdom Egyptian horses and chariots have been studied in numerous scholarly publications.[42] Until Howard Carter's and Lord Carnarvon's (Figures 2a–2b) opening of the antechamber of Tutʿankhamūn's tomb (KV 62), on November, 26–27, 1922, and the sensational discovery of six original chariots[43] (including related equipment), all studies had to rely mainlyon depictions of horse–drawn vehicles on wall reliefs in temples and tombs.[44] The two previously discovered and preserved chariots—the so–called 'Florence Chariot', named after the National Archaeological Museum of Florence where it is on display since the 1830s[45] and the vehicle from the tomb of Yuya and Thuyu[46]—raised several problems regarding their technological and, especially, functional interpretation. The 'Florence Chariot' was found at Thebes in 1829 during the joint Franco–Tuscan expedition under Jean–François Champollion (1790–1832), the decipherer of the Egyptian hieroglyphs, and Ippolito Rosellini (1790–1832), the founding father of Italian Egyptology.[47] Being the earliest chariot discovered at the beginning of modern scholarly research in the 19th century, the 'Florence Chariot' has its own unique history and reception. Mary Littauer and Joost Crouwel published a full technical description of the 'Florence Chariot'.[48] This description is based on a "Rapport sur le CHAR DE FLORENCE" by Jean Spruytte (1919–2007)—colleague and friend of Mary Littauer—sent to Mrs. Littauer in 1981 after examining the chariot

Figure 2a. George Herbert, 5th Earl of Carnarvon (1866–1923), excavator, collector and financial patron of the Tutʿankhamūn excavation at Carter's house in Thebes West. Copyright Griffith Institute, Oxford.

Figure 2b. Howard Carter (1874–1937) in 1924 painted by his brother William (1863–1939). Copyright Griffith Institute, Oxford.

[40] Containing the studies of Abbas 2013; Brock 2013; Calvert 2013; Crouwel 2013; Genz 2013; Gökce, Işık and Değirmencioğu 2013; Herslund 2013; Köpp–Junk 2013; Manassa 2013; Prell 2013; Raccidi 2013; Sabbahy 2013; Sacco 2013; Sandor 2013; Sasada 2013 and Veldmeijer, Ikram and Skinner 2013.

[41] 17th and 18th century studies on wheeled vehicles, such as those of the classicist Johannes Scheffer (1621–1679), or Schefferus as he was known: *De re vehiculari veterum [...]* (1671) and the Dominician Gabriel Fabricy (1725–1800), "de l'ordre des. FF. Prêcheurs Docteur Théologien de Casanate, de l'Académie des Arcades": *Recherches sur l'époque de l'équitation et de l'usage des chars équestres chez les anciens* (1764) as well as the Göttingen orientalist and Old Testament scholar Johann David Michaelis (1717–1791): *Etwas von der ältesten Geschichte der Pferde und Pferdezucht in Palästina und den benachbarten Ländern, sonderlich Aegypten und Arabien* (1776) which was translated by Alexander Smith: *A Dissertation on the Most Ancient History of Horses and Horse–Breeding, in Palestine and the Neighbouring Countries, especially Egypt and Arabia* (1814) and the Bavarian Royal Inspector of Carriage Building Johann Christian Ginzrot (1764–1731), *Die Wagen und Fahrwerke der Griechen und Römer* (1817) naturally completely focused on Greek and Roman authors and mainly the Old Testament when referring to or interpreting "Egyptian" sources.

[42] Hofmann 1989; Littauer and Crouwel 1985; Decker 1984; 1986; Rommelaere 1991; Decker and Herb 1994, I: 209–63 and pls. 97–119; 169–85; Herold 1999; 2004; 2006a; 2006b; with further bibliography. See also Feldman and Sauvage 2010.

[43] Littauer and Crouwel 1985; **A1–A4** were deposited in the *Antechamber*, **A5–A6** in the *Treasury* (see Littauer and Crouwel 1985: pl. I, *in situ*: **A5** = pl. LII [top and bottom] and pl. LIII [right]; **A6** = pl. LII [top and bottom] and pl. LIII [left]).

[44] See Hofmann, 1989, "Katalog" (with pls. 1–115), Rommelaere 1991, "Catalogue" (with pls. 1–113; Planche 1–8) and Decker and Herb 1994, I, "Dokumentation I: Ausgewählte Szenen zum Wagenfahren und Pferdetraining" (pp. 209–63 with pls. XCVII–CXXIX and CLXIX–CLXXXV). On warfare see Shaw 1991, Spalinger 1995, Gnirs 1999, Morkot 2003, Partridge 2003, McDermott 2005 in general; Darnell and Manassa 2007 on "Tutʿankhamūn's Armies".

[45] Littauer and Crouwel 1979, 85[52]. See e.g. Mötefindt 1917, 212 "Florentiner Wagen (wie wir ihn der Kürze wegen nennen wollen)".

[46] The rendering of both names varies in scholarly studies.

[47] See Siliotti 1998, chapter "Jean Francois Champollion, Ippolito Rosellini and the Franco–Tuscan expedition", and Betrò 2011 with further bibliography.

[48] See also Dittmann 1934; Botti 1951; Spruytte 1989; Del Francia 2002b.

Figure 3a. The so–called 'Florence Chariot.' Museo Archeologico di Firence. The yoke saddles are hanging upside–down, the object leaning on the chariot frame is a bow found together with this chariot which—among other arguments—had led the excavator Ippolito Rosellini to the conclusion that this vehicle was booty from a "Scythian country" north of Egypt (Raulwing 1993, 75 ff.). See Guidotti 2010, 47: "[...] the so–called 'Scythian war chariot', one of the most prestigious at the Egyptian Museum in Florence. [...] Rosellini believed that the chariot was a foreign produce because it was made from trees that don't grow in Egypt, but later it was considered an Egyptian produce, datable to the XVIII dynasty, thanks to the comparisons with the chariots from Tutankhamon's equipment".

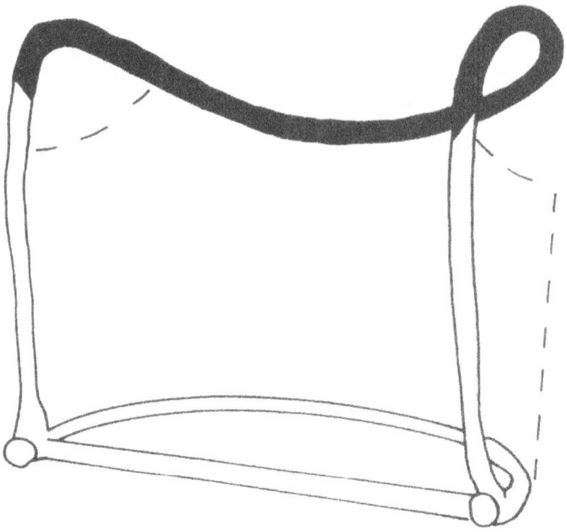

Figure 3b. Correct reconstruction of the railing of the 'Florence Chariot'. After Herold 2006a, 263.

in the Archaeological Museum in Florence.[49] Indicated by its two four–spoked wheels, the 'Florence Chariot' is the oldest of the eight (almost) fully preserved light horse–drawn vehicles from the New Kingdom. It has survived the ravages of over 3000 years due to optimal climatic conditions and burial practices in Egypt. The 'Florence Chariot' represents the basic form of the Egyptian chariot in the early Eighteenth Dynasty: two spoked wheels attached to a non–revolving axle located at the rear end of a D–shaped floor frame underneath two axle blocks (wide enough to carry a charioteer and a chariot driver), a pole, and two yoke saddles for a pair of horses (Figures 3a–3b).

However, the archaeological context of the 'Florence Chariot', found during Rosellini's excavations in 1828–1829, is not entirely clear,[50] partly because Rosellini was not present when his workmen discovered the vehicle in a tomb in Thebes/West Bank, and partly because the chariot as badly damaged and several parts of it were incorrectly assembled after it was shipped to Florence.[51] The tomb was probably not built for royalty, but rather for a private individual, such as an officer in the Egyptian chariot troops.

The chariot of Yuya and Thuyu, the parents–in–law of Amenophis' III (KV 46) is smaller in most dimensions than the 'Florence Chariot' (Decker and Herb 1994, I, 203), and its rear is partially closed[52] (Figures 4a–4b). Due to its dimensions and proportions, the chariot of Yuya and Thuyu was interpreted as a chariot model. However, it seems more likely that this chariot actually might have been used by a juvenile or an elderly person.[53]

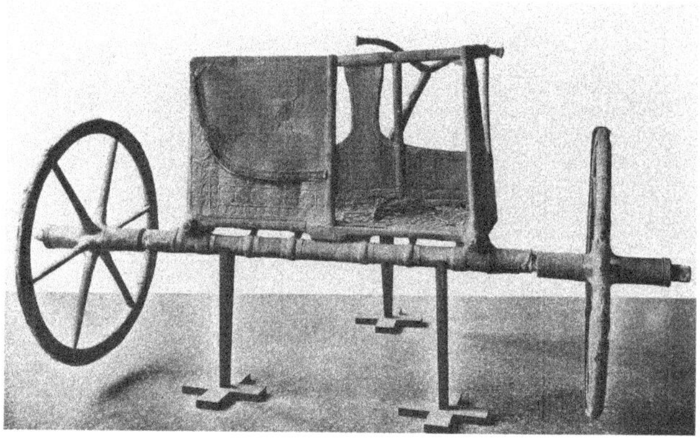

Figure 4a. The chariot from the tomb of Yuya and Thuyu (KV 46) discovered in February 1905 by James Edward Quibell (1867–1935) during the excavation conducted by the American businessman and lawyer Theodore M. Davis (1837–1915). Quibell 1908, Pl. LI = Littauer/Crouwel 1985, Pl. LXVIII bottom. (Yadin 1963, 190 f. illustrates a color plate of "a model of the chariot in the Metropolitan Museum of Art, patterned on the original in the Cairo Museum").

[49] Littauer and Crouwel 1985, Acknowledgements; on Jean Spruytte see the appreciation by Gail Brownrigg (http://antiquity.ac.uk/tributes/spruytte.html).

[50] Del Francia 2002a.

[51] Del Francia 2002b, 21 ff.; see also Herold 2006a, 1[18]; 2006b, 19[16] who uses the phrase "'kaputtrestauriert'", and Spruytte 1989, especially fig. 4. Pier Robert Del Francia (2002a) also documented the way of this unique chariot into the Egyptian Museum of Florence in 1830 and its conservation since. On the technological differences of the »Florence Chariot« compared to those from the tomb of Tut'ankhamūn see Del Francia 2002b, table on p. 36

[52] Thebes/West, Biban el–Muluk No. 58 [Decker and Herb 1994, I, **H 3** and **H 12**].

[53] Herold 2006: 260–65 with further bibliography.

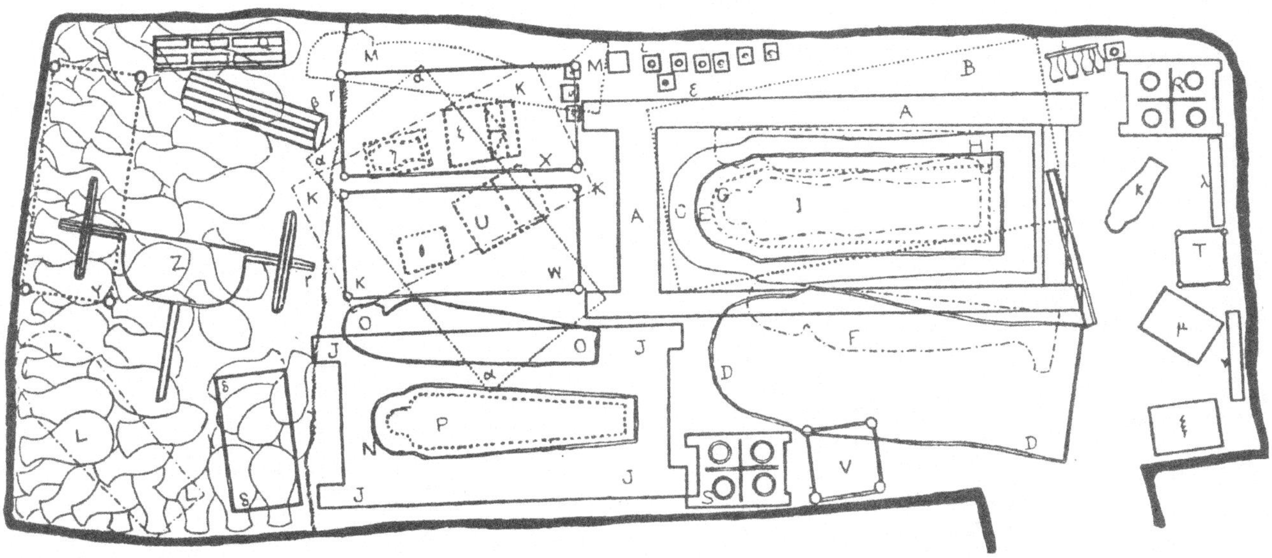

Figure 4b. Drawing of the tomb of Yuya and Thuyu. The placement of the chariot is to the left. After Herold 2006, Fig. 24 b, p. 260 after Quibell 1908, IV with a legend that is not illustrated here, p. 65–67 sub o. 5188 on the chariot with dimensions and Pl. LI–LVI.

Figure 5a. Yoke saddle finials classified as Type c from Pi–Ramesse (the modern Qantir), Ramses' capital in the eastern Delta built in the first quarter of the 13th century BCE. Detail from Herold 2006a, Pl. 5 c.

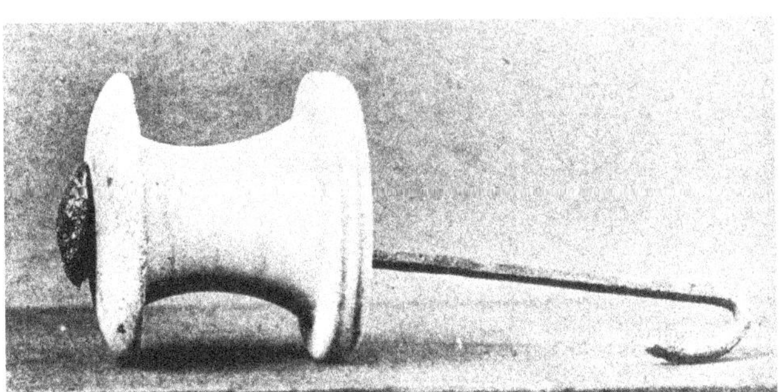

Figure 5b. Yoke saddle finials from KV 58 (left and right of the vessel). In the publication of the French Egyptologist George Daressy (1864–1938) these yoke saddle finials are labeled as material belonging to alabaster vases. After Daressy 1912, Pl. XC and p. 108 sub No. 38 ff. (corrected in Littauer/Crouwel 1985, Pl. LXXIV, shown here, and Herold 2006a, 259 ff. with discussion). KV 58 is also known as the "Chariot Tomb", because gold foils, most probably used as part of harness decoration (Littauer/ Crouwel 1985, Pl. XLV, HH = G13 on p. 38; Decker/Herb 1994, 208 sub H 12 and E 7 with Pl. LXXI) were found bearing the names of Tutʿankhamūn and Ay, successor of Tutʿankhamūn at the end of the 14th century BCE (http:// www.thebanmappingproject.com/sites/browse_tomb_872. html). Nicholas Reeves proposed that the deposits from this tomb were placed here from WV 23, the tomb of Ay (http:// www.thebanmappingproject.com/sites/browse_tomb_837. html) (Reeves 1981 ff.)

Figure 5c. Yoke saddles with yoke saddle finials from the tomb of Tutʿankhamūn. Copyright Griffith Institute, Oxford (= Littauer/Crouwel 1985, Pl. XXXIV and XXXV right).

To the corpus of preserved Egyptian chariots we can, fortunately, add other parts of chariots and their equipment such as blinkers, bridle bosses, and bridle parts, yoke saddles, leather harness, parts of a whip, a part of a wheel, linch pins, a nave with parts of spokes still attached, etc.,[54] and—most prominent—the chariot body from the tomb of

[54] Reeves 181 ff; Littauer and Crouwel 1985; Herold 2006: 1 ff.

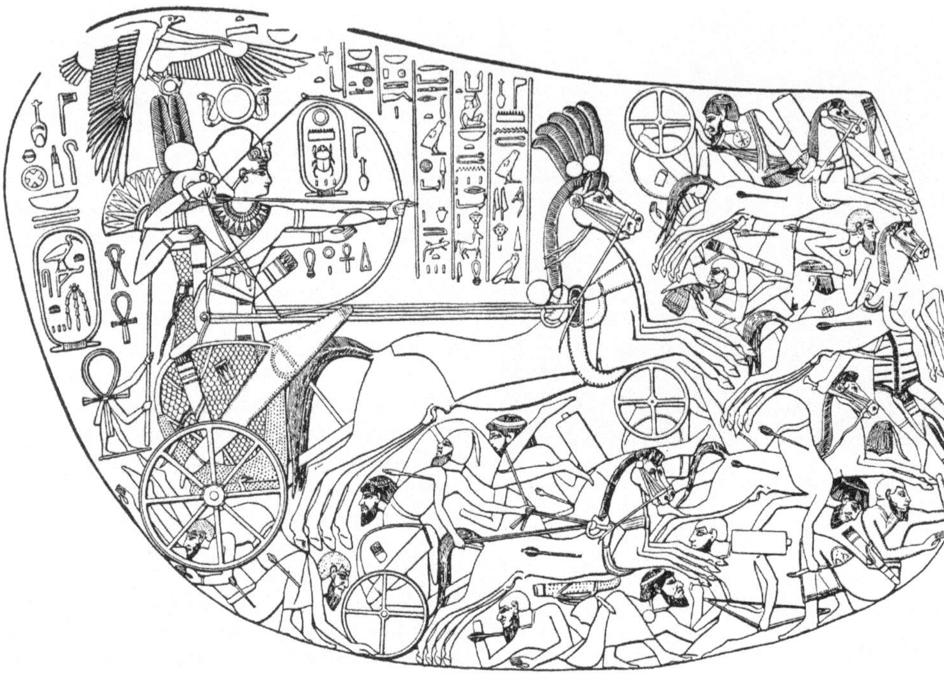

Figure 6. Drawing of the decoration on the left and right side of chariot body of Tuthmosis IV (KV 43, CG 46097) discovered in 1903 by Howard Carter during the excavation of Theodore M. Davis. The scenes show pharaoh smiting his Syrian enemies with bow and arrow and his mace. After Carter/Newberry 1904, plates IX and X (= Hofmann 1979, Plates 002a und 002b; see also Yadin 1963 pp. 192 f. and Littauer/Crouwel 1985, Pl. LXVII).

Tuthmosis IV[55] as well as the above mentioned six–spoked chariot wheel from the first millennium BCE published by Mary Littauer and Joost Crouwel.[56] In addition, yoke saddle finials (Figures 5a and 5b), although prima vista less spectacular or attractive, are invaluable for chariotry research in so far as they serve as a *Leitfossil*, since they were exclusively used for chariots.[57]

Due to a combination of ancient Egyptian burial practices and optimal climate conditions preserving organic materials, we owe most of our technical understanding of how New Kingdom chariots were built, how their parts were assembled, which materials were used, how and in which proportions they were manufactured, etc., almost entirely to those six chariots given to Tut'ankhamūn for his afterlife.[58] Wolfgang Decker, who has studied the preserved Egyptian chariots, has pointed out that the cultural–historical and technological value of Tut'ankhamūn's chariots can hardly be overestimated, because these findings are, so far, unparalleled in antiquity.[59] Mary Littauer and Joost Crouwel elaborated this argument even further:[60]

The six, actual, full–sized chariots of Tut'ankhamūn are of singular importance, not only to the history of chariotry in general, but to the understanding of Late Bronze–Age warfare, and to an intelligent interpretation of what is seen in representations (Figure 6). After documenting and photographing all objects in the tomb of Tut'ankhamūn (Figures 7a–7b), as well as plotting their exact location on a floor plan (Figures 8a–8b), the objects were carefully padded and secured before they could be safely transported out of the tomb. Howard Carter was directly involved in every aspect of this delicate work. The normal, strictly profile character of the latter makes it impossible for them to render many proportions and details accurately. The tomb of Tut'ankhamūn is unique in the ancient orient in offering the possibility of comparing six almost complete vehicles of this period, not only with one another, but with chariot evidence of all categories—actual, representational, and textual—both from Egypt and from other lands.

The technological features of the six chariots from the tomb of Tut'ankhamūn (Figures 9a–9b) are similar in so far as they all follow the same construction principles of a D–shaped light, spoke–wheeled chariot drawn by a pair

[55] From the Tomb of Thutmose IV. See Appendix.

[56] Littauer and Crouwel 1979b.

[57] Reeves 1981 ff. Littauer and Crouwel 1985; Herold 2006.

[58] "The [state] chariots particularly intrigued [James Henry Breasted (1865–1935) who was invited with his son Charles by Lord Carnarvon and Howard Carter to study the Tomb of Tut'ankhamūn], for although they were totally covered with gold leaf and plate, hammered into reliefs of the most accomplished style, and were encrusted with a host of semiprecious stones an inlays, the wheels bore distinct trace of use. They had certainly been driven 'over the rough streets of Thebes' and were therefore not just ornamental pieces prepared for the king's afterlife, something hollow and lifeless." Quoted after Hoving 1978: 121 f.; see also Abt 2011: 303–17.

[59] Decker and Herb 1994, I, 204 sub **H 6**: "Dieser aus unterschiedlichen Typen bestehende Wagenpark, dessen einzelne Fahrzeuge sämtlich zerlegt und dann teils hinter dem Eingang des Vorraums (H 6–8, 10), teils in der sog. Schatzkammer (H 9, 11) deponiert worden waren, kann in seiner Bedeutung für die Ägyptologie, aber auch für die allgemeine Technik– und Sportgeschichte kaum überschätzt werden, da vergleichbare Funde aus anderen antiken Kulturen nicht vorliegen."

[60] Littauer and Crouwel 1985: 96.

Figure 7a. Carter and his friend Arthur Callender (1876–1936), a retired British architect and engineer who was responsible for the successful dismantling of the shrine and many other objects, remove one of the two chariot bodies from the tomb (photographer not known, TAA ii.4.37). Copyright Griffith Institute, University of Oxford.

Figure 7b. The archaeologist Arthur C. Mace (1874–1928), and the chemist Alfred Lucas (1867–1945), seated, examining the beautifully decorated chariot body of *A2* in front of the field laboratory set up in the nearby tomb of Seti II (1203–1197 BCE). Copyright Griffith Institute, University of Oxford (= Littauer/Crouwel 1985, Pl. XV) (= Burton photograph: p0517, http://www.griffith.ox.ac.uk/gri/carter/120-p0517.html). It is quite amazing, as Anja Herold—one of the foremost experts on Egyptian chariots—pointed out (2006a, 377942), that with the rediscovery of bentwood technique by the German–Austrian cabinet maker Michael Thonet (1796–1871) there is a fascinating juxtaposition in this particular picture resembling Thonet's production model "Chair no. 14" ("Konsumstuhl Nr. 14" or "Kaffeehausstuhl Nr. 14") from 1859 and the chariot body of *A2* dated in the late 14th century BCE over 3000 years earlier (Herold 2006b, 6). See also Raulwing 1993, 76 f.; Killen 2000; Herold 2006a, 379.

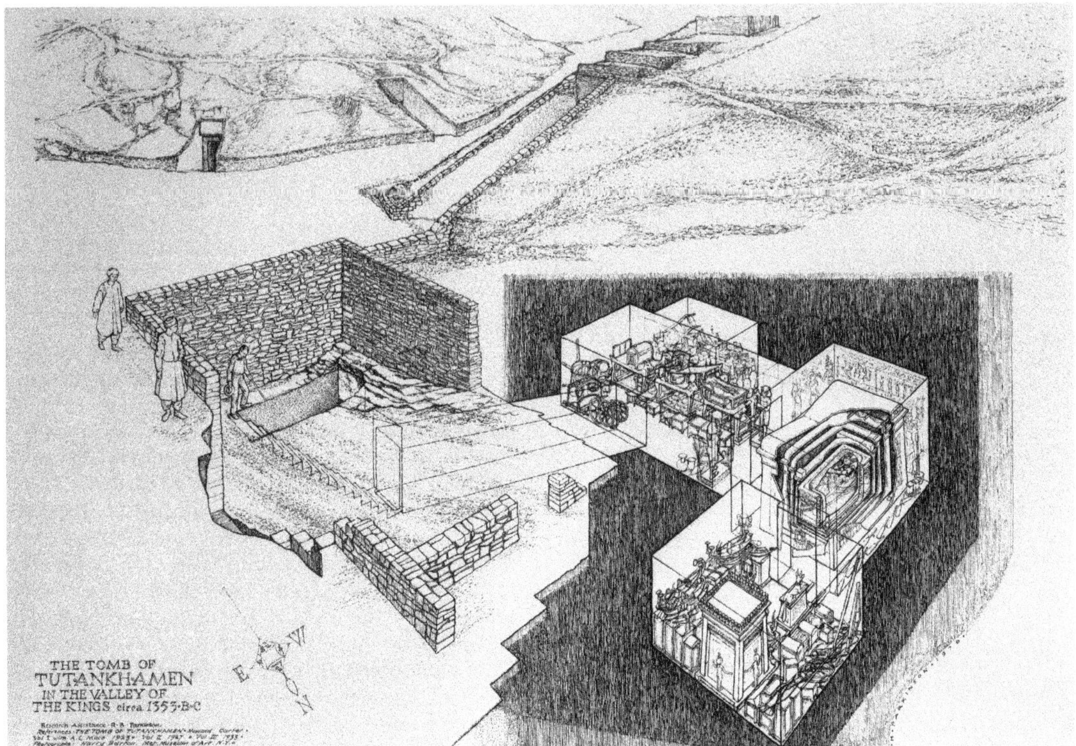

Figure 8a. Isometric drawing of the tomb of Tutʿankhamūn with Antechamber, Annexe right behind, Burial Chamber to the north and Treasury to the East (KV 62) by H. Parkinson from 1978. Copyright Griffith Institute, University of Oxford. (After Littauer/Crouwel 1985, Pl. I; see also James 2000, 51–54 and Reeves 1990a, 50; 70; 78; 80).

Figure 8b. Scale plan of the Antechamber by the American draftsman Lindsey F. Hall (1883–1969) and the American archaeologist and architect Walter Hauser (1893–1959), members of the Tutʿankhamūn excavation and known for their meticulous drawings, showing the placements of all objects *in situ*. (Carter MSS. i.G.11). Copyright Griffith Institute, University of Oxford (= Hoving 1978, p. 17 of the plates; see also Reeves 1990, 80 and James 2000, 51 ff.) (See http://www.griffith.ox.ac.uk/gri/4cartmap.html and http://www.griffith.ox.ac.uk/gri/gif-files/Carter_i_G_11.jpg).

Figure 9a. General view towards the south wall of the Antechamber showing the dismantled chariots *in situ* along the east and south walls. Copyright Griffith Institute, University of Oxford (= Littauer/Crouwel 1985, Pl. II; see also Carter/Mace 1923, Pl. XX) (= Burton photograph p0012, http://www.griffith.ox.ac.uk/gri/carter/143-p0012.html).

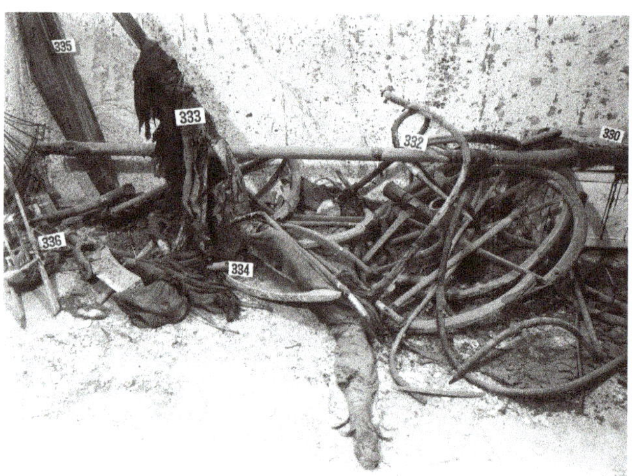

Figure 9b. Chariot material deposited in the Treasury in situ. Close–up of dismantled Chariots *A5* and *A6* in the northwest corner along the north wall. Copyright Griffith Institute, University of Oxford (= Littauer/Crouwel 1985, Pl. LII) (= Burton photograph p1090a. http://www.griffith.ox.ac.uk/gri/carter/332-p1090a.html). The meticulous photographs produced by Harry Burton (1879–1940) of the Metropolitan Museum *in situ* not only provide a glimpse into the overwhelming richness of the tomb (see also the splendid edition by Allen 2006 and the article by Johnson 1997), they also illustrate how difficult it was to identify and allocate all chariot pieces to individual vehicles. As Thomas Hoving describes: The chariots were removed from the tomb "in pieces: a wheel, then another, then the chassis and the [...] poles and at last the bodies themselves. Their removal had been the most difficult work of all. [Arthur Mace] has confided to Carter that the vehicles might prove to be so dilapidated that he was not at all certain they could ever be restored. For weeks he had thought that it would not be worth even a try and wanted at first simply to remove the gold reliefs and mount them in frames. But after months of backbreaking labor with [Alfred Lucas], he succeeded in a complete restoration. Today they are among the most impressive objects" in Egyptian Museums (Hoving 1978, 185).

of horses under a yoke.[61] However, they differ in various details—meticulously elaborated and documented by Mary Littauer and Joost Crouwel—so the assumption can be substantiated that these chariots had been individually designed and crafted; most probably for different purposes.

Carter distinguished two "state chariots" (numbered in Littauer/Crouwel 1985 as **A1** and **A2**) because of their extraordinary decoration—inscriptions and precious materials such as gold and inlay—which were most likely used in parades or for ceremonial purposes (p. 99; Table 1)

[61] Littauer and Crouwel 1979: 4 f.; 2002: 7. See also endnote 31.

Table 1. Concordance of the numbering system of the chariots from the Tomb of Tutʿankhamūn (KV 62) used by the Egyptian Museum Cairo, Littauer/Crouwel 1985 and Decker/Herb 1994, I, 200–208

Chariots from the tomb of Tutʿankhamūn (KV 62)	**Antechambers**				**Treasury**	
Egyptian Museum Cairo, Obj. No. [1]	**122**	**120**	**121**	**161**	**332**	**333**
Journal d'Entrée du Musée du Caire	**JE 61990**	**JE 61989**	**JE 61991**	**JE 61993**	**JE 61994**	**JE 61992**
Littauer/Crouwel 1985	**A 1**	**A 2**	**A 3**	**A 4**	**A 5**	**A 6**
Decker/Herb 1994	**H 7**	**H 6**	**H 8**	**H 10**	**H 11**	**H 9**
Room U 8, Case	**104**	**99**	**100**	—[2]	—[3]	**115**
Howard Carter's description	state chariot	state chariot	open, heavy chariot	open, heavy chariot	light, smaller chariot	light, smaller chariot
Assumed purpose after Littauer/ Crouwel 1985, 99 f.[4]	parade and ceremonial use	parade and ceremonial use	parade	travelling chariot on campaign	hunting	hunting

[1] Carter assigned numbers 1 to 620 to the over 5000 objects found in the tomb. Nos. 14 to 170 were from the Antechamber, nos. 261 to 336 from the Treasury.
[2] Chariot A4 was displayed in case 114 in the Egyptian Museum Cairo. As Robert Hurford and Gail Brownrigg kindly informed me, Chariot A4 is now in the Luxor Museum.
[3] Chariot A5 is on display in the National Military Museum on the Cairo citadel (Herold 2006, 115). Information on Chariot A4 and A5 have been kindly confirmed by Dr. Nadia Lokma.
[4] See Carter/Mace 1923, 60: "The four chariots discovered in the Antechamber (there are others in the Store-room of the Burial Chamber not yet touched) can be divided into two categories: State chariots and curricles. Of these, the latter were more open, of lighter construction probably for hunting or exercising purposes." See also Plates XVIII–XX, XXXVII–XLVI.

Figure 10a. One of the early Egyptian publications in Arabic and English printed immediately after the discovery of the tomb of Tutʿankhamūn is a 19–leaves booklet by Saleh Lutfi titled *Views of the tomb of King Tut–Ankh–Amen at Thebes. Discovered by Mr. Howard Carter on behalf of Lord Carnavon from the Sketches of Hamza Abdalla Carr* [Ḥamzah ʿAbd Ullāh Kār] dated "12–1–23." This illustration shows the copy in the collection of the Cecil H. Green Library at Stanford University from James Heyworth–Dunne (d. 1974), a former senior reader in Arabic at the School of Oriental and African Studies (SOAS) at the University of London from 1928–1948, and later a member of the staff at the Middle East Institute Washington, D.C., whose papers are archived in the Hoover Institution.

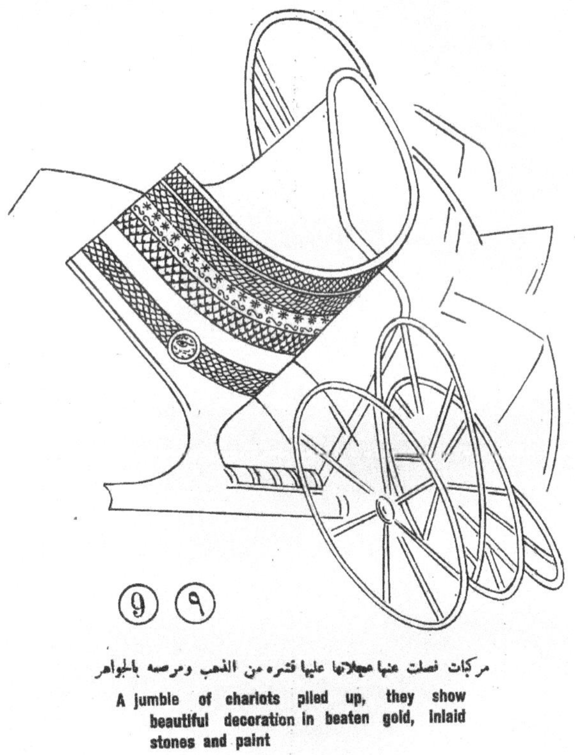

Figure 10b. The booklet contains stylized colored sketches and line drawings of objects found in the Antechamber, among them parts of the dismantled chariot bodies of *A1–A4* (see Figure 9a). The English caption reads: "A jumble of chariots piled up, they show beautiful decoration in beaten gold, inlaid stones and paint."

(Figures 10a–10b). Although **A3** is less elaborate, it still might have been intended for parade purposes, **A4** for traveling or campaigning, **A5** and **A6** for hunting as Littauer and Crouwel propose in their concluding remarks (p. 100).

Of the six chariots filed in Howard Carter's excavation notes,[62] only five vehicles (**A1**, **A2**, **A3**, **A4**, and **A6**) were on display in the Egyptian Museum at Cairo in 1973 when Mary Littauer and Joost Crouwel studied the chariots and related equipment from the tomb of Tutʿankhamūn for their book, published in the *Tutʿankhamūn Tomb Series* edited by the Griffith Institute at Oxford University in 1985 (Figures 11a–11c). When trying to locate the missing sixth vehicle, documented by Carter, Mary Littauer and Joost Crouwel identified warped pieces of wood from chariot parts in the Antechamber, stored in the magazine of the Egyptian Museum, as parts of Chariot **A5**.[63] Chariots **A5** and **A6** had been completely dismantled—most probably because it was decided to store them in the Treasury. Therefore, they must have been transported through the Antechamber and the Burial Chamber.[64] Chariot **A5** was later restored by Nadia Lokma of the Egyptian Museum at Cairo, one of the world's foremost experts on conservation of Egyptian artefacts; a laborious work which took her several years to preserve, to reshape and to assemble the over 200, partly twisted fragments. This Herculean task led Mike Murphy to title an article in the London *Times* on Nadia Lokma's successful restoration efforts "For eight years shall ye labour to restore the king's bespoke vehicle".[65]

Sixty–three years after the unexpected discovery of half a dozen chariots in the tomb of Tutʿankhamūn by Howard Carter and Lord Carnarvon, together with other studies on Egyptian chariots, harness and related equipment[66] Mary Littauer and Joost Crouwel published their marvelous book and "reference work of lasting importance",[67] which would be complemented with the publication of Chariot **A5** by Nadia Lokma (Figure 12).

Addendum

I would like to finish my *hommage* to Mary Littauer with a brief note on her German tutor in the early 1930's—also mentioned in *The New Yorker* article—when she still lived at Upper East Side on Madison Avenue and 72nd St. in New York, before moving with her husband Vladimir Littauer

Figure 11a. Chariot *A1* from the tomb of Tutʿankhamūn, Egyptian Museum Cairo. Copyright Griffith Institute, University of Oxford (= Littauer/Crouwel 1985, Pl. V) (= Burton photograph p0540a, http://www.griffith.ox.ac.uk/gri/carter/122-p0540a.html).

Figure 11b. Chariot *A2* from the tomb of Tutʿankhamūn, Egyptian Museum Cairo. Copyright Griffith Institute, University of Oxford (= Littauer/Crouwel 1985, Pl. XVI bottom) (= Burton photograph p0532a, http://www.griffith.ox.ac.uk/gri/carter/120-p0532a.html).

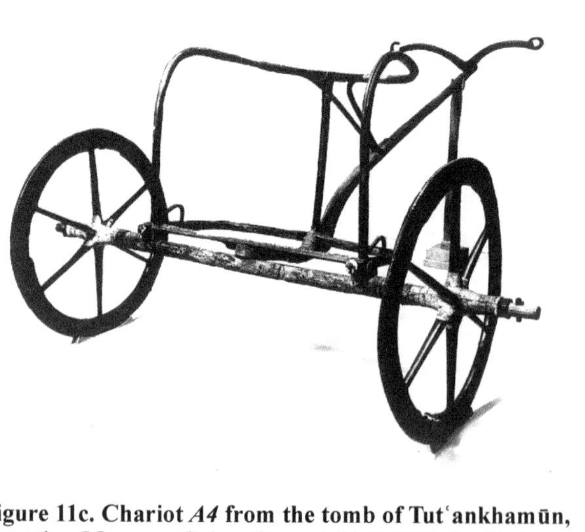

Figure 11c. Chariot *A4* from the tomb of Tutʿankhamūn, Egyptian Museum Cairo. Copyright Griffith Institute, University of Oxford (= Littauer/Crouwel 1985, Pl. XXX top) (= Burton photograph p0532b, http://www.griffith.ox.ac.uk/gri/carter/120-p0532b.html). This chariot went to the Discovery Times Square Exposition in 2010.

[62] Accessible under http://www.griffith.ox.ac.uk/gri/4tutchar.html; see also the "Essay on Tutankhamun's chariots, TAA i.3.8.10–17, based on Howard Carter's notes, probably edited by [Mrs. Jane Waley (who passed away at the age of 83 in 2008) in 1946–1947; Mrs. Waley catalogued the finds from tomb of Tutʿankhamūn]": compare TAA i.3.8.10, page 1: "Six chariots were found in the Tomb of Tut'ankhamûn. Two were highly decorated and appear hardly to have been used before they were placed in the tomb. In the Antechamber with these decorated chariots were two others, of heavier make and plainer construction and decoration. There were two more in the 'Treasury,' slightly smaller and lighter; these were found in a broken and incomplete condition."

[63] Littauer and Crouwel 1985: 53: "[…] chariot **A5** remained dismantled in the Tut. magazine, where the authors were able to inspect the parts" which are described on pp. 53–55. See also Decker and Herb 1994, I: 208 sub **H 11**.

[64] Littauer and Crouwel 1985: 98.

[65] Lokma 1986. See also Murphy 2007. Dr. Lokma visited Mary Littauer in the early 1990s (M.A. Littauer, pers. comm.).

[66] See the bibliography of the authors in Littauer and Crouwel 2002, xxxvii–xlvi.

[67] Reeves 1990b: 236.

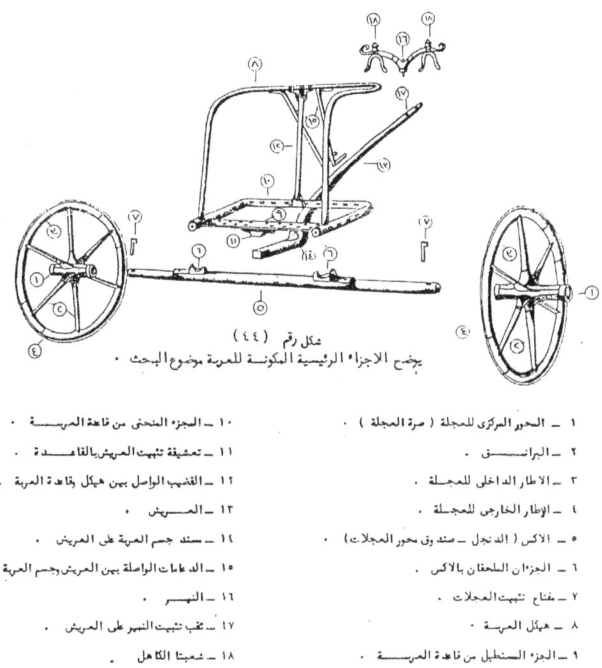

Figure 12. Detail form a drawing of parts of the dismantled Chariot *A5* (KV62) from Nadia Lokma's unpublished thesis (1986). After a Xerox sent to M.A. Littauer reproduced here with kind permission by Dr. Nadia Lokma, Egyptian Museum, Cairo. See also Littauer/Crouwel 1985, Pls. LIV and LV; Herold 2006a, 378, Fig.32) and Carter No. 332. Handlist description: Chariot. Note for Chariots 332 & 333. "Sketch of body to show (approximately) the details of the leather covering that were upon the bodies of the Chariots Nos. 332 & 333 [see Table 1]. From the few fragments that remain of this leather covering show that it was highly coloured & decorated. The sheet gold & bark decoration upon the parts of the upper frame work, suggest that the above was the form the leather took" (http://www.griffith.ox.ac.uk/gri/carter/332-c333-02.html).

to Long Island in early 1940s. One of Mary Littauer's two tutors was Frida Hahn, a German student of the eminent German–American anthropologist at Columbia, Franz Boas (1858–1942).[68] Many years ago, Mary Littauer mentioned to me that she had lost contact to Frida Hahn since the 1930s, and that she was wondering if it would be possible to find out which scholarly career Hahn had pursued, and if she, after all, might still be alive. A first start was, naturally, a verification of studies published under Hahn's maiden name, taking into consideration a change of her name as the result of a possible marriage. At the time, my initial research on Frida Hahn, of whom Mary Littauer always spoke with a certain fondness, ebbed away due to lack of accessible and reliable sources. However, in the course of writing this *hommage*, I came across a few, partly sad pieces of biographical information on Frida Hahn, which I would like to share here.

Frida Hahn studied Native American languages at Columbia since the late 1920s, and specialized—like her mentor—in Ponca,[69] a Dhegihan branch of the Siouan-language group, formerly spoken across an area from South Dakota and Nebraska to Kansas, parts of Iowa and western parts of Missouri in the early 1930s. Hahn mainly worked with the Ponca Tribe of Indians of Oklahoma where the Ponca were relocated by the US government in the late 19th century under great stresses and the loss of many members of their tribe.[70] As Jürgen Langenkämper, news editor of the *Mindener Tageblatt* in Franz Boas' hometown Minden (some 50 miles west of Hanover in Germany) found out in his meticulous research,[71]

> "Frida K. Hahn was [...b]orn in 1903 near Hamburg, she came to Columbia at the end of the 1920s. She followed her husband Dietrich Husemann whom she had met in New York, to China. In fall 1937 the couple fled from Nanking shortly before the massacre by Japanese troops and settled in Berlin (s. Frida K. Hahn–Husemann to FB, APS, February 10, 1938).[72] In the 1940s Katrin Husemann as she preferred now to be called, wrote a dissertation about Liberia[73] under [the Austrian anthropologist Richard Thurnwald (1869–1954) and German anthropologist and scholar in African studies, Diedrich Westermann (1875–1956) in Berlin].[74] Her husband was killed at the end of the war in Berlin. She fled to her family with her newborn daughter, worked as a librarian[75] and died in 1985 near Hamburg [...]."[76]

For Mary Littauer, as with any other students, learning a new language, unsurprisingly, included studying grammar and vocabulary. German has, one almost might concede, more exceptions than rules and vocabulary with nouns belonging to either one of three genders which are syntactically applied to four cases in singular and plural. However, harvesting her efforts at an advanced stage, Mary

[68] M.A. Littauer, personal communication.

[69] See Boas 1907 and Bulletin American Council of Learned Societies 21–25 (1934–36) Index which lists Frida Hahn for pp. 242 and 455.

[70] A type-written "Extensive manuscript grammar of Ponca" is preserved which is attributed to Hahn. See and Park and Rankin 2001: 100 and Eschenbach 2005: 223 sub "Anonymous" with further bibliography). The correspondence between Boas and Hahn is kept in the inventory of Franz Boas papers at the American Philosophical Society (http://www.amphilsoc.org/mole/view?docId=ead/Mss.B.B61.inventory06-ead.xml).

[71] I owe a great depth of gratitude to Jürgen Langenkämper for kindly sharing the sparse unpublished information on Frida Hahn from his research and manuscripts which I am allowed to quote here. Langenkämper 2015: fn. 22. See also Langenkämper 2011.

[72] FB = Franz Boas. http://www.amphilsoc.org/mole/view?docId=ead/Mss.B.B61.inventory06-ead.xml.

[73] Husemann 1949 on the republic of Liberia and her relations to the live of indigenous people (*Die Republik Liberia und ihre Beziehungen zum Eingeborenenleben*). See also Stoecker 2008: 117. J. Langenkämper's research will also shed more light on the years *after* World War II.

[74] On Thurnwald and Westermann see Stoecker 2008: Index p. 358 f. s.v. with multiple entries.

[75] See also *Buchhandel in Hamburg*: 64 listing "Dr. Katrin Husemann Hamburg 24, Quellenweg 17" under the members of the Northern German Publisher and Bookdealer Organization ("Die Mitglieder des Norddeutschen Verleger- und Buchhändler-Verbandes").

[76] Langenkämper, *submitted* in sub-chapter "The fate of Boas' Protégés: Günter Wagner and Frida Hahn" notes that former Boas students had difficulties finding positions as anthropologists after World War II. Since Franz Boas died in 1942, he was not able to help his former students in the new era abroad.

Littauer was also rewarded by enjoying German poetry; among others poems and lyrics by Heinrich Heine (1797–1856) such as "Auf dem Brocken" from his *Harzreise* made in 1824, as she liked to recite. It also enabled Mary Littauer to refresh her acquired German language skills in the late 1950s and early 1960s at a time when she became interested in the history of horses and chariots. While already fluent in French since her childhood, which enabled her to translate Jean Spuytte's book on *Early Harness Systems* (1983), she quickly realized that a scholarly study of horses and chariots requires being able (at least) to read German. Of all studies available to her, she chose the monumental book of the Austrian–Hungarian prehistorian Franz Hančar (1893–1968) on the horse in prehistoric and early historic times published under the German title *Das Pferd in prähistorischer und früher historischer Zeit* (1956). Hančar's German style was much elaborated involving lengthy syntactical constructions, the creation of new compounds and sub–clauses which often expanded over half a page. Immediately, one is reminded of Mark Twain's characterization of "The Awful German Language" in his book *A Tramp Abroad* (1880), here p. 254:

> "There are ten parts of speech, and they are all troublesome. An average sentence, in a German newspaper, is a sublime and impressive curiosity; it occupies a quarter of a column; it contains all the ten parts of speech—not in regular order, but mixed; it is built mainly of compound words constructed by the writer on the spot, and not to be found in any dictionary—six or seven words compacted into one, without joint or seam—that is, without hyphens; it treats of fourteen or fifteen different subjects, each inclosed in a parenthesis of its own, with here and there extra parentheses which re–inclose three or four of the minor parentheses, making pens within pens: finally, all the parentheses and re–parentheses are massed together between a couple of king–parentheses, one of which is placed in the first line of the majestic sentence and the other in the middle of the last line of it—*after which comes the* VERB, and you find out for the first time what the man has been talking about; and after the verb—merely by way of ornament, as far as I can make out—the writer shovels in '*haben sind gewesen gehabt haben geworden sein,*' or words to that effect, and the monument is finished. I suppose that this closing hurrah is in the nature of the flourish to a man's signature—not necessary, but pretty. German books are easy enough to read when you hold them before the looking–glass or stand on your head—so as to reverse the construction—but I think that to learn to read and understand a German newspaper is a thing which must always remain an impossibility to a foreigner."

And Hančar, whom Twain would have certainly incorporated in addition to a German newspaper writer of the Kaiserreich in 1870s, exceeded it all. However, as the hand–written comments and marginalia made by pencil in Mary's copy of Hančar's monograph show,[77] she even mastered Hančar's complex German syntax—across all 651 pages, and many more German, French and even Russian studies to follow which have helped to lay the foundation of more successful work to come.

Acknowledgements

My paper "A Brief History of Chariotry Research" given on June 1st, 2010 at the "Mary Aiken Littauer Symposium" in Lexington (Kentucky) on the occasion of the exhibition A Gift from the Desert (Olsen and Culbertson 2010) has grown into a larger study. I would like to thank Bill Cooke (former Director of the International Museum of the Horse) and his staff, as well Sandra Olsen (at the time of the conference Curator of Anthropology, The Carnegie Museum of Natural History) for their kind invitation and warm hospitality during the symposium; the latter would not have been possible without the inspiring participants with whom I had the honor to spend four unforgettable days in Lexington. Joost Crouwel and Gail Brownrigg provided valuable feedback and corrections, for which I owe them a depth of gratitude. Joost Crouwel kindly granted permission to reproduce figure 1b, the Griffith Institute, University of Oxford, figures 2a, 2b, 4a, 5c, 7a, 7b, 8a, 8b, 9a, 9b, 11a-11c, the Museo Archeologico di Firence (Stefano Casciu and Maria Cristina Guidotti) figure 3a, Anja Herold figure 3b and 5a, Nadia Lokma, Egyptian Museum at Cairo, generously granted permission to publish the drawing of Chariot A5 from her thesis (figure 12a). André J. Veldmeijer kindly sent me a PDF of his excellent proceedings on the Egyptian chariots (Veldmeijer and Ikram 2013).

Abbreviations

KV	Valley of the Kings
Ramsès le Grand	[Exhibition catalogue] *Ramsès le Grand* Paris: Galeries Nationales du Grand Palais, 1976.
TAA	TutAnkhAmun http://www.griffith.ox.ac.uk/gri/4taa.html.

Bibliography

Abbas, M. R. 2013. A survey of the of the charioteers in the Ramesside Period. In *Chasing chariots. Proceedings of the First International Chariot Conference (Cairo 2012),* edited by A. J. Veldmeijer and S. Ikram, 17–27. Leiden, Sidestone Press.

Abt, J. 2011. *American Egyptologist. The Life of James Henry Breasted and the Creation of his Oriental Institute*. University of Chicago Press, Chicago.

Allen, S. J. 2006. *Tutankhamun's Tomb. The Thrill of Discovery*. The Metropolitan Museum of Art, New York. [Photographs by Harry Burton].

Ap–Thomas, D. R. 1981. Review of Littauer and Crouwel 1979. In *Book List. Society for Old Testament Study 1981*, 29–30.

Baer, C. 1832. Review of von Minutoli 1831. *Heidelbergische Jahrbücher der Litteratur* 25. (Erste Hälfte Januar bis Juni), 195–205.

[77] Generously handed on to me by Andrew and Anya Littauer, Princeton.

Betrò, M. (ed.) 2011. *Ippolito Rosellini e gli inizi dell'egittologia. Disegni e manoscritti originali della spedizione franco–toscana in Egitto, 1828–29, dalla Biblioteca universitaria di Pisa [= Ippolito Rosellini and the Dawn of Egyptology. Original Drawings and Manuscripts of the Franco–Tuscan Expedition to Egypt, 1828–29], from the Biblioteca universitaria di Pisa*. Golden House Publications, London. [First published Agenzia italiana, Cairo 2010].

Boas, F. 1907. Notes on the Ponka Grammar. In *Congrès des Américanistes XVe session. Tenu à Québec en 1906*. Vol. II, 217–237. Dussault & Proulx, Quebec.

Botti, G. 1951. Il carro del sogno. *Aegyptus* 31, 190–198 [Reprinted in: (Istituto Editoriale) *Assiciazione Amici collaborati de Museo Egizio di Torino, Omaggio a Giuseppe Botti*, 105–113. Cisalpino–La Goliardica, Milano 1984].

Brock, E. C. 2013. A possible chariot canopy for Tutankhamun. In *Chasing chariots. Proceedings of the First International Chariot Conference (Cairo 2012)*, edited by A. J. Veldmeijer and S. Ikram, 29–43. Leiden, Sidestone Press.

Brownrigg, G. 2006a. Obituary—Mary Aiken Littauer. *International Council for Archaeozoology Newsletter* 7/1, 15. http://www.alexandriaarchive.org/icaz/icaz_website_formembers/pdf/nspring06.pdf

Brownrigg, G. 2006b. Mary Littauer. Expert on the chariots of the ancient world who wrote authoritative reference works on their use in warfare. *The Times* (London), 19 January 2006. http://www.thetimes.co.uk/tto/public/sitesearch.do?querystring=mary+littauer&p=tto&pf=all&bl=on

Buchhandel in Hamburg. 1960. *Buchhandel in Hamburg. Eine Festschrift anläßlich der Gründung des Hamburg–Altonaer Buchhändler–Vereins vor hundert Jahren*. Edited by Norddeutscher Verleger– und Buchhändler–Verband. Broschek, Hamburg.

Calvert, A. M. 2013. Vehicle of the Sun: The Royal Chariot in the New Kingdom. In *Chasing chariots. Proceedings of the First International Chariot Conference (Cairo 2012)*, edited by A. J. Veldmeijer and S. Ikram, 45–71. Leiden, Sidestone Press.

Carter, H. and A. C. Mace. 1922–33. *The Tomb of Tut.ankh. Amen*. Vols. I–III. London: Cassell. [Reprint *Duckworth Egyptology*. London: Duckworth, 2000–03].

Carter, H. and P. E. Newberry. 1904. *The Tomb of Thoutmosis IV*. (*Catalogue général des antiquités égyptiennes du Musée du Caire*, 15). Westminster: Constable. [Reprint *Duckworth Egyptology*. London: Duckworth, 2002].

Carus, C. G. 1848. *Mnemosyne. Blätter aus Gedenk– und Tagebüchern*. Vol. II. *Erinnerungen an Florenz aus dem Jahr 1841*. Flammer & Hoffmann, Pforzheim.

Cooney, J. D. 1965. *Amarna Reliefs from Hermopolis in American Collections*. Mainz: von Zabern.

Crouwel, J. H. 1981. *Chariots and Other Means of Land Transport in Bronze Age Greece*. Allard Pierson Series 3. Allard Pierson Museum, Amsterdam.

Crouwel, J. H. 1992. *Chariots and Other Wheeled Vehicles in Iron Age Greece*. Allard Pierson Series 9. Allard Pierson Museum, Amsterdam.

Crouwel, J. H. 2012. *Chariots and Other Wheeled Vehicles in Italy Before the Roman Empire*. Oxbow, Oxford.

Crouwel, J. H. 2013. Studying the six chariots from the of Tutankhamun—An update. In *Chasing chariots. Proceedings of the First International Chariot Conference (Cairo 2012)*, edited by A. J. Veldmeijer and S. Ikram, 73–93. Leiden, Sidestone Press.

Daressy, G. 1912. Catalogue of the objects found in an unknown tomb, supposed to be Toutânkhamanou's. In *The Tombs of Harmhabi and Toutânkhamanou*, edited by T.M. Davis, 125–135. Constable, London. [Reprint *Duckworth Egyptology*. London, 2001].

Darnell, J. C. and C. Manassa. 2007. *Tutankhamun's Armies. Battle and Conquest During Ancient Egypt's Late Eighteenth Dynasty*. Wiley & Sons, Hoboken, NJ.

Davis, T. M. 1907. *The Tomb of Iouiya and Touiyou*. Constable, London. [Reprint *Duckworth Egyptology*. London, 2000].

Decker, W. 1981. Review of Littauer and Crouwel 1979. *Stadion. Zeitschrift für Geschichte des Sports und der Körperkultur* 7/2, 285–286.

Decker, W. 1986. Der Wagen im Alten Ägypten. In *Achse, Rad und Wagen. Fünftausend Jahre Kultur– und Technikgeschichte*, edited by W. Treue, 35–51; 360–368 [endnotes]. Vandenhoeck & Ruprecht, Göttingen.

Decker, W. 1988. Review of Littauer and Crouwel 1985. *Bibiotheca Orientalis* 45, 128–134.

Decker, W. and W. Herb. 1994. *Bildatlas zum Sport im Alten Ägypten. Corpus der bildlichen Quellen zu Leibesübungen, Spiel, Jagd, Tanz und verwandten Themen*. Handbuch der Orientalistik, Erste Abteilung: *Der Nahe und Mittlere Osten* 14. Part 1. *Text*, Part 2. *Abbildungen*. Brill, Leiden.

Decker, W. and J.–P. Thuillier. 2004. *Le Sport dans l'antiquité. Egypte, Grèce et Rome*. Picard, Paris.

Decker, W. and J.–P. Thuillier. 2006. *Pharao und Sport*. Zaberns Bildbände zur Archäologie / Sonderbände der Antiken Welt. von Zabern, Mainz.

Del Francia, P. R. 2002a. La raccolta di armi del Museo Egizio di Firenze. In *Il carro e le armi di Museo Egizio di Firenze*. edited by M. C. Guidotti, 6–10. MAAT—Materiali del Museo Egizio di Firenze 2.

Del Francia, P. R. 2002b. Il carro di Firenze. In *Il carro e le armi di Museo Egizio di Firenze*. edited by M. C. Guidotti, 216–237. MAAT—Materiali del Museo Egizio di Firenze 2.

Dittmann, K. H. 1934. Die Herkunft des altägyptischen Streitwagens in Florenz. *Germania* 18, 249–252.

Drews, R. 2004. *Early Riders. The Beginning of Mounted Warfare in Asia and Europe*. Routledge, New York and London.

von den Driesch, A. and P. Raulwing. 2004. Pferd, (Esel*, Halbesel, Maulesel, Maultier*). D. Archäozoologisch. *Reallexikon der Assyriologie* 10, 493–503.

Dunlap, D. W. 2011. 'Gay Synagogue' Finds a Home, Full of Ancient Assyrians. *The New York Times*, August 8, 2011. http://cityroom.blogs.nytimes.com/2011/08/08/gay-synagogue-finds-a-home-full-of-ancient-assyrians/

Eaton–Krauss, M. 1990. Review of Littauer and Crouwel 1985. *Orientalistische Literaturzeitung* 85, 530–532.

Emiliozzi, A. 2011. The Etruscan chariot from Monteleone di Spoleti. *Metropolitan Museum Journal* 46, 9–132.

Eschenberg, A. 2005. *The Article System of Umonhon (Omaha)*. PhD Dissertation, State University of New York at Buffalo. http://www.base-search.net/Search/Results?q=id:d348934594e0b3a7c15f4bbb4b2df196ef6cb6a87903c647886b8adc5e33350e

Fabricy, G. 1764. *Recherches sur l'époque de l'équitation et de l'usage des chars équestres chez les anciens* [...]. Vols. I–II. Jean Mossy, Marseille and Pierre Durand, Rome.

Feldman, M. H. and C. Sauvage. 2010. Objects of prestige? Chariots in the Late Bronze Age Eastern Mediterranean and Near East. *Ägypten und Levante* 20, 67–181.

Genz, H. 2013. The introduction of the light, horse–drawn chariot and the role of archery in the Near East at the transition from the Middle to the Late Bronze Ages: Is there a Connection? In *Chasing chariots. Proceedings of the First International Chariot Conference (Cairo 2012),* edited by A. J. Veldmeijer and S. Ikram, 95–105. Leiden, Sidestone Press.

Germer, R. 1985. *Flora des pharaonischen Ägypten*. Deutsches Archäologisches Institut / Sonderschrift 14. Mainz: von Zabern.

Ginzrot, J. C. 1817. *Die Wagen und Fahrwerke der Griechen und Römer und anderer alten* [sic] *Völker nebst der Bespannung, Zäumung und Verzierung ihrer Zug–, Reit– und Lastthiere*. Bey dem Verfasser, München and J. Lentner, Stuttgart und Tübingen, Cotta'sche Buchhandlung. [Reprinted Olms, Hildesheim, 1992; partly reprinted see also as *Die Wagen und Fuhrwerke von der Antike bis zum 19. Jahrhundert, nebst Bespannung, Zäumung und Verzierung der Zug–, Reit– und Lasttiere*. Prisma, Gütersloh 1981].

Gnirs, A. M. 1999. Ancient Egypt. In *War and Society in the Ancient and Medieval Worlds. Asia, the Mediterranean, Europe, and Mesoamerica*, edited by K. Raaflaub and N. Rosenstein, 71–104. Center for Hellenic Studies Colloquia 3. Cambridge/MA.

Gökce, B., K. Işık and H. Değirmencioğlu. 2013. On Urartian chariots. In *Chasing chariots. Proceedings of the First International Chariot Conference (Cairo 2012),* edited by A. J. Veldmeijer and S. Ikram, 107–122. Leiden, Sidestone Press.

Guidotti, M. C. 2002a. L'iconografia della battaglia di Qadesh. In *La battaglia di Qadesh. Ramesse II contro gli Ittiti per la conquista della Siria*, edited by M.C. Guidotti and F. Pecchiolo Daddi, 170–175. Ministero per i Beni e le Attività Culturali [...], Livorno.

Guidotti, M. C. 2002b (ed.) *Il carro e le armi di Museo Egizio di Firenze.* MAAT—Materiali del Museo Egizio di Firenze 2. Ministero per i Beni e le Attività Culturali [...], Firenze.

Guidotti, M. C. 2010. Gli oggetti della spedizione di Rosellini nel Museo Egizio di Firenze. In *Ippolito Rosellini e gli inizi dell'egittologia. Disegni e manoscritti originali della spedizione franco–toscana in Egitto, 1828–29* edited by Betrò, 44–49. Agenzia italiana, Cairo. [Italian and English].

Hančar, F. 1955. *Das Pferd in prähistorischer und früher historischer Zeit.* Wiener Beiträge zur Kulturgeschichte und Linguistik 11. Herold, München and Wien.

Hankey, J. 2001. *A Passion for Egypt. Arthur Weigall, Tutankhamun, and the 'Curse of the Pharaohs'*. Tauris, London and New York.

Häusler, A. 1985. Review of Littauer and Crouwel 1979. In *Ethnographisch–Archäologische Zeitschrift* 26, 534–536.

Häusler, A. 1994. Archäologische Zeugnisse für Pferd und Wagen in Ost– und Mitteleuropa. In *Die Indogermanen und das Pferd*, edited by B. Hänsel and S. Zimmer, 217–257. Archaeolingua 4. Archaeolingua Alapítvány, Budapest.

Hawkins, J. D. 1999. *Corpus of Hieroglyphic Luwian Inscriptions*. Vol. I/Part 1. *Inscriptions of the Iron Age. Introduction, Karatepe, Karkamiš, Tell Ahmar, Maraş, Malatya, Commagene* Part 3. *Plates*. Untersuchungen zur indogermanischen Sprach– und Kulturwissenschaft N.F. 8. de Gruyter, Berlin.

Herold, A. 1999. *Streitwagentechnologie in der Ramses–Stadt. Bronze an Pferd und Wagen.* Die Grabungen des Pelizaeus–Museums Hildesheim in Qantir–Piramesse, 2 / Forschungen in der Ramses–Stadt [...]. Von Zabern, Mainz.

Herold, A. 2004. Funde und Funktionen—Streitwagentechnologie im Alten Ägypten. In *Rad und Wagen. Der Ursprung einer Innovation. Wagen im Vorderen Orient und Europa*, edited by M. Fansa and

S. Burmeister, 123–142. Beiheft der Archäologischen Mitteilungen aus Nordwestdeutschland 40. Von Zabern, Mainz.

Herold, A. 2006a. *Streitwagentechnologie in der Ramses–Stadt. Knäufe, Knöpfe und Scheiben aus Stein.* Die Grabungen des Pelizaeus–Museums Hildesheim in Qantir–Piramesse 3 / Forschungen in der Ramses–Stadt. [...]. Von Zabern, Mainz.

Herold, A. 2006b. Auf Biegen und Brechen ... Zum Nachbau eines altägyptischen Streitwagens. *Achse, Rad und Wagen. Beiträge zur Geschichte der Landfahrzeuge* 14, 4–19.

Herslund, O. 2013. Chariots in the daily life of New Kingdom Egypt: A survey of production, distribution and use in texts. In *Chasing chariots. Proceedings of the First International Chariot Conference (Cairo 2012),* edited by A. J. Veldmeijer and S. Ikram, 123–129. Leiden, Sidestone Press.

Hofmann, U. 1989. *Fuhrwesen und Pferdehaltung im Alten Ägypten.* PhD thesis, University of Bonn.

l'Horte, N. 1993. Champollion, Ippolito Rosellini and the Franco–Tuscan expedition. In *Egypt Lost and Found. Explorers and Travellers on the Nile*, edited by A. Siliotti. Thames & Hudson, London.

van den Hout. T. 2004. Pferd. A. II. in Anatolien. In *Reallexikon der Assyriologie* 10, 481–490.

Hoving, T. 1978. *Tutankhamun. The Untold Story*. Simon & Schuster, New York.

Huot, J.–L. 1981. Review of Littauer and Crouwel 1979. In *Revue Archéologique 1981*, 287–288.

Husemann, K. [formerly Frida K. Hahn]. 1949. *Die Republik Liberia und ihre Beziehungen zum Eingeborenenleben.* PhD thesis, University Berlin. [Typescript, 108 pp. and 3 illustrations, German National Library Leipzig, Signature Di 1949 B 1852].

James, T. G. H. 1992. *Howard Carter. The Path to Tutankhamun.* Thamses & Hudson, London.

James, T. G. H. 2000. *Tutankhamun.* White Star, Vercelli. [Edited by V. Manferto de Fabianis and L. Accomazzo. Text by T. G. H. James].

Johnson, G. B. 1997. Painting with light: The work of archaeology photographer Harry Burton. *KMT. A Modern Journal of Ancient Egypt* 8/2, 58–77. [With a biographical essay by M. Hill].

Kawami, T. 1981. Review of Littauer and Crouwel 1979. *American Journal of Archaeology* 85/2, 227–228.

Kennedy, R. 2010. King Tut's Chariot Arrives in Times Square. *The New York Times*, 2 August, 2010. http://www.nytimes.com/2010/08/03/arts/design/03chariot.html?scp=1&sq=King Tut's Chariot Arrives in Times Square&st=cse

Killen, G. 2000. Wood [Technology]. In *Ancient Egyptian Materials and Technology*, edited by. P. T. Nicholson and I. Shaw 353–371. Cambridge University Press, Cambridge.

Kinkead, E. 1980. The Early Horse. *The New Yorker*, August 18, 24–27.

Klengel–Brandt, E. 1981. "Review of Littauer and Crouwel 1979". *Orientalistische Literaturzeitung* 77/5, 454–456.

Köpp–Junk, H. 2013. The chariot as a mode of locomotion in civil contexts. In *Chasing chariots. Proceedings of the First International Chariot Conference (Cairo 2012),* edited by A. J. Veldmeijer and S. Ikram, 131–142. Leiden, Sidestone Press.

Langenkämper, J. 2011. "Ich fuerchte nur, wir verstehen einander nicht". In *Franz Boas—Kultur, Sprache, Rasse. Wege einer antirassistischen Anthropologie*, edited by F. Pöhl and B. Tilg, 131–149. Ethnologie: Forschung und Wissenschaft 19. 2nd edition, LIT, Wien and Münster. [2009].

Langenkämper, J. 2015. Franz Boas' correspondence with German friends and colleagues in the Early 1930s. In *Franz Boas Papers Vol. I. Franz Boas as Public Intellectual—Theory, Ethnography, Activism*, edited by R. Darnell et al. 277–292. University of Nebraska Press, Lincoln.

Langenkämper, J. *submitted.* "A Prophet Hath No Honor in His Own Country: The fate of Franz Boas, his German protégés, and the Westphalian town of Minden". To be published in the proceedings of the centennial Franz Boas conference "Indigenous Visions: Rediscovering the World of Franz Boas" held at Yale 15–17 September 2011.

Littauer, M. A. and J. H. Crouwel. 1979a. *Wheeled Vehicles and Ridden Animals in the Ancient Near East.* Handbuch der Orientalistik Siebente Abteilung. Kunst und Archäologie. Erster Band. *Der Alte Vordere Orient.* Zweiter Abschnitt. *Die Denkmäler*. Brill, Leiden and Köln.

Littauer, M. A. and J. H. Crouwel. 1979b. An Egyptian wheel from Brooklyn. *Journal of Egyptian Archaeology* 65, 107–120 [Reprinted in Littauer and Crouwel 2002, 296–313].

Littauer, M. A. and J. H. Crouwel. 1985. *Chariots and Related Equipment from the Tomb of Tutʿankhamūn.* Tutʿankhamūn Tomb Series 8. Griffith Institute, Oxford.

Littauer, M. A. and J. H. Crouwel. 1987. 'Unrecognized linch pins from the tomb of Tutankhamon and Amenhotep II': A reply. *Göttinger Miszellen* 100, 57–61.

Littauer, M. A. and J. H. Crouwel. 2001. The earliest evidence for metal bits. *Oxford Journal of Archaeology* 20, 329–338.

Littauer, M. A. and J. H. Crouwel. 2002. *Selected Writings on Chariots, Other Early Vehicles, Riding and Harness*, edited by P. Raulwing. Culture and History of the Ancient Near East 6. Brill, Leiden.

Littauer, V. S. 1965. *Russian Hussar*. London: Allen. [2nd print with suppl. materials with the title *Russian Hussar. A Story of the Imperial Cavalry, 1911–1920.* White Mane, Shippensburg/PA 1993. See also the translations into Russian by E. Izbitser with the title *Russkii gusar. Rasskaz ob Imperatorskoi kavalerii, 1911–1920.* Sankt Petersburg: Russika, 2005 and L. A. Igorevskogo with the title *Russkie gusary. Memuary ofitsera Imperatorskoi kavalerii 1911–1920.* Tsentropoligraf, Moskva 2006].

Lokma, N. I. A. 1986. *Treatment and Conservation of Warped Wood Applied on one of the Chariots of King Tutankhamun.* Cairo University, Faculty of Archaeology. Conservation Treatment. [Unpublished Paper in Arabic for Fulfilment of the Degree of Diploma (Master) in Conservation].

Lorenzi, R. 2010. King Tut's Chariots Marvels of Engineering. *Discovery News*, Aug 3, 2010. http://news.discovery.com/history/king-tut-chariot.html

Lutfi, S. [n.d.] *Views of the tomb of King Tut–Ankh–Amen at Thebes. Discovered by Mr. Howard Carter on Behalf of Lord Carnavon from the Sketches of Hamza Abdalla Carr.* Cairo, Typo Dawawin. [ca. 1923].

Manassa, C. 2013. The chariot that plunders foreign lands: 'The Hymn to the King in His Chariot'. In *Chasing chariots. Proceedings of the First International Chariot Conference (Cairo 2012),* edited by A. J. Veldmeijer and S. Ikram, 143–156. Leiden, Sidestone Press.

Martin, D. 2005. Obituary: Mary Aiken Littauer, historian. *The New York Times*, 19 December 2005. http://www.nytimes.com/2005/12/19/world/americas/19iht-web.1219obitlitt.html?_r=1

McDermott, B. 2004. *Warfare in Ancient Egypt.* Sutton, Stroud.

Michaelis, J. D. 1776. *Etwas von der ältesten Geschichte der Pferde und Pferdezucht in Palästina und den benachbarten Ländern, sonderlich Aegypten und Arabien.* In, *Mosaisches Recht*, by J.D. Michaelis, 257–359. Third Part, 2nd revised and enlarged edition. Garbe, Frankfurt/M.

The German text was also published separately by Johann Gottlieb Garbe in Frankfurt am Main (1776).

Michaelis, J. D. 1814 [posthumous ("APPENDIX TO ART. CLXVI.")] *A Dissertation on the Most Ancient History of Horses and Horse–Breeding, in Palestine and the Neighbouring Countries, especially Egypt and Arabia.* In *Commentaries on the Laws of Moses.* Translated by A. Smith. Vol. II. London: Printed for F. C. and J. Rivington [etc.], 431–514; summary pp. XXV–XXVII.

Minichiello, J. K. and A. W. White (eds.) 1997. *From Blue Ridge to Barrier Islands. An Audubon Naturalist Reader.* Johns Hopkins University Press, Baltimore.

von Minutoli, J. H. K. 1831. *Abhandlungen vermischten Inhalts.* Zweiter Cyklus. Erstes Bändchen. Nicolai, Berlin and Stettin.

von Minutoli, J. H. K. 1838. Der Wehrstand in dem alten Aegypten, durch Stellen aus den Klassikern und Abbildungen auf den alten Denkmälern jenes Landes erläutert und belegt. *Zeitschrift für Kunst, Wissenschaft, und Geschichte des Krieges* 44/9, 189–238.

Moore, M. B. 1971. *Horses on Black–Figured Greek Vases of the Archaic Period, ca. 620–480 B.C.* Ph.D. thesis, New York University.

Moorey, P. R. S. 1980. Review of Littauer and Crouwel 1979. *Antiquity* 54/212, 247–248.

Morkot, R. 2003. *Historical Dictionary of Ancient Egyptian Warfare.* Scarecrow Press, Lanham, ML and Oxford.

Mötefindt, H. 1917. Der Wagen im nordischen Kulturkreise zur vor– und frühgeschichtlichen Zeit. In *Festschrift Eduard Hahn zum LX. Geburtstag dargebracht von Freunden und Schülern*, 209–240. Studien und Forschungen zur Menschen– und Völkerkunde 14. Strecker & Schröder, Stuttgart.

Murphy, M. 2007. For eight years shall ye labour to restore the king's bespoke vehicle. *The Times* (London) 15 November 2007. http://www.thetimes.co.uk/tto/arts/visualarts/article1890690.ece

Nibbi, A. 1990. Review of Littauer and Crouwel 1985. *Discussions in Egyptology* 18, 97–99.

North, A. 1984. Review of Littauer and Crouwel 1979. *Orientalia* 153, 125–126.

Olsen, S. L. 2006. Early horse domestication. Weighing the evidence. In *Horses and Humans. The Evolution of Human–Equine Relationships*, edited by S. L. Olsen et al., 81–113. BAR International Series 1560. BAR Publishing, Oxford.

Olsen, S. L. and C. Culbertson eds. 2010. *A Gift from the Desert. The Art, History and Culture of the Arabian Horse.* International Museum of the Horse, Kentucky Horse Park. Lexington/KY.

Parks, D. R. and R. L. Rankin. 2001. Siouan Languages. In *Handbook of North American Indians*, edited by W. C. Sturtevant. Vol. 13: *Plains*, edited by R. J. DeMallie, 94–114. Smithsonian Institution, Washington.

Partridge, R. B. 1996. *Transport in Ancient Egypt.* London: Rubicon.

Partridge, R. B. 2002. *Fighting Pharaohs. Weapons and Warfare in Ancient Egypt.* Peartree, Manchester.

Petschel, S. and M. von Falck (eds.) 2004. *Pharao siegt immer. Krieg und Frieden im Alten Ägypten.* Kettler, Bönen.

Prell, S. 2013. A glimpse into the workshops of the chariotry of Qantir–Piramesse—Stone and tools of Site Q I. In *Chasing chariots. Proceedings of the First International Chariot Conference (Cairo 2012),* edited by A. J. Veldmeijer and S. Ikram, 157–174. Leiden, Sidestone Press.

Quibell, J. E. 1908. *Tomb of Yuaa and Thuiu.* Nos *CG 51001–51191.* (Catalogue général des antiquités égyptiennes du Musée du Caire / Service des antiquités de l'Égypte. Catalogue général des antiquités égyptien du musée du Caire, 31). Institut français d'archéologie orientale, Le Caire.

Raccidi, M. 2013. Wagons and carts in the 3rd millennium BC Syrian Jazirah: A study through the documentation. In *Chasing chariots. Proceedings of the First International Chariot Conference (Cairo 2012),* edited by A. J. Veldmeijer and S. Ikram, 175–190. Leiden, Sidestone Press.

Raulwing, P. 1993. Pferd und Wagen im Alten Ägypten Forschungsstand. Beziehungen zu Vorderasien und methodenkritische Aspekte. *Göttinger Miszellen* 136, 71–83.

Raulwing, P. 2009. Kleines Florilegium zur Skythen–Rezeption in Kunst und Literatur. [Österreichische Akademie der Wissenschaften] *Anzeiger der philosophisch–historischen Klasse* 144/1, 93–128.

Raulwing, P. and J. Clutton–Brock. 2008. The Buhen horse: Fifty years after its discovery (1958–2008). *Journal of Egyptian History* 2/1, 1–106.

Reeves, C. N. 1981. A State Chariot from the Tomb of Ay? *Göttinger Miszellen* 46, 11–19.

Reeves, C. N. 1982. The discovery and clearance of KV58. *Göttinger Miszellen* 53, 33–45.

Reeves, C. N. 1990a. *The Complete Tutankhamun. The King, The Tomb, The Royal Treasure.* London: Thames and Hudson.

Reeves, C. N. 1990b. Review of Littauer and Crouwel 1985. *Journal of Egyptian Archaeology* 76, 232–236.

Ridgeway, W. 1905. *The Origin and Influence of the Thoroughbred–Horse.* Cambridge University Press, Cambridge.

Röllig, W. 1981. Review of Littauer and Crouwel 1979. *Zeitschrift der Deutschen Morgenländischen Gesellschaft* 131, 207.

Rommelaere, C. 1991. *Les chevaux du Nouvel Empire égyptien. Origines, races, harnachement.* Connaissance de l'Egypte ancienne 3. Bruxelles: Connaissance de l'Egypte ancienne.

Rosellini, I. 1830. *Breve notizia degli oggetti di antichità egiziani. Riportati dalla Spedizione letteraria toscana in Egitto e in Nubia eseguita negli anni 1828 e 1829,* ed esposti al pubblico nell'Accademia delle arti e mestieri in S. Caterina. Firenze: Piatti.

Rosellini, I. 1836. *I monumenti dell'Egitto e della Nubia.* [...]. *Monumenti Civili.* Vol. III. Pisa: Capurro.

Rosellini, I. 1994. *Ippolito Rosellini e il suo Giornale della spedizione letteraria toscana in Egitto negli anni 1828–1829,* edited by G. Gabrieli. Pisa: ETS. [Reprint of the 1st edition, Roma: Tipografia Befani 1925].

Rovetta, A. 2002. Cocchi reali e cocchi virtuali. In *Il carro e le armi di Museo Egizio di Firenze,* edited by M. C. Guidotti, 11–15. MAAT—Materiali del Museo Egizio di Firenze 2. Ministero per i Beni e le Attività Culturali [...], Firenze.

Rovetta, A., I. Nasry and A. Helmi. 2000a. The chariots of the Egyptian Pharaoh Tut Ankh Amun in 1337 B.C.: kinematics and dynamics. *Mechanism and Machine Theory* 35/7, 1013–1031.

Rovetta, A., I. Nasry and A. Helmi. 2000b. Erratum to "The chariots of the Egyptian Pharaoh Tut Ankh Amun in 1337 B.C.: kinematics and dynamics". *Mechanism and Machine Theory* 35/11, 1651–1653.

Rovetta, A., A. Emanuaeli, I. Nasry and A. Helmi. 2000c. Ancient Egyptian chariots. Design and functionality. In *International Symposium on History of Machines and Mechanisms. Proceedings HMM 2000,* edited by M. Ceccarelli, 149–154. Kluwer, Dordrecht and Boston.

Sabbahy, L. 2013. Depictional study of chariot use in New Kingdom Egypt. In *Chasing chariots. Proceedings of the First International Chariot Conference (Cairo 2012),* edited by A. J. Veldmeijer and S. Ikram, 191–202. Leiden, Sidestone Press.

Sacco, A. 2013. Art and imperial ideology: Remarks on the depiction of Royal Chariots on wall reliefs in New–Kingdom Egypt and the Neo–Assyrian Empire. In *Chasing chariots. Proceedings of the First International Chariot Conference (Cairo 2012),* edited by A. J. Veldmeijer and S. Ikram, 203–215. Leiden, Sidestone Press.

Sandor, B. I. 2004a. The rise and decline of the Tutankhamun–class chariot. *Oxford Journal of Archaeology* 23, 153–175.

Sandor, B. I. 2004b. Tutankhamun's chariots: secret treasures of engineering mechanics. *Fatigue & Fracture of Engineering Materials & Structures* 27/7 (July), 637–646.

Sandor, B. I. 2013. Chariots' inner dynamics: Springs and rotational inertias". In *Chasing chariots. Proceedings of the First International Chariot Conference (Cairo 2012),* edited by A. J. Veldmeijer and S. Ikram, 217–228. Leiden, Sidestone Press.

Sasada, Y. 2013. An alternative theory for found on the lower second of the Buhen Horse". In *Chasing chariots. Proceedings of the First International Chariot*

Conference (Cairo 2012), edited by A. J. Veldmeijer and S. Ikram, 229–236. Leiden, Sidestone Press.

Schäfer, H. 1931. Armenisches Holz in altägyptischen Wagnereien. *Sitzungsberichte der Preußischen Akademie der Wissenschaften* 25, 1–11.

Scheffer, J. 1671. *De re vehiculari veterum, libri duo. Accedit Pyrrhi Ligorii. V.C. de vehiculis fragmentum nunquam ante publicatum ex Bibliotheca Sereniss. Reg. Christinae cum ejusquem J. Schefferi Arg. annotationibus*. Francoforti: Zunner. [Reprinted by Giovanni Poleni (1683–1761), *Utriusque thesauri antiquitatum romanarum graecarumque. Nova Supplementa. Congesta Ab Joanne Poleno*. Venetiis: Pasquali, 1737. Vol. V, 1045–1502 (http://digi.ub.uni-heidelberg.de/diglit/poleni1737bd5/0558?sid=77bea91bba8d2979317eafb1e54a5010). Also available as an online resource: Farmington Hills, MI: Thomson Gale, 2005 in the series "The Making of the Modern World"]. http://digi.ub.uni-heidelberg.de/diglit/poleni1737bd5/0558?sid=77bea91bba8d2979317eafb1e54a5010

Shaw, I. 1991. *Egyptian Warfare and Weapons*. Shire Egyptology 16. Shire Publications, Princes Risborough/Buckinghamshire.

Siliotti, A. 1998. *Egypt Lost and Found. Explorers and Travellers on the Nile*. Thames & Hudson, London.

Solomon, J. 2001. *The Ancient World in the Cinema*. Yale University Press, New Haven.

Spalinger, A. J. 2004. *Warfare in New Kingdom Egypt*. Blackwell, Malden/MA.

Spalinger, A. J. 2013. Egyptian chariots: Departing for war. In *Chasing chariots. Proceedings of the First International Chariot Conference (Cairo 2012),* edited by A. J. Veldmeijer and S. Ikram, 237–256. Leiden, Sidestone Press.

Spruytte, J. 1977. *Études expérimentales sur l'attelage. Contribution à l'histoire du cheval.* Crépin–Leblond, Paris.

Spruytte, J. 1983. *Early Harness Systems.* Allen, London. [Translation by M. A. Littauer of Spruytte 1977].

Spruytte, J. 1989. Le char antique du musée archéologique de Florence. *Traveaux du LAPMO 1989*, 185–190. [LAPMO: Laboratoire d'Anthropologie et de Préhistoire et d'Ethnologie des Pays de la Méditerranée Occidentale. Université de Provence, Aix–en–Provence].

Spruytte, J. 1999. L'attelage égyptien sous la VIIIe Dynastie: étude technique et technologique. *Kyphi. Bulletin du Cercle Lyonnais d'egyptologie Victor Loret* 2, 77–87.

Stoecker, H. 2008. *Afrikawissenschaften in Berlin von 1919 bis 1945. Zur Geschichte und Topographie eines wissenschaftlichen Netzwerke*s. Pallas Athene 25. Stuttgart: Steiner.

Telegin, D. Y. ed. 1986. *Dereivka. A Settlement and Cemetery of Copper Age Horse Keepers on the Middle Dnjepr*. BAR International Series 287. BAR Publishing, Oxford.

Twain, M. 1880. The Awful German Language = Appendix D. In *A Tramp Abroad*, vol. I, 252–273. Collection of British Authors. Tauchnitz Edition 1899. Tauchnitz, Leipzig.

Veldmeijer, A. J. and S. Ikram (eds.) 2013. *Chasing chariots. Proceedings of the First International Chariot Conference (Cairo 2012)*. Leiden, Sidestone Press.

Veldmeijer, A. J., S. Ikram, and L. Skinner. 2013. Charging chariots: Progress report on the Tano Chariot in the Egyptian Museum Cairo. In *Chasing chariots. Proceedings of the First International Chariot Conference (Cairo 2012),* edited by A. J. Veldmeijer and S. Ikram, 257–271. Leiden, Sidestone Press.

Vogel, C. 2007. A More Precise Version of Your Chariot Awaits. *The New York Times*, March 29, 2007. http://www.nytimes.com/2007/03/29/arts/design/29char.html?pagewanted=all

Weszeli, M. 2004. Pferd. A. I. Mesopotamien. *Reallexikon der Assyriologie* 10, 469–481.

Yadin, Y. 1963. *The Art of Warfare in Biblical Lands in the Light of Archaeological Discovery*. Weidenfeld & Nicolson, London.

Zaccagnini, C. 1982. Review of Littauer and Crouwel 1979. *Bibliotheca Orientalis* 39, 425–428.

Zarins, J. 1976. *The Domestication of Equidae in Third Millennium B.C. Mesopotamia*. Vols. I–II. PhD Dissertation, University of Chicago.

Zarins, J. 1978. The Domesticated Equidae of Third Millennium B.C. Mesopotamia. *Journal of Cuneiform Studies* 30, 3–17.

Zarins, J. 1986. Equids Associated with Human Burials in Third Millennium B.C. Mesopotamia: Two Complementary Facets. In *Equids in the Ancient World*, edited by R. H. Meadow and H.–P. Uerpmann, 164–193. Beihefte zum Tübinger Atlas des Vorderen Orients, Reihe A—Naturwissenschaften 19/1. Wiesbaden: Reichert.

Zarins, J. 2014. [with the assistance of R. Hauser]. *The Domestication of Equidae in Third Millennium B.C. Mesopotamia.* Cornell University Studies in Assyriology and Sumerology 24. CDL Press, Bethesda, Maryland [update of *The Domestication of Equidae in Third Millennium B.C. Mesopotamia,* PhD Thesis University of Chicago 1976].

Appendix

Florence Chariot

The so–called **Florence Chariot** (from an unknown private tomb, Thebes/West). Found in 1828/29. 4 spokes. Wooden frame construction. Today on display in the *Museo Egizio di Firenze* (acc. no. 2678). Parts replaced with modern wood and other material; certain parts wrongly assembled.

Descriptions and illustrations of the Florence Chariot can be traced back to 1830s. Already in 1830, Rosellini organized an exhibition of the artefacts shipped to Florence from the Franco–Tuscan Expedition in the Accademia di Arti e Mestieri in S. Caterina at the Piazza San Marco (Guidotti 2010, 45). In addition to the following bibliography of commonly quoted studies, I only list here as an example: First, two publications by the Prussian army officer, explorer and archaeologist who excavated in Saqqara and Hermopolis, Johann Heinrich Karl Freiherr von Minutoli (1772–1846). And second, by the eminent German physiologist, natural philosopher and painter Carl Gustav Carus (1789–1869), one of the most prominent intellectuals of the 19th century Germany in the tradition of the Romance period: von Minutoli 1831 (see also Baer 1832 and the review of von Minutoli in *Göttingische Gelehrte Anzeigen*. 10. Stück. Den 19. Januar 1832. Berlin und Stettin, 89–96, here p. 92), von Minutoli 1938, Carus 1848: 211 f. (with a hand–drawn, but partly inaccurate illustration). The Griffith Institute at the Oxford University holds an early illustration of the Florence Chariot by the British antiquarian, traveler and amateur artist George Alexander Hoskins (1802–1863) from 1832–33. [http://www.griffithinstituteprints.com/image/670965/george-alexander-hoskins-ancient-egyptian-chariot].

Selected Bibliography

Rosellini 1836, 263–71 (and vol. II, pl. CXXII); Schäfer 1931: pl. VII, 1; Dittmann 1934; Botti 1951; Littauer and Crouwel 1985: 105–108 and pls. LXXII–LXIII; Decker, 1986, 38–42 (ill. p. 38f.); Spruytte 1989 (with various illustrations); Raulwing 1993: 75–79 (after Littauer and Crouwel 1985, 68f.; 105–108. pl. LXXII); Decker and Herb 1994, I, 200 f. **H 1** and pl. LXXXIVf.; Partridge 1996, 113, fig. 93. Del Francia 2002b, ills. pp. 26–34; McDermott 2004: 133, fig. 85; Decker 2006: fig. 29; Herold 2006a, 259–65 with further bibliography; Index p. 369 s.v. "Florenz", "Streitwagen in".

Chariot from the tomb of Yuya and Thuyu

Chariot from the tomb of Yuya and Tjuyu. Thebes/West = Biban el–Muluk, No. 46). Discovered by Theodore M. Davis (1837–1915) in 1905. 6 spokes. Decorated chariot body, partly closed at the read. On display in Cairo, *Egyptian Museum*, U 13, Case H. CG 51188.

Selected Bibliography

Carter & Newberry in Davies 1907, 35f., frontispiece a. pl. XXXII; Quibell 1908, 65–67; pl. 51 = Littauer and Crouwel 1985, 67; pl. LXVIII–LXX; Decker and Herb 1994, I, 200 f. **H 5** and pl. LXXXVII; Partridge 1996, 117, figs. 100 (side view), 101 (floor of the body from above), 102 (rear view), 103 (dismantled chariot body) = Partridge 2002, 71, fig. 111 (side view) and 112 (rear view), p. 27, fig. 30 (detail of chariot body, part of the pole and axle from the front left side). See also Herold 2006, 259–65, and http://www.griffith.ox.ac.uk/gri/tut-scans/TAA_i_3_8_15.jpg

Chariots from the tomb of Tut'ankhamūn (KV 62)

Antechamber

Chariot A1. Gold–decorated chariot body with inscriptions. 6 spokes. Chariots **A1** and **A2** were called "state chariots" by Carter (Carter and Mace 1922–33, II: 54 ff.). Thebes/West = Biban el–Muluk, No. 62, Obj. No. 122, JE 61990, Exhib. nos. [*Descr. Sommaire*] 97; 100; 101; 104. On display in Cairo, *Egyptian Museum*, room U 8, case 104. There has been some confusion which of the two "state chariots" in the Antechamber is "the first" and which is "the second". According to James 2000: 274 f., A 2 must be identified with Carter's "first state chariot".

Selected Bibliography

Carter and Mace 1922–33, II: 54–63; pls. 17b; 21a; c; 38a; b, 39a; b (above), 41a (above); Littauer and Crouwel 1985: 9–17 and pl. VII–XIV; Decker 1986: 42–46 (colour, p. 43); Reeves 1990a: 170; Decker and Herb 1994, I: 205, sub **H 7** and pl. XCI–XCII; Partridge 1996: 225, fig. 108 f. (after Harry Burton).

Chariot A2. Gold–decorated chariot body. 6 spokes. Thebes/West = Biban el–Muluk, No. 62, Obj. No. 120, JE 61989, Exhib. nos. [*Descr. Sommaire*] 98; 99; 102; 103. On display in Cairo, *Egyptian Museum*, room U 8, case 99.

Selected Bibliography

Carter and Mace 1922–33, II: 54–63, pls. 17a; 18–20; 21b; 37a; b; 39a; b (below), 40,]; 41a (below). *Ramses le Grand*: pl. LI a–m (details in color photographs); p. 242 (part of the left side of A2); b/w photographs of further details and related equipment on pp. 244–63; p. 256 "Schéma « en éclaté » montrant la structure du char"; Littauer and Crouwel 1985: 18–23 and pl. XV–XXV; Decker and Herb 1994, I: 203–205, sub **H 6** and pl. LXXXVIII–XC; Partridge 1996: 225, fig. 108 f. (after Harry Burton); p. 129, fig. 113 (pair of wheels); James 2000: 274 f. (colour); Decker and Thuillier 2004: 52, pl. VI (color); Decker 2006, fig. 30 (color).

Chariot A3. Wooden frame construction. 6 spokes. Thebes/West = Biban el–Muluk, No. 62, Obj. No. 121, JE

61991, Exhib. nos. [*Descr. Sommaire*] 1325. On display in Cairo, *Egyptian Museum*, room U 8, case 100.

Selected Bibliography

Carter and Mace 1922–33, II: 54–63; Decker 1984: pls. 1, 3, 4; Littauer and Crouwel 1985: 23–26; pls. XXVI–XXVIII; XXIX (drawings); Decker and Herb 1994, I: 205f., sub **H 8** and pl. XCIII–XCIV.

Chariot A4. Wooden frame construction with gold application. 6 spokes. Thebes/West = Biban el–Muluk, No. 62, Obj. No. 161, JE 61993, Exhib. nos. [*Descr. Sommaire*] 1324. It was on display in Cairo, *Egyptian Museum*, room U 8, case 114, and is now shown in the Luxor Museum (kindly confirmed by Dr. Nadia Lokma, personal communication).

Selected Bibliography

Littauer and Crouwel 1985: 26f.; pl. XXX–XXXII; Reeves 1990a: 172; Decker and Herb 1994, I: 207 f., sub **H 10** and pl. XCIV–XCVI; Decker and Thuillier 2004: 32, fig. 15 (left wheel). Decker 2006: fig. 31 and fig. 32 (left wheel, nave and parts of the axle and chariot body; color). http://news.discovery.com/history/zooms/king-tut-chariot.html; http://www.nytimes.com/2010/08/03/arts/design/03chariot.html?scp=1&sq=King Tut's Chariot Arrives in Times Square&st=cse

Treasury

Chariot A5. Wooden frame construction with gold application. 6 spokes. Thebes/West = Biban el–Muluk, No. 62, Obj. No. 332, JE 61994, Exhib. nos. [*Descr. Sommaire*] 3231. (Decker and Herb 1994, I: 208 sub **H 11**). Restoration by Nadia Lokma (1986). It took Dr. Lokma eight years to preserve, reshape and assemble the over 200, partly twisted fragments (Murphy 2007). On display in the Military Museum on the Citadel (kindly confirmed by Dr. Nadia Lokma, personal communication).

Selected Bibliography

Carter and Mace 1922–33, III: 34; 95; Littauer and Crouwel 1985: 53–55; pl. XXX–XXXII; Lokma 1986 (unpublished); Decker and Herb 1994, I: 207f., sub **H 10** and pls. XCIV–XCVI.

Chariot A6. Wooden frame construction with gold application. 6 spokes. Thebes/West = Biban el–Muluk, No. 62, Obj. No. 333, JE 61992, Exhib. nos. [*Descr. Sommaire*] 1701. It was on display in Cairo, Egyptian Museum, room U 8, case 115 and has been moved to its final resting place at the Grand Egyptian Museum in spring of 2018 (https://www.independent.ie/world-news/egypt-moves-last-chariot-of-king-tut-to-new-museum-36876776.html). Gail Brownrigg kindly shared this information.

Selected Bibliography

Littauer and Crouwel 1985: 55–57; pl. LVII; Decker and Herb 1994, I: 206f., sub **H 11** and pl. XCVI.

Chariot Body of Thutmose IV

Chariot Body of Thutmose IV. Thebes/West = Biban el–Muluk, No. 35 (KV 43): CG 46097, Exh. No. 3000. On display in Cairo, *Egyptian Museum*, room U 12, case P.

Selected Bibliography

Carter and Newberry 1904, 24–33, pls. 9–12; figs. 1–20; Littauer and Crouwel 1985, 67, No. 2 and pl. LXVII; Decker and Herb 1994, I., 201 f., sub **H 4**. Decoration: Hofmann 1989, pl. 002a; 002b; Rommelaere 1991, 176 f., cat. No. 40; Partridge 1996, 114, fig. 94 (front), fig. 95 (rear view), fig. 96 (detail of the decoration), fig. 97 and fig. 98: Howard Carter's drawing of the exterior decoration (left side = Partridge 2002, 69, fig. 108 [see also p. 70, fig. 109 and p. 219, fig. 285]). Detail (photograph) of the decoration: Partridge 2002, p. 68, fig. 107.

Part II

Equines and Wheeled Vehicles in the Near East

3

Wheeled Vehicles and their Draught Animals in the Ancient Near East—an Update

Joost H. Crouwel

This paper presents a brief overview of the history and development of wheeled vehicles in the Near East—from their first appearance in the later fourth millennium to the earlier first millennium B.C. The text, which is largely based on many years of collaboration with Mary Littauer, concentrates on the construction of the various vehicles—two–wheeled carts and chariots and four–wheeled wagons—the ways in which their draught animals were harnessed and controlled, and the use to which the equipages were put.

The first evidence for the use of wheeled vehicles in the Near East comes from southern Iraq and dates to the Late Uruk period (ca. 3400–3100 B.C.).[1] The evidence consists of simple pictographic signs on inscribed clay tablets, mostly from the site of Uruk.[2] One tablet with similar pictographs, probably comes from Tell Uqair, not far from Baghdad, and dates to the Jamdat Nasr period of ca. 3100–3000 B.C.[3] The signs appear to represent sledges with runners curving upwards at one end, raised over what may either be two (captive) rollers or four solid disk wheels (Figure 1, left). Wheels are more likely than rollers.[4] These vehicles, which may be called 'sledge cars,' carry roofed structures, exactly like the sledges represented by other, more numerous pictographs of the Late Uruk period from Uruk (Figure 1, right) and a few other sites.[5] The interpretation of the ideograms as (wheeled) sledges is supported by linguists such as Falkenstein (1936: 56), Steinkeller (1990: 22) and Burmeister, Krispijn and Raulwing (in this volume). The roofed structures may well be covered litters intended to carry important seated persons. This is suggested by contemporary representations on a small stone plaque, without provenance, but bought in Baghdad (Figure 2), and on impressions of a cylinder seal from excavations at Arslantepe in southeastern Turkey (Figure 3), both in the Late Uruk style, as well as on an unprovenanced contemporary cylinder seal cut in the Jamdat Nasr style.[6] Two of these documents show wheelless sledges involved in threshing, which may well have been of a ceremonial nature in view of the royal or priestly figures (or perhaps statues of them?) seated inside the roofed litters with legs and carrying poles (for threshing sledges in third millennium cuneiform texts: Steinkeller 1990). Here it may be noted that threshing sledges, with flints attached underneath, are still used in parts of the the Near East today.[7] Also of note is the fact that remains of an actual ceremonial sledge, without flints but with paired bovid draught, were found among the funerary assemblage of 'queen' Pu–abi in the Royal Cemetery of Ur in southern Iraq and which dates to Early Dynastic III, ca. 2600–2350 BCE.[8]

Figure 1. 'Sledge cars' and sledges. Pictographs on clay tablets, mostly from Uruk (after Littauer and Crouwel 1979: fig. 1).

Figure 2. Sledge. Stone plaque. London, British Museum 128858 (after Littauer and Crouwel 1979: fig. 2).

[1] For what follows: Burmeister, Krispijn and Raulwing (in this volume); see already Burmeister 2011: 211–14.
[2] Falkenstein 1939: 56 and signs 743–44; Green and Nissen 1986, 220, sign 248; Englund 1994: 131 and 1998: 69, fig. 22; Steinkeller 1999: 22, note 33; Burmeister 2011: 211–14 with note 9.
[3] Falkenstein 1936, 56, sign 745; Green 1986; Burmeister 2011: 214 and note 9: fig. 1: 2.
[4] Littauer and Crouwel 1979: 12–13; Sherratt 2003: 242 f.; Burmeister 2011: 213 f.; but cf. Nagel and Strommenger 1994; Jacobs 2005: 435.
[5] Falkenstein 1936, signs 741–42, Green and Nissen 1986, 220, sign 247; Steinkeller 1990: 22; Englund 1994: 131 and 1998: 69, fig. 22.
[6] Littauer and Crouwel 1990 = 2002: 329–35.
[7] D. Potts (2001) compared the ancient depictions to recent sherpherd huts set on runners in Bulgaria, thus casting doubt on the 'ceremonial threshing' interpretation.
[8] Littauer and Crouwel 1990: 18, pl. IIIA = 2002: 333, pl. 152.

Figure 3. Sledge. Cylinder seal impression from Arslantepe (after Littauer and Crouwel 1990: fig. 1).

In the representations, the sledges appear to be pulled by single bovids, harnessed between ropes fastened to the runners, and presumably running to the animal's horns. (Shafts are less likely than ropes, the harnessing of a single animal between shafts not being otherwise attested for another 3,000 years.) In contrast, bovids and equids pulling (wheeled) vehicles from the third millennium onwards were harnessed in pairs, under a yoke attached to a central pole which transmitted the pull. The animals pulling the early sledges and 'sledge cars' were controlled by a single line to a nose ring, the traditional manner of controlling bovids and the one also employed for equids in the third millennium BCE.[9]

Figured, textual, and also material evidence for both four– and two–wheeled vehicles becomes quite plentiful during the Early Dynastic II and III periods (ca. 2750–2350 BCE) in Iraq and other parts of the Near East.[10] For recent additions to vehicle representations on cylinder seals or seal impressions: Matthews 1981: 152 no. 11: fig. 2; Jans 1993; Martin and Matthews 1993: 60f. no. 71: fig. 2: 81; Dohmann–Pfälzer 2000: fig. 29; Oates 2003: 117, figs. 9.1–9.2; Zarins 2014: 138, figs. 70a–c). In many instances, the wheels were tripartite disks made of planks. There was a central plank with a nave through which the axle passes, being lentoid in shape or straight–sided where the flanking planks joined it. Ancient and modern disk wheels alike are not made of cross–sections of tree trunk, but are of wood cut lengthwise of the grain, in order to avoid the weakness of the pithy heartwood. The tripartite construction permitted the use of timber of smaller girth than a single plank which, even if cut lengthwise, would be limited in diameter to something under the maximum girth of the trunk. It seems likely that the first experiments in making disk wheels were with the single plank, hence, in an area of small timber, the results would be strictly limited in size.[11] The tripartite disk is a fairly sophisticated solution to this problem, and implies that the single–piece disk wheel must have been in use long enough for greater size to be desired. The three sections of the wheel were held together by external bonding slats.

The disk wheels have striking parallels in actual finds from different parts of Europe. Recent calibrated radio–carbon dates assign some of these surviving wheels to as early as the middle of the fourth millennium BCE Prior to the calibrations, a commonly held view was that the (disk) wheel—and the animal–drawn wheeled vehicle—were first developed in the proto–urban centres in southern Iraq, largely on the basis of the evidence of the Uruk pictographs (see above). From there, knowledge and use of the wheel would have spread to other parts of the ancient world. This is currently subject to debate.[12] Suffice it to say, here, that a 'monocentric' rather than a 'polycentric' origin of the (disk) wheel and vehicles with such wheels still seems more likely. Whatever the role of parts of the Near East and Europe, this form of transport could only have originated—or been adapted—in societies with an adequate level of technology, suitable trees, terrain, and domestic animals.[13]

Returning to the the development of wheeled vehicles in the Near East, remains of actual tripartite disk wheels from tombs at Ur and Kish in southern Iraq, and from Susa in Elam in southwest Iran, yielded diameters of 0.50–0.83 m. and evidence that they revolved on fixed axles, with linch pins to prevent them from slipping off. It is not clear whether the naves were raised parts of the central wheel plank or separately made and inserted tubular elements. The latter would have considerably helped to keep the wheel perpendicalular to the axle and prevent it from wobbling. On the other hand, there is material and figured evidence (from the Near East, but not from Europe) for a tyre, made of rawhide, wood or metal, that was sometimes pierced by metal hobnails (Figures 4–5, 12). This outer element of the wheel would have played an important role, not only in protecting the tread, but in consolidating the three wheel planks. Hobnails, when present, added depth to the protection and also increased the traction in certain types of terrain.

[9] A wall painting from public Building IV of the Late Uruk period (ca. 3374 BCE, according to a series of calibrated radio–carbon dates; see Sherratt 2004: 418) at Arslantepe in south–eastern Turkey was described by the excavator as showing "two figures of stylized bulls facing one another [...] which seem to be pulling what appears to be a monumental cart driven by a coachman [...] On the 'bulls' horns are attached reins which end in a ring held by the coachman" (Frangipane 1997: 64). Unfortunately, this description cannot be verified with the help of the one published photograph (Frangipane 2004: fig. 15). While the use of the word 'cart' suggests a (two–)wheeled vehicle, the comparison made by Frangipane (2004: 64 f.) between this scene and those on the Late Uruk seal impressions from the same site would point to a sledge. However this may be, the wall painting from Arslantepe may well provide the earliest illustration of *paired* (bovid) traction (so Sherratt 2003: 243, and 2004: 418). Knowledge of disk–wheeled vehicles in the Near East in the later fourth millennium BC may perhaps be inferred from two small perforated disks, one from Arslantepe, the other from Gebel Aruda in Syria (Fragipane and Palmieri 1983: 407: fig. 64;4; Bakker *et al.* 1999: 782: fig. 4; Burmeister 2011: 214 f.: fig. 3: 1–2). Made of terracotta and limestone respectively, these objects have been interpreted as model wheels.

[10] For the various types of vehicle: Littauer and Crouwel 1979: chapter 5; Nagel 1966: 1–10; 1984/1985; 1986; 1992a; Messerschmidt 1992; Bollweg 1999; Nagel and Strommenger 2001; Eder and Nagel 2006: 46–51.

[11] Littauer and Crouwel 1979: 18–19; Piggott 1983: 17–21, 24–26; also Spruytte 1983: pl. 33.

[12] See specially Burmeister 2004, 2011 and 2017; Maran 2004a and b; Sherratt 2003: 242 f.; also 2004; also Bakker 2004; Schlichtherle 2004; Jacobs 2005: 430–34; Drenth and Bakker 2005–2006: 6f., 29–37; Kohl 2007: 85; Bulliet 2016, chapters. 3-4.

[13] Burmeister 2004: 23.

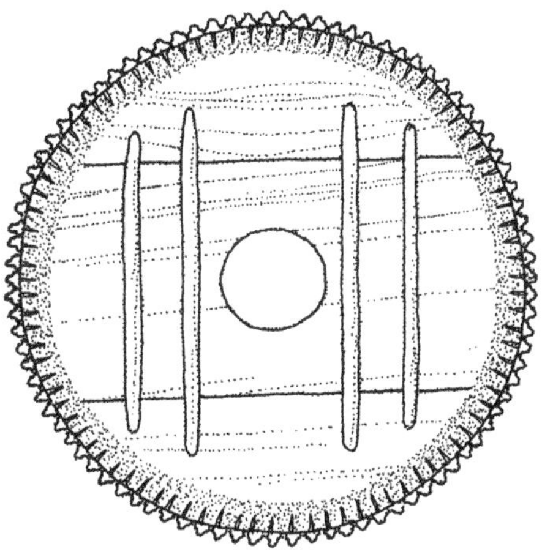

Figure 4. Wheel from Susa (after Littauer and Crouwel 1979: fig. 5).

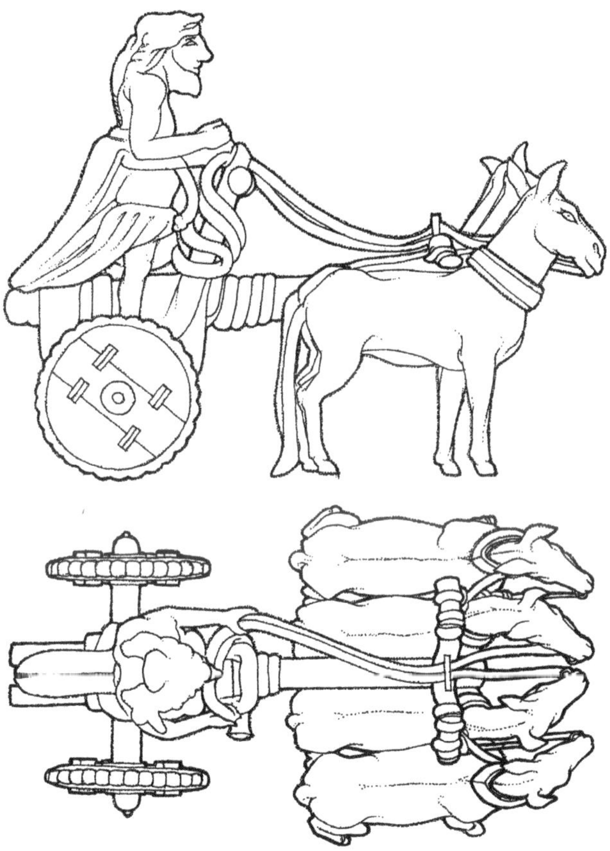

Figure 5. 'Straddle car.' Copper model from Tel Agrab. Bagdad, Iraq Museum 31389 (after Littauer and Crouwel 1979: fig. 7).

Bovids—in teams of two—appear to have been the draught animals used in funerary and other ritual, as sanctified by tradition. Oxen, castrated males that are slow but strong, docile, and steady, would also have been employed in transporting agricultural or other material too heavy and/or bulky for human porters or pack animals to carry. (The use of pack animals, presumably asses, is first documented in the Near East by terracotta figurines of the fourth and third millennia BCE).[14]

[14] See Moorey 2001: 345; Oates 2003: 115; Benecke 2004: 458–60, for the ass in the Near East and elsewhere.

At the same time, livelier and faster teams were now composed of (usually) four equids, either asses or hybrids of ass and hemione or ass and horse. In contrast to what has often been claimed, the native hemione itself appears never to have been fully domesticated and used only in cross–breeding.[15]

All draught was by the pole–and–yoke system. The single, central draught pole would have originated with paired bovids under yoke, on either side of the plough beam that helped to keep the share straight.[16] Domestic cattle had long been known in Iraq and other parts of the Near East, as well as in parts of Europe.[17] The plough beam must have been converted into a pole of four– or two–wheeled vehicles when the time came.[18] The bovids were harnessed under a yoke either attached to their horns, or laid on their necks in front of the withers which pushed against the neck yoke to produce the tractive power. A very simple yoke may have been used, with rods or straps to keep it in position. When equids first began to be used in draught, they were harnessed by the same system, which is less suitable for equine anatomy than for bovine. Equids have longer, slenderer necks and low withers. This would have reduced the pushing power of the withers and permit the yoke, unless it was held firmly forward, to ride back over the lower withers, bruising and chafing them, especially in fast going. It has often been assumed that the strap that went across the equid's neck in order to hold the yoke forward would rise against the animal's trachea, thereby impairing its breathing and greatly reducing tractive power. (Modern equid shaft–and–breastcollar harness permits the pressure to be taken by the shoulders or breast, the most suitable areas.) However, experiments have shown that the ancient method cannot have been quite as inefficient as it has often been made out to be, or modifications would have taken place sooner.[19]

The harnessing of the teams of four equids is well illustrated by a copper model of a type of two–wheeled vehicle known as the 'straddle car' (see below) from Tell Agrab in Iraq (Figure 5). Here, only the two inner animals carry a yoke, slightly shaped to fit their necks and attached on top of the pole near its end, while the other animals are more loosely attached outriggers. All four animals wear thick (padded?) neck straps, those of the inner animals being fastened directly to the yoke, those of the outer ones being attached to the yoke on one side only and with a little play. Thus, the outriggers would have provided very little pulling power since they were not under yoke. Equids are herd animals and, as such, are not only more willing to go in company than alone, but also competitive at speed.

[15] Postgate 1986; Zarins 1986: 180–89 and 2014: chapter II; von den Driesch 1993; Oates 2003: 115, 117; but cf. Bollweg and Nagel 1992; Strommenger and Bollweg 1996; Nagel, Bollweg and Strommenger 1999.

[16] Littauer and Crouwel 1979: 9f.; Sherratt 2004: 416–19.

[17] Benecke 2004: 455–57.

[18] See the Uruk IV pictograph for plough: Falkenstein 1936, sign 214; Green and Nissen 1986, sign 33; also Moorey 1994: 2 f.

[19] Spruytte 1983.

The outer animals, feeling less yoke pressure on their necks than the inner ones, may well have been inclined to set a faster pace, which would then stimulate the yoked animals, and help to increase the speed. They could also serve as reserve animals, while at the same time adding an element of display to the equipage.

As evidenced on the Tell Agrab model and elsewhere, the equids were controlled by single lines to nose rings, a circumstance that also points to original bovid draught. The lines passed through a double terret which, in the case of the Tell Agrab model, is positioned on the yoke (Figure 5), but is shown in two–dimensional representations as placed farther back, on the pole (Figures 8, 12); the latter position is also indicated by the shape of the bases of actual metal terrets found at various sites. A well–known, costly example is the terret crowned by an equid figure that was associated with Pu–abi's sledge in the Royal Cemetery at Ur (Figure 6). The nose rings of the outer animals were each connected by a line only to the yoke, so that when the driver pulled on the lines of the central animals and these slowed the equipage, the yoke itself would check the outer ones. This method of control was limited to braking. Directional control, if attempted, must have been by voice, goad, or whip, or a combination thereof—as oxen may still be directed. Goads and whips are indeed shown with equine harness teams (Figure 12). Although a nose ring does not require a headstall of leather straps or rope to hold it, the combination is repeatedly shown with equids. The headstall includes a basket muzzle and/or a tight–fitting noseband so placed as to press on the sensitive tissue below the equids' nasal bones (Figure 8). Such arrangements may well have been to discourage the draught teams (in the most explicit renderings shown as four ungelded males) from bickering among themselves, and to protect the grooms when harnessing them.

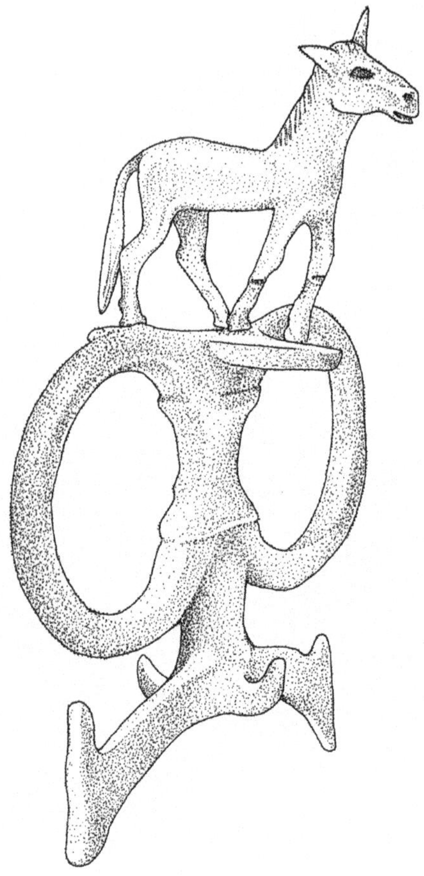

Figure 6. Silver and electrum terret from Ur. London, British Museum 121438 (after Littauer and Crouwel 1979: fig. 10).

The four–wheelers seen in two– and three–dimensional figured documents differ markedly from those of the earlier pictographs. Many representations show a single type which may be called 'battle car' since—during the Early Dynastic III period—it is depicted primarily in military contexts. The 'battle car' was narrow, with room for a complement of two—a driver and a passenger standing behind him. The fairly high front breastwork of bent wood is topped by an open rail by which the driver may steady himself in the springless vehicle. The sides are low and panelled. Sometimes a seat for the driver is shown.

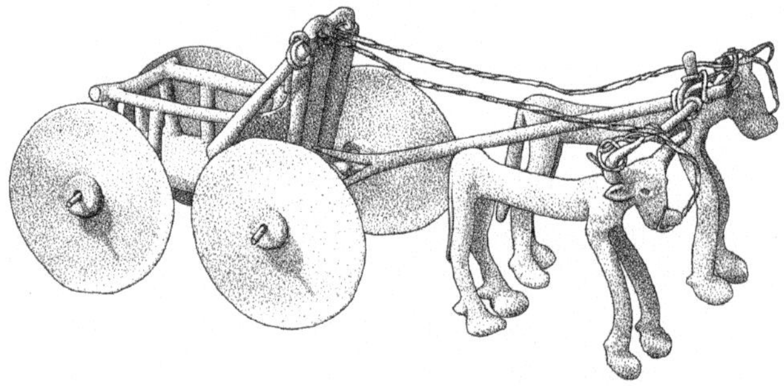

Figure 7. 'Battle car.' Copper/bronze model. New York, Metropolitan Museum of Art 66.15 (after Littauer and Crouwel 1979: fig. 15).

The draught pole may be of two types: one is straight, rising obliquely from the level of the vehicle floor to the yoke area; the other type is high and arching. Since the poles of four–wheelers have no supportive functions, unlike those of two–wheelers, they need not be attached rigidly. Indeed, a rigid attachment to the vehicle would be disadvantageous in anything but completely level terrain. Several metal models, most of them unprovenanced but including some from a cemetery near Abamor in south–eastern Turkey and dating to the later third or early second millennium BCE, testify to the presence of a vertically articulating pole which would have reduced the strain on draught team, pole and wagon (Figure 7).[20] There is no evidence to suggest that the poles of the third–millennium 'battle cars', like those of wagons in the much later Hallstatt period in Europe, had any horizontal movement through being combined with an undercarriage and a swiveling front axle. This means that the 'battle cars' could turn only in wide arcs.

[20] Littauer and Crouwel 1973 = 2002: 336–79; Nagel 1984–85; Kulakoğlu 2003 (model no. 1 stands out in having an arched tilt).

Figure 8. 'Battle cars.' Detail of the 'Standard of Ur'. London, British Museum 121201 (after Littauer and Crouwel 1979: fig. 3).

In Early Dynastic representations, this type of vehicle is often depicted with a sheath of throwing spears attached to the front edge of the breastwork, suggesting a use in warfare or hunting. Several times, the 'battle car' is shown in an explicitly military context, the most graphic illustration of this being a scene on the so–called Standard of Ur (Figure 8). It shows a row of four wagons, manned by helmeted drivers and warriors; the latter steadying themselves on the shoulders of the drivers in front and carrying a throwing spear or axe. The vehicles are moving at varying speed across a battlefield, symbolised by the stripped and bleeding human bodies of enemies prostrate beneath the draught teams. The position of the bodies need not be read literally since, if at all possible, equids avoid stepping on bodies for fear of injuring their own limbs. We have here, in fact, the earliest appearance of the motif of 'the enemy beneath the team's hooves' that is to return early in the next millennium and later with another type of vehicle, the two–wheeled chariot. Wherever the motif appears, it must be taken as merely symbolic of victory.

Although the appearance of the 'battle car' in warfare may originally have had as daunting a psychological effect on third millennium foot soldiers, as had that of the tank on those in the First World War, this often drawn comparison is fallacious. It is most unlikely that the wagons, costly to make and maintain, and with their poor manoeuvrability, were used in head–on charges against each other or against unbroken infantry lines. In contrast, tanks could do that, while also being able to negotiate many kinds of terrain that are inaccessible to draught animals and wheels, and offering protection to both motive power and crew. However, the ancient 'battle cars' could have functioned as mobile arsenals and firing platforms from which javelins could have been thrown in warfare or hunting. Indeed, modern experiments have shown that an expert javelin thrower can throw 30 javelins a minute over a distance of up to 60 meters from a reconstruction of such a wagon moving at a speed of 10–12 miles an hour.[21] The 'battle cars' may also have been used as a prestigious means of transport for kings and important officers to and from the field of battle. The axe seen in one warrior's hand on the 'Standard of Ur' and his position at the rear of the vehicle, convenient for mounting and dismounting, also suggests fighting on the ground.

After the Early Dynastic period there is no longer evidence for the use of these four–wheeled vehicles in warfare. Indeed, the various disadvantages noted above may have led to their abandonment for this purpose. A use in cult may be suggested by representations on cylinder seals of the Akkadian period (ca. 2350–2100 BCE) showing 'battle cars' driven by deities and pulled by winged monsters. Cult use may also be implied by the group of metal models mentioned above, which show bovid–drawn four–wheelers with a high front breastwork clearly deriving from that of the traditional 'battle car' (Figure 7).

A civil use can be assumed for the wagons represented by various terracotta models which show no features typical of the 'battle car.' Among them, four–wheelers with arched tilts figure promimently (Figure 9). These vehicles were presumably bovid–drawn, like the actual wagons, with and without tilts, from Transcaucasia and the Caspian steppe and were of similar as well as later date.[22] Of special interest is also the four–wheeler represented in detail by a terracotta model from Tulul el–Khattab near Baghdad, with its floor of bent wood and osier or reed, held together by bindings (Figure 10).[23]

As to the two–wheeled vehicles of the third millennium BCE, one type, the so–called platform car, appears to be a two–wheeled version of the 'battle car,' including the typical high, bent–wood front breastwork. This is most clearly seen in terracotta models, such as one from Kish (Figure 11). Other such models, mostly dating to the later third millennium like those from Ashur, show no side breastwork but a seat for the single occupant (Figure 12). The second type of two–wheeler is the so–called straddle car, in which the standing or seated occupant straddled the end of the straight or arching draught pole, with his feet

[21] Littauer and Crouwel 1979: 33, based on unpublished information from D. Noble. Nobel's full–scale (very heavy) reconstruction of a 'battle car', as depicted on the 'Standard of Ur', was recently exhibited at the International Museum of the Horse in the Kentucky Horse Park (Olsen and Culbertson 2010: no. 2).

[22] Littauer and Crouwel 1974, 2002: 380–402; Strommenger 1990; Moorey 2001: 345–47. But cf. Özgen 1986 (three 'battle car'–like models).

[23] Postgate 1992: fig. 10: 1.

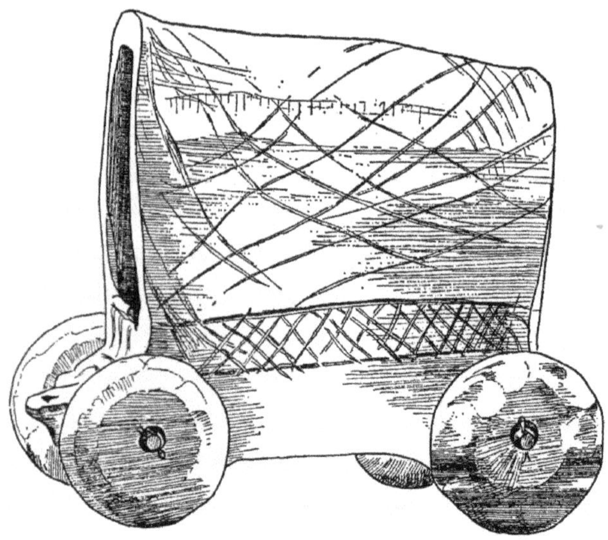

Figure 9. Tilt wagon. Terracotta model from Hammam. Oxford, Ahsmolean Museum 1913.183 (after Littauer and Crouwel 1974: fig. 1).

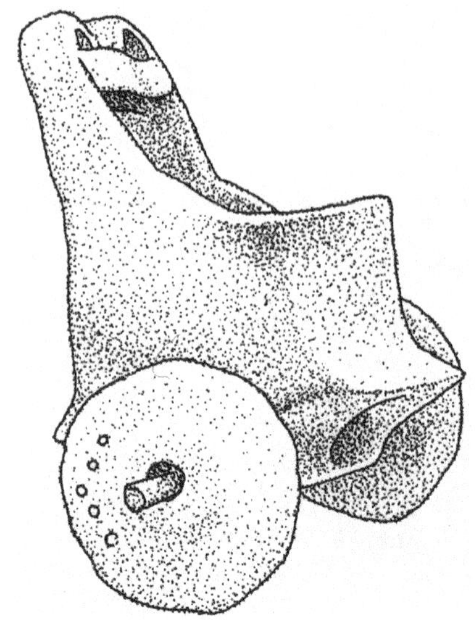

Figure 11. 'Platform car.' Terracotta model, from Kish. Chicago, Field Museum of Natural History FM 229170 (after Littauer and Crouwel 1979: fig. 9).

Figure 10. Wagon. Terracotta model from Tulul el–Khattab. Baghdad, Iraq Museum 87213 (after Postgate 1992: fig. 10: 1).

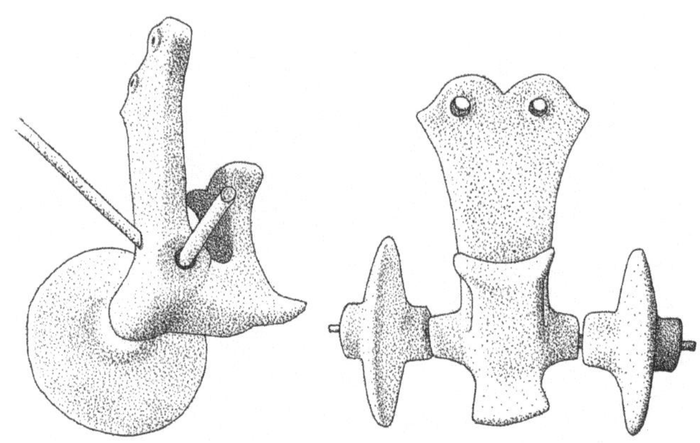

Figure 12. 'Platform car.' Terracotta model from Ashur (after Littauer and Crouwel 1979: fig. 20).

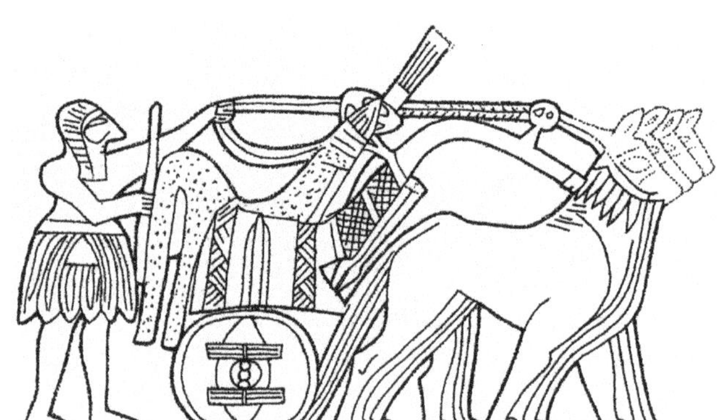

Figure 13. 'Straddle car.' Detail of stone plaque from Ur. Philadelphia, University Museum CBS 17086 (after Littauer and Crouwel 1979: fig. 8).

placed on the axle or a bracket, much as a later rider would place his feet in the stirrups of a saddle (Figures 5, 13). The saddle seat may be shown with a high cantle behind and a leopard skin to increase the rider's comfort. There appear to have been variations on these two basic types. As two–wheelers, both must have been more manoeuvrable than the four–wheeler, but only the light 'straddle car' would have been suitable at speed, since the rider could make use of the spring in his hips, knees and ankles to absorb the jolts. The 'straddle car' sometimes also carries javelins, but is never shown in an explicitly military setting. The necessity for the driver to control the team of four equids and to use weapons at the same time would have limited its usefulness in battle. Such a vehicle could, however, have been driven at speed in prepared hunts, where game would be chased across its arranged path over appropriate terrain and, perhaps, even in a confined area, as happened later with Egyptian and Assyrian chariots. The 'straddle car' may also have served as a fast courier's transport.

All in all, various types of wheeled vehicles were used for a variety of purposes in Early Dynastic times. According to textual sources, some were sacred vehicles to be employed in processions, while others figure among grave goods.[24] The latter use is confirmed by the discovery of vehicles, along with their draught teams, in the Royal Cemetery at Ur and other richly furnished tombs at Kish and Susa, where they must also have functioned as status symbols.[25]

In the Near East by the later third millennium BCE, fast, single–person, equid–drawn two–wheelers with solid wheels had been in use for centuries. It is at this time and including the early second millennium that we can observe a progressive series of modifications in the construction of the body and wheels of these vehicles and in the methods by which the draught teams were controlled (for what follows: Littauer and Crouwel 1979, chapters 6 and 7). These modifications resulted in the establishment in the Near East—in the eighteenth to seventeenth centuries at the latest—of the chariot, i.e. a light vehicle with two spoked wheels, usually drawn by horses which were controlled by bitted bridles, and carrying one or more standing riders.

The horse is documented sparingly in the Near East in the later third millennium BCE, among the faunal remains[26] and in cuneiform texts. The latter refer to the horse either as a breeder to get hybrids, or as a mount or draught animal.[27] Although his superior speed, strength, and handsomeness made him desirable, the horse could not be made efficient use of until suitably harnessed and bridled.

In the later third millennium, (indirect) evidence for the use of a bridle bit—an element for control of the equid by mouth, composed of a mouthpiece and cheekpieces, and essential for driving at speed—becomes available (for what follows: Littauer and Crouwel 2001; Brownrigg with Dietz 2004). It consists of wear, possibly left by metal bridle bits, on the teeth of five asses buried in a temple at Tell Brak in Syria, belonging to the Akkadian period and dated ca. 2200 BCE.[28] Bit wear has also been reported in connection with the remains of a 'probable mule or possible horse' from Malyan in Elam and attributed to the Kaftari phase (ca. 2200–1800 BCE). Other, again indirect indications of an early use of (metal) bits in the Near East, may be yielded by terracotta figurines. An explicitly male caballine figurine (viz. the falling mane) from a context dated ca. 2300–2100 BCE at Tell Es–Sweyhat in Syria has a transverse hole through its muzzle (Figure 14), as has a broken–off equine head of Akkadian date from Tell Taya in northern Iraq (Figure 15) and a roughly contemporary head from Selenkahiyeh in Syria. The holes would have been for paired reins belonging to a (bitted?) bridle, and not a nose ring. In addition, the Tell Taya head preserves part of the headstall, made of straps or rope and designed to hold a controlling bit or noseband in place. Complete headstalls, but no pierced muzzles, can be seen on several equid figurines from Selenkahiyeh and other sites.

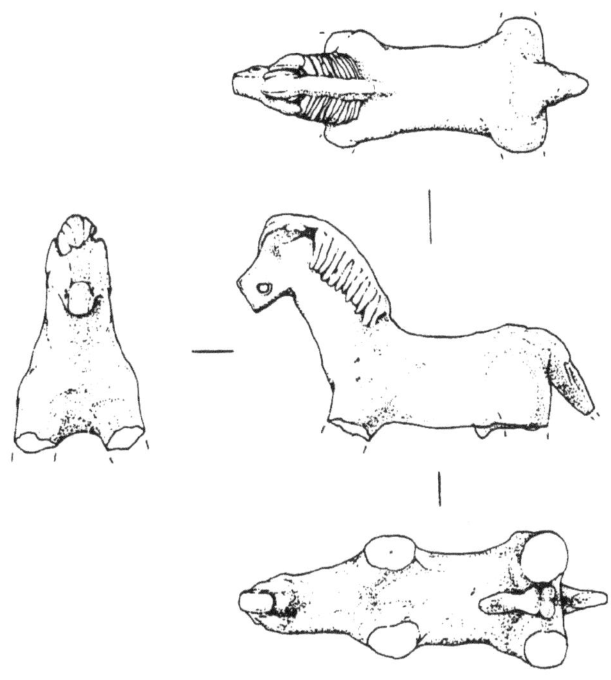

Figure 14. Stallion. Terracotta figurine from Tell Es–Sweyhat (after Holland 1993–94: fig. 11).

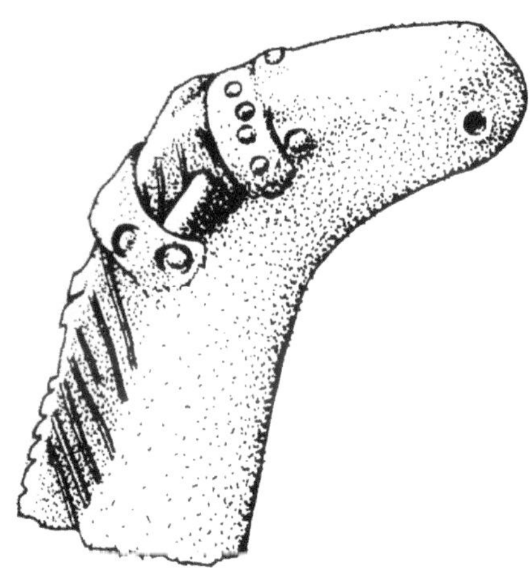

Figure 15. Bridled equid. Incomplete terracotta figurine from Tell Taya (after Littauer and Crouwel 2001: fig. 3).

Whoever thought of eliminating the nose ring and attaching a rein to either side of the low noseband, already seen in figured documents of the earlier third millennium, was on the way to achieving the graded braking effect and, at least partial, directional control impossible with a single nose ring. One of the drawbacks of the noseband type of control is its weak lateral effect. It would be natural to place a metal reinforcement in the areas where the reins were attached. If these two reinforced sections were joined by a bar mouthpiece through the mouth (in

[24] Civil 1968; Zarins 1986: 180–88. For textual evidence for wheeled vehicles in general: Salonen 1951; Zarins 2014: chapter III.
[25] Zarins 1986: 165–80, including burials of (paired) equids without traces of vehicles at various sites.
[26] Becker 1994: 156–170; Oates 2003: 117, 119 f.; Benecke 2004: 460–62; Vila 2006: 117–20; Zarins 2014: 83–87.
[27] Postgate 1986; Zarins 2014: chapter III; also J. Marzahn in this edition.
[28] Clutton–Brock and Davies 1993: 214 f.; Clutton–Brock 2001; but cf. Clutton–Brock 2003.

this area, equids' teeth are conveniently missing), their lateral effectiveness would be further increased. A pair of reins attached to either side of the bit would give the driver not only braking power, but directional control, thereby greatly increasing the manoeuvrability, hence the usefulness of the vehicle, something particularly desired in battle. The presence of a bit would also eliminate the tendency of the entire headstall to skew around when one rein was pulled strongly.[29] Interestingly, the osteological evidence from Tell Brak raises the possibility that in the Near East bit control was initially not exclusively associated with horse–drawn chariots, as has often been assumed.

There are also indications of developments in wheel construction.[30] Actual metal tyres, made in segments and held on by clamps, have been found in different parts of the Near East, as well as in Turkmenistan and possibly Afghanistan (Figure 16).[31] Such tyres, studded with hobnails, belong to the 'straddle car' depicted on a fragmentary stone stela of Ur III date from Telloh in southern Iraq (Figure 17). While such clamped tyres would have been an improvement on the plain metal hoop ones that are sometimes apparently illustrated in Early Dynastic times and which would have easily worked loose, the amount of metal required still made them costly as well as adding considerably to the weight of the wheels.

The so–called cross–bar wheel represents an attempt to lighten the disk wheel (for what follows: Littauer and Crouwel 1977 = 2002: 272–88). It is first seen, with a crudely rendered two–wheeled vehicle, on a cylinder seal from level IIIb at Tepe Hissar in northern Iran and usually dated to ca. 2250–2000 BCE (Figure 18). In this type of wheel, the central plank of the wheel, through which the axle passes, is narrowed to a diametral bar; the flanking planks of the tripartite wheel are eliminated, and the former bonding slats are turned into sturdy transverse bars between the diametral bar and the felloe. This cross–bar wheel is also illustrated in the second millennium BCE and later. Fixed on a revolving axle, it has remained in use with simple carts in various parts of the world. A more important innovation was the much lighter, spoked wheel which is first attested early in the second millennium BCE (see below).

The traditional four–wheeled 'battle car' and two–wheeled 'straddle' and 'platform cars' continue to be depicted in figured documents of the earlier second millennium BCE (for what follows: Littauer and Crouwel 1979: chapter 7). Vehicles deriving from the 'battle car' appear on cylinder seals and impressions thereof from the time of the Old Assyrian trade colony at Kültepe (Kârum Kaneš, level II) in central Turkey, often dated to the later 20th and earlier nineteenth century BCE. The vehicles, with a seated

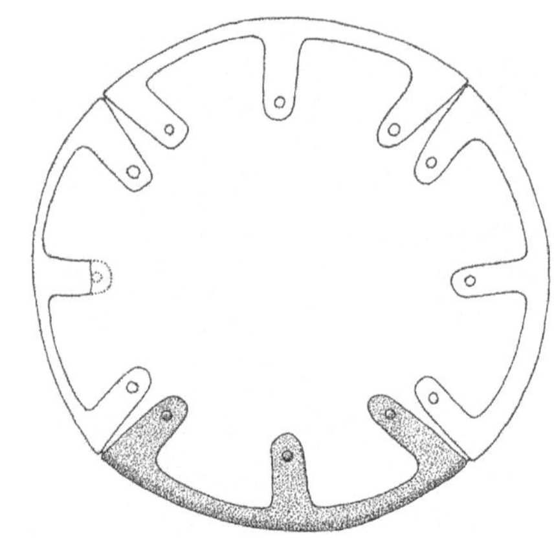

Figure 16. Bronze tyre in segments, from Susa. Teheran, Iran Bastan Museum (after Littauer and Crouwel 1979: fig. 1).

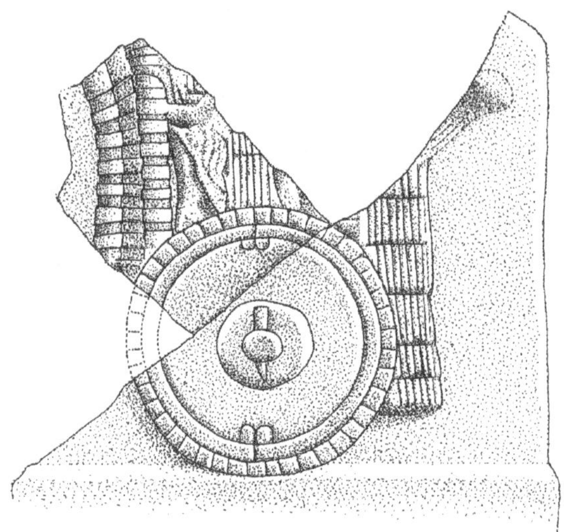

Figure 17. 'Straddle car.' Fragments of stone relief from Telloh. Berlin, Vorderasiatisches Museum 2904 (after Littauer and Crouwel 1979: figs. 18a–b).

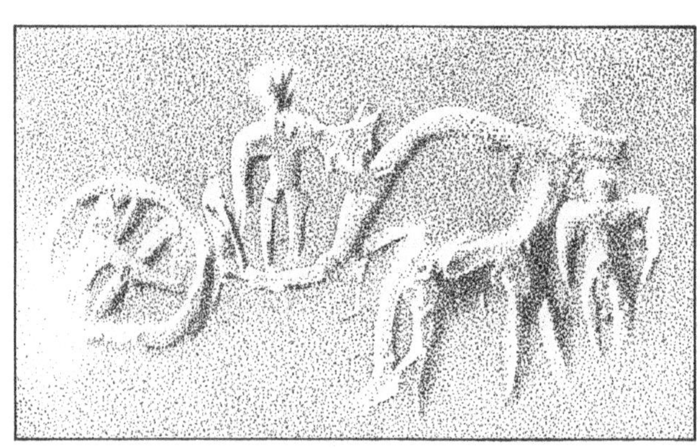

Figure 18. Two–wheeler with 'cross–bar' wheel. Cylinder seal from Tepe Hissar. Teheran, Iran Bastan Museum (after Littauer and Crouwel 1979: fig. 21).

[29] Littauer 1969 = Littauer and Crouwel 2002: 487–504; Littauer and Crouwel 2001: 335.

[30] For what follows: Littauer and Crouwel 1989 = 2002: 261–71; Teufer 1999: 73, 75, and 2012, 291; Crouwel 2012.

[31] *In situ* remains of actual wagons with such tyres were quite recently found in the 'Royal Cemetery' of ca. 2100–1800 BCE at Gonur–depe in south–eastern Turkmenistan (Sarianidi 2006: 252 f. with figs. 72 and 99; Kohl 2007: 195–99, 209, 244 with figs. 5.9–11; Crouwel 2012: 92 f.; Teufer 2012: 291, fig. 15). This and other sites of the so–called Bactrian–Margiana Archaeological Complex (BMAC) were clearly in contact with Iran and other parts of the Near East. I am grateful to T. Kawami for further information.

Figure 19. 'Battle car' with spoked wheels. Cylinder seal. New York, Morgan Collection 893 (after Littauer and Crouwel 1979: fig. 25).

Figure 20. Proto–chariot. Detail of cylinder seal impression. New York, Metropolitan Museum of Art 66.245.17b (after Littauer and Crouwel 1979: fig. 29).

Figure 21. Proto–chariot. Front screen of fragmentary terracotta vehicle model from Uruk. Berlin, Vorderasiatisches Museum 11576 (after Littauer and Crouwel 1979: t).

Figure 22. Copper/bronze model spoked wheel from Acemhöyük (after Littauer and Crouwel 1986, pl. 23: fig. 2).

driver behind the characteristic high front breastwork, are drawn by a team of four equids. The wheels are of the cross–bar or new spoked types (Figure 19). On other seal impressions spoked wheels are shown with a new form of two–wheeled vehicle, like the four–wheelers shown in a peaceful, parade setting (Figure 20). This two–wheeler is light, with only a hip–high railing to protect the single, standing occupant who is clearly a deity. There is no seat anymore. Instead, the rail probably extended along the front and sides but leaving the back open for quick (dis)mounting. This interpretation is supported by the evidence of a terracotta model, possibly of slightly later date, from Uruk in southern Iraq (Figure 21). As on the four–wheelers, the old nose ring is still in evidence, but the team now consists of only two equids. Like the four draught animals illustrated with the traditional 'battle cars', the equids are too schematically rendered to permit more precise identification. This spoke–wheeled vehicle and those shown in Anatolian glyptics may be called proto–chariots, as they still lack yoke saddles and bridle bits with reins attached to either side. At the same time, the two–wheeled vehicles represent an important step towards the true chariot.

The earliest three–dimensional evidence for spoked wheels consists of the copper/bronze four–spoked wheels (0.17–0.177 m. in diameter) belonging to a large model or trolley from the so–called Burnt Palace at Acemhöyük in central Turkey, usually dated to the first half of the eighteenth century BCE (Figure 22).[32] Rather than four spokes inserted into a separately made barrel nave, these

[32] Littauer and Crouwel 1986 = 2002: 289–95.

wheels may suggest the sophisticated nave–and–spoke construction materially documented in the second half of the second millennium (see below).

Syrian cylinder seals of the 18th and 17th centuries BCE illustrate spoke–wheeled two–wheelers with a variety of body forms and drawn by teams of two equids.[33] Some of the vehicles appear to be traditional high–fronted 'platform cars' or modifications thereof, in which the seat is removed, thereby permitting rapid access from the rear (Figure 23). Others may be called chariots, with either a light, open siding of bent rails or a variety of closed, box–like superstructures of equal height at front and sides (Figures 24–25). The traditional high, arching pole is usually reduced to a lower, more gently curving one (Figures 23–25). Some of the chariots seem to have the breastwork brace and/or pole support dropping from the top of the siding to the pole a short distance in front that is explicitly attested in the later second millennium BCE (Figure 24). Most vehicles still carry only a single occupant, but one chariot explicitly has a two–man crew, depicted standing one close behind the other, but in reality probably actually standing abreast, as did the crews on many later chariots shown in this convention (Figure 24). This implies a wider floor and a lengthened axle, which would provide a wider wheelbase and increase lateral stability on rapid turns.

The draught animals may well be exclusively horses, which are now amply documented in osteological and textual sources, the latter often referring to their use with vehicles.

Syrian seals may show rods with curved ends over the yoke area (Figures 24–25). These rods may be interpreted as the (exaggeratedly long) 'handles' of so–called yoke saddles that are well documented in the later second millennium BCE (see below). These wooden Y–shaped objects, one to each animal and with their 'handle' attached to the yoke, were a means of adapting the yoke better to horses' anatomy: by transferring the pull from the neck to the shoulders they increased the animals' pulling power and stamina.[34]

Some of the Syrian seals depicting two–wheelers with teams of two equids still show single lines to nose rings. The others indicate two pairs of lines running to the heads of the animals, suggesting the use of a bridle with or without a bit (Figures 23–25). As we have seen, there is indirect evidence for the use of bridle bits by the later third millennium BCE. Now we have direct testimony of a copper/bronze bit, dated to ca. 1700/1650–1550 BC (Middle Bronze Age III) from Haror in the northern Negev.[35] This bit was actually found in the mouth of an equid, identified as an approximately four years–old ass, that was buried in a cultic context (Figure 26). The Haror bit belongs to a

Figure 23. 'Platform car'. Detail of cylinder seal (after Littauer and Crouwel 1979: fig. 36).

Figure 24. Chariot. Detail of cylinder seal. Yale, Babylonian Collection (Newell Collection no. 343; after Littauer and Crouwel 1979: fig. 33).

Figure 25. Chariot. Detail of cylinder seal. Oxford, Ashmolean Museum 1920.25 (after Littauer and Crouwel 1979: fig. 35).

[33] Littauer and Crouwel 1979: 68–70; Moorey 1986; Nagel 1992: 72–74.
[34] Littauer 1968 = Littauer and Crouwel 2002: 479–86.
[35] Littauer and Crouwel 2001; Bar–Oz et al. 2013.

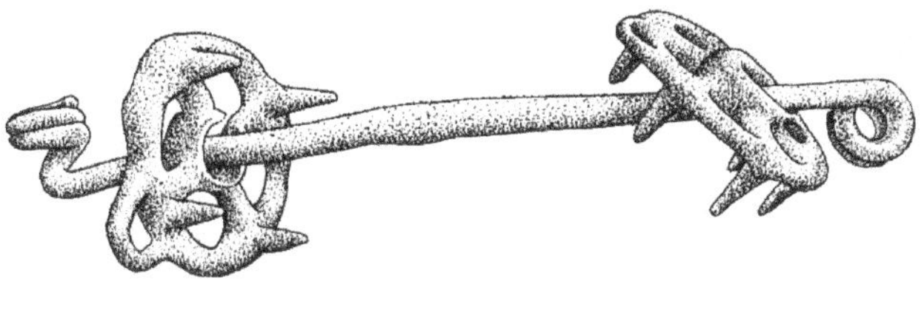

Figure 26. Copper/bronze bridle bit from Tel Haror (after Littauer and Crouwel 2001: fig. 1).

well–known type with circular cheekpieces, often studded on the inner faces, and a single bar mouthpiece. Most, if not all of these cast–metal bits, which are known not only from different parts of the Near East, but also from Egypt and mainland Greece, date within the second half of the second millennium BCE (see below). They are driving bits, with the emphasis on enforcing directional control from a vehicle—something further improved by studs on the inside of the cheekpieces. Mention may be made here of a horse skeleton from the Egyptian fortress at Buhen in Nubia, possibly dating to some time in the seventeenth century.[36] It has been proposed that the wear on one tooth was caused by a metal bit.[37]

One Syrian seal (Figure 23) provides the first evidence for the use of the bow from a vehicle, here in a hunting setting. The driver is using his bow, a quiver hanging at his back, while at the same time controlling his team by reins tied around his hips (for what follows: Littauer and Crouwel 1979: 63). This is a theme that is to become popular in the Near East, and particularly in Egypt, in the later second millennium—a royal or high–ranking personnage alone in his chariot, attacking the enemy or game, usually using the bow, the reins tied around his hips. This ostentatious feat was indeed accomplished by much later Etruscan and Roman racing charioteers. While it might have been practical with the hunting chariot in prepared hunts where game was driven across the raised hunter's chosen path, it would have been both too dangerous and too uncontrolled for a head of state to have taken such a risk under the uncertainty of battle conditions.

The use of a bow from a wheeled vehicle is implicit in some other Syrian seals where the single driver carries a quiver or bowcase on his back (Figure 25). The use of the bow in warfare would be most effective when the archer did not need to concern himself with driving, but had a charioteer. This is precisely the practice illustrated in many battle scenes of the later second and earlier first millennium BCE. Appropriate conditions may already have existed by the earlier second millennium, with the chariot presumably wide enough to permit a second person abreast of the driver and the new low front breastwork not interfering with the use of the bow. Improved stability and manoeuvrability, thanks to a wide wheel base and control by bitted bridle, would also have increased the suitability of the vehicle for warfare.

Several of the Syrian seals show the vehicles, with one or two occupants, accompanied by a file of marching unarmed attendants, and sometimes a prone human figure 'beneath the team's hooves'.[38] The latter is a symbol of military victory, already encountered with the four–wheeled 'battle cars' depicted on the 'Standard of Ur' and also illustrated with chariots in the later second and earlier first millennium BCE.

Firmer evidence for a role of chariots in warfare at this time is found in cuneiform texts of the Hittite Old Kingdom, which refer to chariotry forming part of the armies of the Hittites and their Anatolian and Levantine enemies. Only in one case—a passage in the Annals of king Hattusilis I (ca.1650–1600 BCE)—are 80 vehicles more specifically reported as being used in surrounding the besieged city of Ursum, somewhere in southeastern Turkey.[39]

In the Near East, the earliest archaeological sources for the light, spoke–wheeled, horse–drawn chariot come from central Turkey, Syria, and Iraq. The first texts refer to the new type of vehicle and its team of horses among the Amorites, Hurrians, and Hittites, the latter speaking an Indo–European language. Despite repeated claims, there is no compelling evidence for a special or unique role of Indo–European or other linguistic groups in the early development of the chariot in the Near East.[40]

In this account, it is argued that the chariot of the Near East is the outcome of a long evolution in the area itself. This stands in contrast to the long–held theory that the vehicle and its draught animals were introduced from the Eurasian steppes, probably by Indo–European–speaking people.[41] The idea of a northern origin for the chariot has recently been revived, chiefly on the basis of calibrated radio–carbon dates of ca. 2000–1800 BCE for tombs containing remains of spoke–wheeled two–wheelers at Sintashta and other sites in northern Khazakstan, just east of the Urals.[42] In most cases, all that remained of the actual two–wheelers were the impressions left in the soil by the lower parts of ten– or twelve–spoked wheels. Horse remains and discoid bone cheekpieces of soft–mouthed bridle bits found with them, show that these early vehicles were horse–drawn

[36] An early date for the Buhen horse receives some support from the recent find of a horse skeleton in a palace of ca. 1640–1530 BCE at Tell el–Dab'a in the Nile delta (Bietak and Förstner–Müller 2009: 98–100 with fig. 8).

[37] Clutton–Brock 1974: 238–39; but cf. Payne 1995: 238f.; Raulwing and Clutton–Brock 2009.

[38] Littauer and Crouwel 1979: 63; Moorey 1986: 205 f.

[39] Littauer and Crouwel 1979: 64–65; Houwink ten Cate 1984: 57–60; Moorey 1986: 198, 204–05; Beal 1992: 143–44, 147–48, 279–82.

[40] Littauer and Crouwel 1979: 69–71; Moorey 1989; extensive discussion and bibliography in Raulwing 2000, and also 2005; Raulwing and Burmeister 2012; Teufer 2012.

[41] Nagel 1987: 170–175; 1992: 67–81; Jacobs 2005: 434–37; Eder and Nagel 2006; Anthony 2005; 2007: chapters 15–17 (cf. the highly critical review by Izbitser 2012).

[42] Gening *et al.* 1992; Anthony 1995: 561 f.; Anthony and Vinogradov 1995; Epimachov and Korjakova 2004; Kuznetsov 2006; Anthony 2007: chapter 15; Kohl 2007: 151–53; Kupriyanova, Epimakhov, Berseneva and Bersenev 2017.

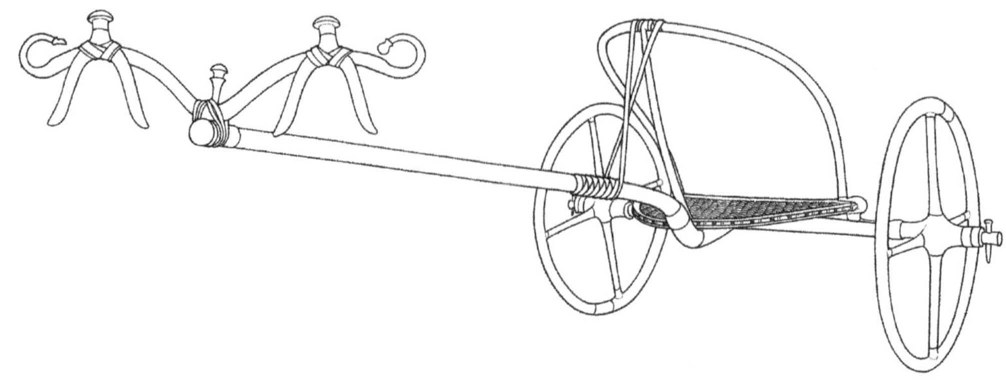

Figure 27. Drawing of the so-called Florence chariot from Thebes, 18th Dynasty (after Littauer and Crouwel 1979: fig. 42).

(for what follows: Littauer and Crouwel 1996b = 2002: 45–52). The steppes have yielded burials of ox–drawn wagons with four disk wheels, going back to the third and late fourth millennium BCE, but no early tradition of fast transport by two–wheeler.[43] No doubt, ridden horses were used for this purpose from early on. It has been calculated that most of the buried two–wheelers from northern Khazakstan had narrow wheel bases of 1.20 m and short naves, of 0.20 m. at most. These dimensions would render the vehicles impractical at speed and limit their manoeuvrability, and hence they cannot yet be called true chariots.[44] Instead, it may be suggested that it was the prestige value of the locally developed Near Eastern two–wheelers that inspired imitations on the steppes.

The true chariot of the Near East was perfected by about the middle of the second millennium BCE, as the culmination of successive developments over a period of nearly 2000 years. Details of construction and dimensions come from New Kingdom Egypt, where the chariot had been introduced from the Levant at some time during the Second Intermediate period (ca. 1650–1550 BCE). Large numbers of Egyptian and Asiatic chariots are illustrated in detail in painting and low relief in tombs and on temple walls, and (all or parts of) eight actual chariots have survived, six of them from the tomb of pharaoh Tutankhamun.[45] The actual chariots are constructed entirely of heat–bent wood and rawhide (Figure 27). Their D–shaped floors (about 1.00 m. wide by 0.50 m. deep) have a flooring of interwoven thongs, which not only help to keep the bent–wood frame in tension, but also provided some springiness. The approximately hip–high siding, in profile shown as rounded or rectangular, extends around the front and sides, leaving the rear open for rapid mounting or dismounting. The siding is solid or fenestrated. Parade and royal chariots might have gilded and decorated sidings.

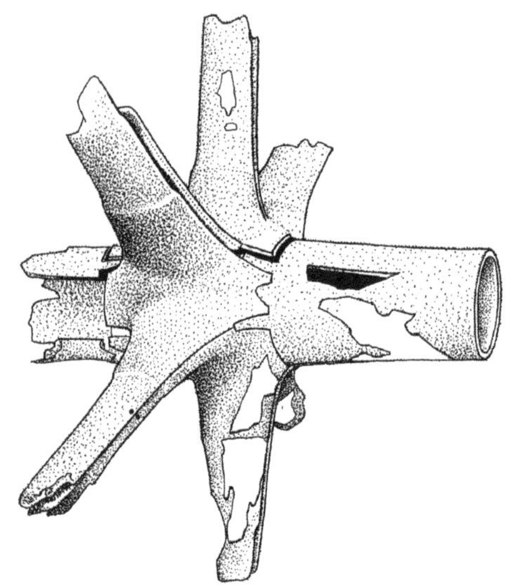

Figure 28. Fragmentary chariot wheel from tomb of Amenophis III, Thebes. Oxford, Ashmolean Museum 1923.663 (after Littauer and Crouwel 1979: fig. 46).

The light, fast vehicle was given stability by a rear axle and a wide wheel base (1.51–1.80 m.). The wheels are 0.90–0.97 m. in diameter and revolve freely on fixed axles, their naves (0.325–0.440 m. in length) elongated to reduce wobbling (Figures 27–28). The interiors of the naves were protected by a leather lining, which would also have reduced the noise produced by wood revolving on wood. Traces of a lubricant were present on the axle of one of the chariots. There are examples of both four– and six–spoked wheels, the latter being more common. All have composite spokes, made from single rods, semi–elliptical in section and heat–bent in the middle to form V's (at an angle of 60 degrees for a six–spoked wheel and of 90 degrees for a four–spoked one). The 'legs' of one V were glued back to back to the 'legs' of the adjacent V's, each whole spoke being fully elliptical in section. The rounded angles of the V's nestle in bays of the nave. The outer ends of the composite spokes are inserted into the felloe which is composed of two lengths of bent wood, overlapping at the ends. Glue and rawhide, applied wet and shrunk as it dried, tightened and consolidated the construction of naves and felloes. In most cases, there is evidence to show that shrunk–on rawhide was also used for tyres, which not only protected the running surface of the wheel, but helped to consolidate its parts.

[43] Izbitser 1993; several contributions to *Rad und Wagen*; Kohl 2007: 84 f., 115, 119 f.

[44] It is unlikely that these vehicles were used on the steppe by warrior–drivers hurling javelins or firing arrows, as has been suggested (Anthony 2007: 399 f., 403, 462; Anthony and Dorcas 2011: 155 f.).

[45] For what follows: Littauer and Crouwel 1979: chapter 8; 1985; Spruytte 1983: 23–52; Sandor 2004a and 2004b; Herold 2004 and 2006; Crouwel 2013.

This construction, based on tension and without metal parts that could be jolted loose, required great skill to make, and was vulnerable to dampness. Elm, tamarisk, and birch bark (which has waterproofing properties) have been identified in the construction of various parts, and particularly the wheels, of the actual chariots found in Egypt. The fact that neither elm nor birch are native to Egypt suggests that the type of chariot and wheel used by the Egyptians originated in the Near East, in parts of which these timbers are indigenous. Indeed, elm wood has been attested on a six–spoked wheel, 0.86 m. in diameter and made according to the same technique of nave–and–spoke construction, from a burnt building dated to the early twelfth century BCE at Lidar Höyük in southeastern Turkey (Figure 29).[46] It may be noted that elm is a hard timber that was also preferred by later wheelwrights, particularly for naves. In the Near East this technique of wheel–making may be traced back to the eighteenth century BCE, if we can judge from the four–spoked metal wheels belonging to a large model or trolley from central Turkey (see above).

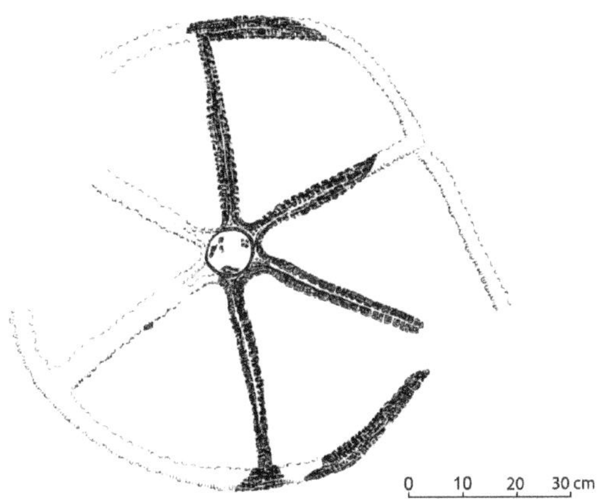

Figure 29. Remains of chariot wheel from Lidar Höyük (after Littauer, Crouwel and Hauptmann 1991: fig. 3).

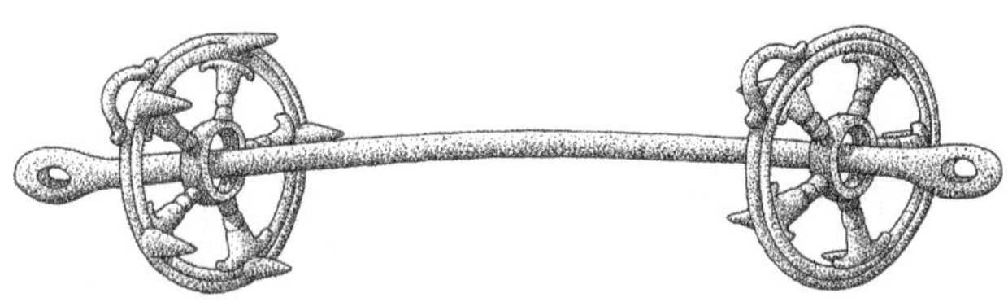

Figure 30. Copper/bronze bridle bit from Tell el Ajjul. Jerusalem, Rockefeller Museum 37.3271 (after Littauer and Crouwel 1979: fig. 48).

Figure 31. Copper/bronze bridle bit from Tell el Amarna. Oxford, Ashmolean Museum 1933.1209 (after Littauer and Crouwel 1979: fig. 49).

The single, long draft pole (2.46–2.60 m.) of the extant chariots from Egypt runs all the way under the floor and helped to support it. This pole, with its end fitted into a socket under the rear floor bar and lashed to the front of the floor frame, rises in front of the chariot in an S–curve to the yoke. A pair of rods, or thongs, drops diagonally from the front railing to the draught pole, restricting the tendency of the floor frame and pole to pull apart in rough going and bracing the front of the chariot. Lashings and a pin secure the top of the pole near its far end to the two–horse yoke. Two thongs (so–called yoke braces) run out from the pole to each arm of the yoke in order to keep the latter at right angles to the pole and to distribute tractive stress. The yoke is adapted to equine conformation by means of a Y–shaped yoke saddle attached to each yoke 'arm'. (As seen above, yoke saddles, one to each animal, are possibly already documented in the earlier second millennium.) The 'legs' of these wooden objects, secured at either end by a crescentic strap across the front of the neck, would have lain along the shoulders of the horse, taking the pull in a manner similar to that of the much later horse collar. There is now evidence for a backing element, i.e. a strap running under the belly which would have prevented the team from breaking out of its harness and transmitted the backward movement to the pole.

The horses used were distinctly small by present–day standards. Their size calculated from the yoke height of extant Egyptian chariots, i.e. ca. 1.35 m. at the withers, agrees quite well with skeletal information from various places in Egypt and the Near East: ranging from 1.34–1.38 to 1.40–1.45 m., thereby falling within the modern classes of 'large pony' or 'small horse.'

The two horses under yoke and—padded—yoke saddles were controlled by a bridle, composed of reins, a bit, and a headstall. The metal bits, with their single–bar or jointed mouthpieces and wheel–shaped or long rectangular cheekpieces, were often studded on their inner faces, exactly as on the Haror example of the earlier second millennium (Figures 30–31).[47] Because the ends of the (often long) solid or jointed mouthpiece were inserted in the cheekpieces, a pull on one rein would press the studs on the opposite side against the horse's lips, thus providing for extra directional control, something particularly welcome in warfare.

Most chariots represented in Egypt and the Near East appear to agree closely with the various extant ones from Egypt (Figures 32–35). The vehicles are usually shown with crews of no more than two men standing side by

[46] Littauer, Crouwel and Hauptmann 1991 = 2002: 314–26.

[47] For the types of bit used: Potratz 1966: 103–16; Littauer and Crouwel 1979: 87–88; 2001.

Figure 32. Egyptian chariot. Detail of stone relief of Ramesses III from Medinet Habu (after Littauer and Crouwel 1979: fig. 44).

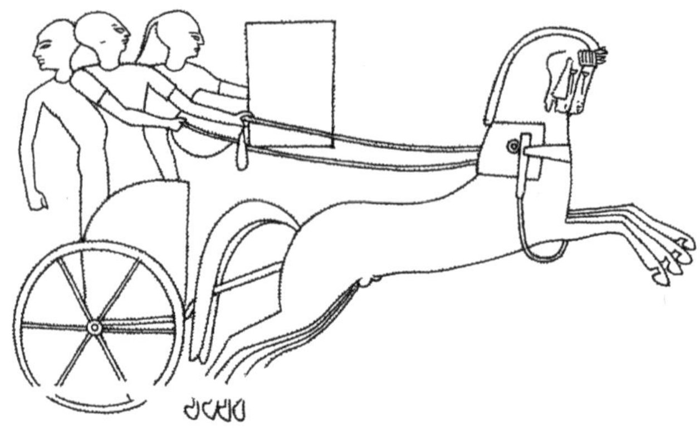

Figure 34. Chariot of the Hittites or their allies. Detail of stone relief of Ramesses II, Abydos (after Littauer and Crouwel 1979: fig. 45).

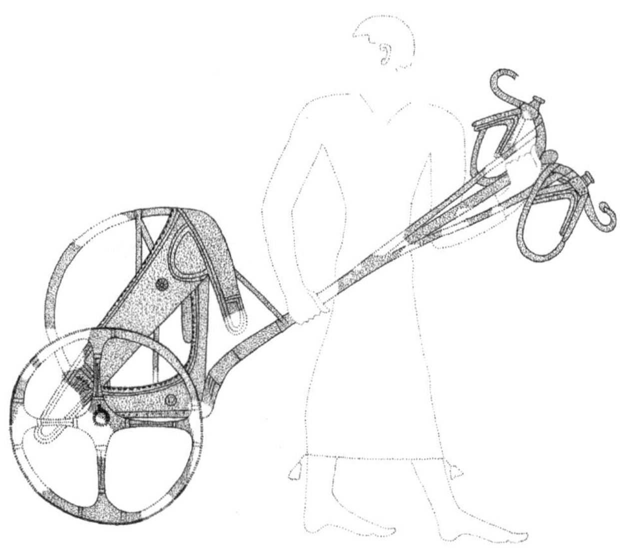

Figure 33. Syrian chariot. Detail of wall painting from tomb of Rekhmire, Thebes (after Littauer and Crouwel 1979: fig. 43).

Figure 35. Assyrian chariot. Detail of stone relief of Ashurnasirpal II, Nimrud (after Littauer and Crouwel 1979: fig. 53).

side. The chariots of the Hittites and their allies depicted on Egptian temple reliefs commemorating the battle at Kadesh in Syria (1275 BCE) are exceptional in having a large enough floor space to permit a third man to stand behind the other two (Figure 34).

From the Near East the use of the chariot had spread to Egypt;[48] and other parts of the ancient world, including Greece.[49] The question then is why was this piece of equipment so widely adopted, and why did it remain in use for so long, in the Near East even until the later first millennium BCE? The answer to this question has mainly to do with the vehicle's military possibilities and its prestige value. In warfare, horse–drawn chariots combined speed with manoeuvrability, provided the terrain of operation was level and open. At the same time, both the chariot and their teams of horses could be lavishly decorated, thus catering to the desire of conspicuous display of kings and an elite class. They could also be used in hunting (with beaters) and for ceremonial purposes. On the other hand, chariots were far too light and small for goods transport, which was served by boats, pack animals or ox– or mule–drawn carts or wagons.

Chariots were costly to make and maintain, and the draught teams had to be especially trained and to be matched in height.[50] In warfare at this time, chariotry and infantry were both important, though chariotry alone was probably not always the deciding factor.[51] There is ample figured and textual documentation from the Near East and Egypt

[48] Shaw 2001; Herold 2004: 123–26; Raulwing and Clutton–Brock 2009: 59–78.

[49] Crouwel 1991; 2004b; 2005. For the opposing, long–held view that chariots were introduced in Greece from the north by Indo–European warriors: a.o. Nagel 1987: 174–77; Messerschmidt 1988; Lichardus and Vladár 1996; Penner 1998; Eder and Nagel 2006: 44f; cf. Kristiansen 2004: 448–52, and 2005. R. Drews (1988: chapter 8) argued for Indo–European conquerors coming from eastern Anatolia.

[50] Piggott 1992: 45–48. Cf. the famous Kikkuli treatise from the Hittite capital Boğazköy for training chariot horses: Raulwing and Meyer 2004; Raulwing 2005.

[51] Yadin 1963: 4 f., 86–90; Kendall 1975; Littauer and Crouwel 1979: 90–95; Schulman 1979; Beal 1992: 141–84; Richter 2004; Vita 2008; Feldman and Sauvage 2010; Crouwel 2013, 85–88.

that—under the right circumstances—chariots, often in large numbers, performed a variety of military functions, primarily as a mobile firing platforms for archers standing beside their drivers and to play a fast flanking and pursuing role, and with bowcases and quivers of arrows or javelins attached outside the chariot body providing reserve arms (Figures 32–33). The three–man chariots of the Hittites and their allies at Kadesh, depicted in Egyptian temple reliefs, carry a driver, shield bearer, and spearman (quivers are absent), and must have served essentially as a means of transport (Figure 34).

The chariots, however, were not only limited in their field of operation, but, despite the use of armour on their bodies, their harness teams, as well as their crews, remained extremely vulnerable.[52] The wounding of one horse, or its laming from an obstacle–strewn battlefield, could bring the entire equipage to a standstill, thereby eliminating most of its effectiveness and making the elevated crew an easy target. Like the third millennium four–wheeled 'battle cars' and other ancient military vehicles, they were not launched in head–on charges against well–prepared enemy lines.[53] The only exception are the scythed chariots of the later first millennium BCE, but even these were strictly limited to level and open terrain for their field of operation.

Chariots survived the upheavals that struck most of the Near East in the late second millennium. When they are illustrated again, on Assyrian reliefs of the ninth century BCE, they are still lightly built, with rear axles, six–spoked wheels, and the pole–and–yoke hitch (Figure 35). Of the three or four draft horses, only two are under yoke, the others outriggers. In military settings, the chariots still function as mobile firing platforms for archers. Their use in prepared hunts and as prestige means of transport for kings and their entourage also continues (Littauer and Crouwel 1979, chapter 9).

This paper concludes with a few words on experiments that were conducted in the winter of 2002 with full–scale reconstructions of ninth–century Assyrian military chariots. The trials, in which I was involved, took place near Harran in south–eastern Turkey, as part of the preparations for a British television documentary (in the series 'Inventing the Past: Machines Time Forgot', 2003). The experiments suggested that Assyrian and other chariots did not, as had previously been thought, run along the enemy front line to soften it up with volleys of arrows.[54] Rather, columns of chariots would have approached the enemy line, at full speed and, at different places, simultaneously, firing, before making a sharp turn and getting away quickly enough so as not to offer easy targets to enemy archers. This manoeuvre would be repeated again and again, until the enemy was sufficiently softened up for the infantry to come up and finish him off (Crouwel 2013: 87 f.).[55]

There is, however, a major problem with these tactics: how to visualise both armies launching their chariotry at the same time in battle? Would there not be chaos and loss of many precious chariots, draught teams and crews?

Acknowledgements

Much of this text goes back to a chapter—in German—in Rad und Wagen (Crouwel 2004), which is likewise largely based on my many years of collaboration with Mary Littauer. Most of the illustrations are by Dr J. Morel. For glossaries of technical terms to do with vehicles and the harnessing and control of their draught animals see Littauer and Crouwel 1979: chapter 2, and 2000, xv–xx. The absolute chronology used in this chapter is taken from van de Mieroop 2007. I wish to thank Th.J.K. Krispijn S. Burmeister and P. Raulwing for contributing the study following this paper, which has been specifically written after our discussion on the tablets from Uruk, and for their expertise and help, G. Skyte–Bradshaw for checking the English text, and A. Dekker and J. Eerbeek for assistance with the preparation of text and illustrations.

Abbreviation

Rad und Wagen *Rad und Wagen. Der Ursprung einer Innovation. Wagen im Vorderen Orient und Europa*, edited by M. Fansa and S. Burmeister. Beiheft der Archäologische Mitteilungen aus Nordwestdeutschland 40. Von Zabern, Mainz am Rhein 2004.

Bibliography

Anthony, D. W. 1995. Horse, wagon and chariot: Indo–European languages and archaeology. *Antiquity* 69, 554–565.

Anthony, D. W. 2007. *The Horse, the Wheel, and Language. How Bronze–Age Riders from the Eurasian Steppes Shaped the Modern World.* Princeton University Press, Princeton.

Anthony, D. W. and D. Brown. 2011. The secondary products revolution, horse–riding and mounted warfare. *Journal of World Prehistory* 24, 131–160.

Anthony, D. W. and N. B. Vinogradov. 1995. Birth of the chariot. *Archaeology* 48, 36–41.

Archer, R. 2010. Developments in Near–Eastern chariotry and chariot warfare in the early first millennium BCE and their contribution to the rise of cavalry. In *New Perspectives on Ancient Warfare*, edited by G.G. Fagan and M. Trundle, 57–79. Brill, Leiden and Boston.

Bakker, J. A. 2004. Die neolithische Wagen im nördlichen Mitteleuropa. In *Rad und Wagen*, 283–294.

Bakker, J. A., J. Kruk, A. E. Lanting and S. Milisauskas. 1999. The earliest evidence of wheeled vehicles in

[52] Littauer and Crouwel 1979: 92; 1996a: 298–99 = 2002: 68–69; Schulman 1979: 114–44.

[53] Drews 1988, chapters 5 and 7; 1993: 10–14, assumed just such a role of chariots in the armies of the Near East, Egypt and Greece in the second half of the second millennium BCE, eventually to be superseded by that of infantry; but cf. Littauer and Crouwel 1996a = 2002: 66–74; Archer 2010: 63–66.

[54] Littauer and Crouwel 1979: 128–30; Spruytte 1993.

[55] As was forcibly argued by Mike Loades who drove the chariots and fired arrows from them in the television documentary.

Europe and the Near East. *Antiquity* 73, 778–790 (reprinted in Drenth and Bakker 2005–06, 10–28).

Bar–Oz, G., P. Nahshoni, H. Motro and E. D. Oren. 2013. Symbolic metal bit and saddlebag fastenings in a Middle Bronze Age donkey burial. *Plos One* 8/3 (March 2013), 1–7.

Beal, R. H. 1992. *The Organisation of the Hittite Military*. Texte der Hethiter 20. Winter, Heidelberg.

Becker, C. 1994. Zur Problematik früher Pferdenachweise im östlichen Mittelmeergebiet. In *Die Indogermanen und das Pferd. Bernfried Schlerath zum 70. Geburtstag gewidmet*, edited by B. Hänsel and S. Zimmer, 145–183. Archaeolingua 4. Archaeolingua Alapítvány, Budapest.

Benecke, N. 2004. Die Domestikation der Zugtiere. In *Rad und Wagen*, 455–466.

Bietak, M. and I. Förstner–Müller. 2009. Der Hyksos–Palast bei Tell el–Dab'a. Zweite und dritte Grabungskampagne (Frühling 2008 und Frühling 2009). *Ägypten und Levante* 19, 91–119.

Bollweg, J. 1999. *Vorderasiatische Wagentypen im Spiegel der Terrakottaplastik bis zur altbabylonischen Zeit*. Orbis Biblicus et Orientalis 167. Universitäts–Verlag, Freiburg (Schweiz) and Vandenhoeck & Ruprecht, Göttingen.

Bollweg, J. and W. Nagel. 1992. Equiden Vorderasiens in sumerisch–akkadischen Schriftquellen und Ausgrabungen. *Acta Praehistorica et Archaeologica* 24, 17–63.

Brownrigg, G. [with the cooperation of U. Dietz]. 2004. Schirrung und Zäumung des Streitwagenpferdes: Funktion und Rekonstruktion. In *Rad und Wagen*, 481–490.

Bulliet, R. 2016. *The Wheel. Inventions and Reinventions*. New York: Columbia University Press.

Burmeister, S. 2004. Der Wagen im Neolithikum und in der Bronzezeit: Erfindung, Ausbreitung und Funktion der ersten Fahrzeuge. In *Rad und Wagen*, 13–40.

Burmeister, S. 2011. Innovationswege—Wege der Kommunikation. Erkenntnisprobleme am Beispiel des Wagens im 4. Jt. v. Chr. In *Sozialarchäologische Perspektiven. Gesellschaftlicher Wandel 5000–1500 v. Chr. zwischen Atlantik und Kaukasus*. Internationale Tagung (Kiel 2007), edited by S. Hansen and J. Müller, 211–240. von Zabern, Darmstadt.

Burmeister, S. 2017. Early Wagons in Eurasia: Disentangling an Enigmatic Innovation. In *Appropriating Innovations Entangled Knowledge in Eurasia, 5000–1500 BCE*, edited by Philipp W. Stockhammer and Joseph Maran, 69–77. Oxford and Philadelphia: Oxbow Books.

Burmeister, S. and P. Raulwing. 2012. Festgefahren. Die Kontroverse um den Ursprung des Streitwagens. Einige Anmerkungen zu Forschung, Quellen und Methodik. In *Archeological, Cultural and Linguistic Heritage. Festschrift fur Erzsébet Jerem in Honour of her 70th Birthday*, edited by P. Anreiter, E. Bánffy, L. Bartosiewicz, W. Meid and C. Metzger–Nebelsick, 93–113. Archaeolingua Alapítvány, Budapest.

Civil, M. 1968. Išme–Dagan and Enlil's chariot. *Journal of the American Oriental Society* 88, 3–13.

Clutton–Brock, J. 1974. The Buhen horse. *Journal of Archaeological Science* 1, 89–100.

Clutton–Brock, J. 2001. Bit wear and evidence for the stabling and the use of donkeys 1–5. In *Excavations at Tell Brak 2. Nagar in the Third Millennium BC*, edited by D. Oates, J. Oates and H. McDonald, 336–338. McDonald Institute for Archaeological Research, Cambridge, and Brith School of Archaeology in Iraq, London.

Clutton–Brock, J. 2003. Were the donkeys at Tell Brak (Syria) harnessed with a bit? In *Prehistoric Steppe Adaptation and the Horse*, edited by M. Levine, C. Renfrew and K. Boyle, 126–127. McDonald Institute Monographs. McDonald Institute for Archaeological Research, Cambridge.

Clutton–Brock, J. and S. Davies. 1993. More donkeys from Tell Brak. *Iraq* 55, 209–221.

Crouwel, J. H. 1991. *Chariots and Other Means of Land Transport in Bronze Age Greece*. Allard Pierson Series 3. Allard Pierson Museum, Amsterdam.

Crouwel, J. H. 2004a. Der alte Orient und seine Rolle in der Entwicklung von Fahrzeugen. In *Rad und Wagen*, 69–86.

Crouwel, J. H. 2004b. Bronzezeitliche Wagen in Griechenland. In *Rad und Wagen*, 341–346.

Crouwel, J. H. 2005. Early chariots in the Aegean and their eastern connections. In *Emporia. Aegeans in the Central and Eastern Mediterranean. Proceedings of the 10th International Aegean Conference (Athens, Italian School of Archaeology, 14–18 April 2004)* vol. I, edited by R. Laffineur and E. Greco, 39–44. *Aegaeum* 25. Liège, Université de Liège Histoire de l'art et archéologie de la Grèce antique and Austin, University of Texas Program in Aegean Scripts and Prehistory.

Crouwel, J. H. 2012. Metal wheel tyres from the ancient Near East and Central Asia. *Iraq* 74, 89–95.

Crouwel, J. H. 2013. Studying the six chariots from the tomb of Tutankhamun. In *Chasing Chariots*. Proceedings of the First International Chariot Conference (Cairo 2012), edited by A.J. Veldmeijer and S. Ikram, 73–93. Sidestone Press, Leiden.

Dohmann–Pfäfzer, H. 2000. Ausgrabungen der Deutschen–Orient–Gesellschaft in der zentralen Oberstadt von Tall Mozan/Urkeš. Bericht über die in Kooperation mit dem IIMAS durchgeführte Kampagne 1999. *Mitteilungen der Deutschen Orient–Gesellschaft* 132, 185–228.

Drenth, E. and J. A. Bakker. 2005–06. In memoriam Albert Lanting (10 April 1941—13 December 2004). *Palaeohistoria* 47–48, 3–37.

Drews, R. 1988. *The Coming of the Greeks. Indo–European Conquests in the Aegean and the Near East.* Princeton University Press, Princeton.

Drews, R. 1993. *The End of the Bronze Age. Changes in Warfare and the Catastrophe ca. 1200 B.C.* Princeton University Press, Princeton.

Eder, C. and W. Nagel 2006. Grundzüge der Streitwagenbewegung zwischen Tiefeurasien, Südwestasien und Ägäis. *Altorientalische Forschungen* 33, 42–93.

Englund, R. K. 1994. *Archaic Administrative Texts from Uruk. The Early Campaigns. Ausgrabungen der Deutschen Forschungsgemeinschaft in Uruk–Warka* 15. *Archaische Texte aus Uruk* 5. Gebr. Mann, Berlin.

Englund, R. K. 1998. Texts from the Late Uruk period. In *Mesopotamien. Späturuk–Zeit und Frühdynastische Zeit. Annäherungen* 1, edited by P. Attinger and M. Wäfler, 15–233. Orbus Biblicus et Orientalis 160/1. Freiburg and Göttingen.

Epimachov, A. and L. Korjakova. 2004. Streitwagen der eurasischen Steppe in der Bronzezeit: das Wolga–Uralgebiet und Kasachstan. In *Rad und Wagen*, 221–336.

Falkenstein, A. 1936. *Archaische Texte aus Uruk.*: Deutsche Forschungsgemeinschaft, Berlin und Harrassowitz, Leipzig.

Feldman, M. H. and C. Sauvage. 2010. Objects of prestige? Chariots in the Late Bronze Age eastern Mediterranean and Near East. Ägypten und Levante 20, 67–181.

Frangipane, M. 1997. A 4th–millennium temple/palace complex at Arslantape/Malatya. North–south relations and the formation of early state societies in the northern regions of Greater Mesopotamia, *Paléorient* 23/1, 45–73.

Frangipane, M. 2004. Dal tempio al palazzo. Espansione del potere. Una stato prima della scritture. In *Alle origini del potere. Arslantepe, la collina dei leoni*, edited by M. Frangipane, 46–75. Electra, Milano.

Frangipane, M. and A. Palmieri. 1983. A proto–urban centre of the Late Uruk period. *Origini* 2, 287–574.

Gening, V. F., G. B. Zsanovič and V. V. Gening. 1992. *Sintašta.* Archeologičeskie pamjatniki arijskich plemen Uralo–Kazachstanskich stepej. [Archeological Sites of Aryan Tribes of the Ural–Kazakh Steppes; Russian]. Južno–Ural'skoe Knižnoe Izdatel'stvo, Čeljabinsk.

Green, M. 1986. Urum and Uqair. *Acta Sumerologica* 8, 77–83.

Green, M. W. and H. J. Nissen. 1987. *Zeichenliste der Archaischen Texte aus Uruk.* Archaische Texte aus Uruk 2; Ausgrabungen der Deutschen Forschungsgemeinschaft in Uruk–Warka 11. Gebr. Mann, Berlin.

Herold, A. 2004. Funde und Funktionen—Streitwagentechnologie im Alten Ägypten. In *Rad und Wagen*, 123–142.

Herold, A. 2006. Auf Biegen und Brechen... Zum Nauchbau eines altägyptischen Streitwagens, *Achse, Rad und Wagen. Beiträge zur Geschichte der Landfahrzeuge* 14, 5–19.

Holland, T. A. 1993–94. Tall as–Swehat 1989–1992. *Archiv für Orientforschung* 40/41, 275–285.

Houwink ten Cate, Ph. H. J. 1984. The history of warfare according to Hittite sources: the Annals of Hattusilis I (Part II). *Anatolica* 11, 47–83.

Izbitser, E. 1993. *Progrebenija povozkami stepnoj polosy Vostočnoj Evropy i Severnogo Kavzaka. III–I tyc. do. n.é* [Vehicle Burials in the East–European Steppe Zone and the North Caucasus. III–II Millennium B.C.; Russian]. PhD Thesis University of St. Petersburg.

Izbitser, E. 2012. The Iron Curtain and Eurasian Archaeology. *Tyragetia. The Journal of the National Museum of History and Archaeology of Moldova*, serie nouă 6 (XXI), nr. 1, Arheologie. Istorie Antică: 65–73. http://www.nationalmuseum.md/en/press_releases/journal_tyragetia/the_iron_curtain_and_eurasian_archaeology/. (Review article of Anthony 2007).

Jacobs, B. 2005. Review of *Rad und Wagen.* In *Orientalistische Literaturzeitung* 100, 429–438.

Jans, G. 1993. Lion hunt by wagon on a Syrian seal impression from the third millennium. *Ugarit–Forschungen* 25, 181–188.

Jones–Bley, K. 2000. The Sintashta "chariots". In *Kurgans, Ritual Sites and Settlements. Eurasian Bronze and Iron Age*, edited by J. Dans–Kimball et al., 135–140. BAR International Series 890. BAR Publishing, Oxford.

Kendall, T. B. 1975. *Warfare and Military Matters in the Nuzi Tablets* (PhD Thesis Brandeis University 1975) University Microfilms International, Ann Arbor/MI.

Kohl, P. L. 2007. *The Making of Bronze Age Eurasia.* Cambridge World Archaeology. Cambridge University Press, Cambridge.

Kristiansen, K. 2004. Kontakte und Reisen im 2. Jahrtausends v. Chr. In *Rad und Wagen*, 443–454.

Kristiansen, K. 2005. What language did Neolithic pots speak? Colin Renfrew's European farming–language–dispersal model challenged. *Antiquity* 79, 679–691 [See also 692–695, Renfrew's critique and Kristiansen's reply].

Kulakoğlu, F. 2003. Recently discovered bronze wagon models in Şanlıurfa, Southeastern Anatolia, *Anadolu / Anatolia* 24: 63-77.

Kupriyanova, E., A. Epimakhov, N. Berseneva and A. Bersenev. 2017. Bronze Age charioteers of the Eurasian steppe: a part-time occupation for select Man? *Praehistorische Zeitschrift* 92, 40–65.

Kuznetsov, P. F. 2006. The emergence of Bronze Age chariots in eastern Europe. *Antiquity* 80, 638–645.

Lichardus, J. and J. Vladár. 1996. Karpathenbecken—Sintašta—Mykene. Ein Beitrag zur Definition der Bronzezeit als historische Epoche. *Slovenská Archeológia* 44, 25–93.

Littauer, M. A. 1968. The function of the yoke saddle in ancient harnessing. *Antiquity* 42, 27–31. [Reprinted in Littauer and Crouwel 2002, 479–486].

Littauer, M. A. 1969. Bits and pieces. *Antiquity* 43, 289–300 [Reprinted in Littauer and Crouwel 2002, 487–504].

Littauer, M. A. and J. H. Crouwel. 1973. Early metal models of wagons from the Levant. *Levant* 5, 102–126. [Reprinted in Littauer and Crouwel 2002, 336–379].

Littauer, M. A. and J. H. Crouwel. 1974. Terracotta models as evidence for vehicles with tilts in the ancient Near East. *Proceedings of the Prehistoric Society* 40, 20–36. [Reprinted in Littauer and Crouwel 2002, 380–402].

Littauer, M. A. and J. H. Crouwel. 1977. The origin and development of the cross–bar wheel? *Antiquity* 51, 95–105. [Reprinted in Littauer and Crouwel 2002, 272–288].

Littauer, M. A. and J. H. Crouwel. 1979. *Wheeled Vehicles and Ridden Animals in the Ancient Near East.* Handbuch der Orientalistik Siebente Abteilung. Kunst und Archäologie. Erster Band. *Der Alte Vordere Orient.* Zweiter Abschnitt. *Die Denkmäler*. Brill, Leiden and Köln.

Littauer, M. A. and J. H. Crouwel. 1985. *Chariots and Related Equipment from the Tomb of Tut'ankhamūn.* Tut'ankhamūn Tomb Series 8. Griffith Institute, Oxford.

Littauer, M. A. and J. H. Crouwel. 1986. The earliest known three–dimensional evidence for spoked wheels. *American Journal of Archaeology* 90, 395–398 [Reprinted in Littauer and Crouwel 2002, 289–295].

Littauer, M. A. and J. H. Crouwel. 1989. Metal wheel tyres from the ancient Near East. In *Archaeologia Iranica et Orientalis. Miscellanea in Honorem Louis Vanden Berghe*, vol. I, edited by L. de Meyer and E. Haerinck, 111–121. Peeters, Gent [Reprinted in Littauer and Crouwel 2002, 261–271].

Littauer, M. A. and J. H. Crouwel. 1990. Ceremonial threshing in the ancient Near East I. Archaeological evidence. *Iraq* 52, 15–19 (reprinted in Littauer and Crouwel 2002, 325–335).

Littauer, M. A. and J. H. Crouwel. 1996a. Robert Drews and the role of chariotry in Bronze Age Greece. *Oxford Journal of Archaeology* 15, 297–305. [Reprinted in Littauer and Crouwel 2002, 66–74].

Littauer, M. A. and J. H. Crouwel. 1996b. The origin of the true chariot. *Antiquity* 70, 934–939. [Reprinted in Littauer and Crouwel 2002, 45–52].

Littauer, M. A. and J. H. Crouwel. 2001. The earliest evidence for metal bridle bits. *Oxford Journal of Archaeology* 20, 229–338.

Littauer, M. A. and J. H. Crouwel. 2002. *Selected Writings on Chariots and other Early Vehicles, Riding and Harness*. Culture and History of the Ancient Near East 6, edited by P. Raulwing. Brill, Leiden and Köln.

Littauer, M. A., J. H. Crouwel and H. Hauptmann 1991. A Late Bronze Age spoked wheel from Lidar Höyük in southeast Turkey. *Archäologischer Anzeiger* 1991, 348–358. [Translated into English in Littauer and Crouwel 2002, 314–326].

Maran, J. 2004a. Die Badener Kultur und ihre Radfahrzeuge. In *Rad und Wagen*, 265–282.

Maran, J. 2004b. Kulturkontakte und Wege der Ausbreitung der Wagentechnologie im 4. Jahrtausend v.Chr. In *Rad und Wagen*, 429–442.

Martin, H. P. and R. J. Matthews. 1993. Seals and sealings. In *Abu Salabikh Excavations* 4. *The 6G Ash-tip and its Contents. Cultic and Administrative Discard from the Temple?*, edited by A. Green, 23–81. British School of Archaeology in Iraq, London.

Matthews, D. 1991. Tell Brak 1990, the glyptic. *Iraq* 53, 147–157.

Messerschmidt, W. 1988. Der ägäische Streitwagen und seine Beziehungen zum nordeurasisch–vorderasiatischen Raum. *Acta Archaeologica et Praehistorica* 20, 31–44.

Messerschmidt, W. 1992. Ein Kanzelbockwagen auf dem Geierstele—Überlegungen zur Typologie frühdynastischer Fahrzeuge. *Acta Praehistorica et Archaeologica* 24, 77–83.

Meyer, J. W., M. Novák and A. Pruss (eds.). 2001. *Beiträge zur vorderasiatischen Archäologie. Winfried Orthmann gewidmet.* Johann–Wolfgang–Goethe–Universität, Archäologisches Institut, Frankfurt/Main.

Moorey, P. R. S. 1986. The emergence of the light, horse–drawn chariot in the Near East, c. 2000–1500 B.C. *World Archaeology* 18, 196–215.

Moorey, P. R. S. 1989. The Hurrians, the Mitanni and technological innovation. In *Archaeologica Iranica et Orientalis. Miscellanea in Honorem Louis Vanden Berghe*, vol. I, edited by L. de Meyer and E. Haerinck, 273–286. Peeters, Gent.

Moorey, P. R. S. 1994. *Ancient Mesopotamian Materials and Industries. The Archaeological Evidence.* Clarendon Press, Oxford.

Moorey, P. R. S. 2001. Clay models and overland mobility in Syria c. 2350–1800 B.C. In Meyer, Novák and Pruss 2001, 344–351.

Nagel, W. 1966. *Die mesopotamische Streitwagen und seine Entwicklung im ostmediterranen Bereich.* Berliner Beiträge zur Vor– und Frühgeschichte 1. Hessling, Berlin.

Nagel, W. 1984–85. Zwei Kupfermodelle eines Kultwagens mit Rinderzweigespann vom zweiachsigen Gatterkanzeltyp aus der Alaçahüyük–Kultur im Museum für Vor– und Frühgeschichte Berlin. *Acta Praehistorica et Archaeologica* 16/17, 143–151.

Nagel, W. 1986. Die Entwicklung des Wagens im frühen Vorderasien. In *Achse, Rad und Wagen. Fünftausend Jahre Kultur– und Technikgeschichte* (2nd revised edition), edited by W. Treue, 7–34. Vandenhoeck & Ruprecht, Göttingen. [^{1}1965 Bruckmann, München].

Nagel, W. 1987. Indogermanen und Alter Orient: Rückblick und Ausblick auf den Stand des Indogermanenproblems. *Mitteilungen des Deutschen Orient–Gesellschaft* 119, 157–213.

Nagel, W. 1992a. Die Wagentypen des zentral–vorderasiatischen Frühdynastikums und der Bremsbügel. *Archäologische Mitteilungen aus Nordwestdeutschland* 15, 167–178.

Nagel, W. 1992b. Das Aufkommen des klassischen Streitwagens *wr̥ta– in Syrien und Ägypten. *Damaszener Mitteilungen* 6: 67–81.

Nagel, W., J. Bollweg and E. Strommenger. 1999. Der 'Onager' in der Antike und die Herkunft des Hausesels. *Altorientalische Forschungen* 26, 154–202.

Nagel, W. and E. Strommenger. 1994. Der frühsumerische Kultschlitten—ein Vorläufer des Wagens? In *Beiträge zur Altorientalischen Archäologie und Altertumskunde. Festschrift für Barthel Hrouda zum 65. Geburtstag*, edited by P. Calmeyer, et al., 202–209. Harrassowitz, Wiesbaden.

Nagel, W. and E. Strommenger. 2001 Drei Kultwagen aus der Gudea– oder der Ur III–Zeit. In Meyer, Novák and Pruss 2001, 352–265.

Oates, J. 2003. A note on the early evidence for the horse in Western Asia. In *Prehistoric Steppe Adaptation and the Horse*, edited by M. Levine, C. Renfrew and K. Boyle, 115–125. McDonald Institute Monographs. McDonald Institute for Archaeological Research, Cambridge.

Olsen, S. L. and C. Culbertson. 2010. *A Gift from the Desert. The Art and Culture of the Arabian Horse* (Catalogue accompanying an exhibition held 29 May—15 October 2010 at the International Museum of the Horse at the Kentucky Horse Park, Lexington, Kentucky, USA), Lexington/KY.

Özgen, E. 1986. A group of terracotta wagon models from Southeastern Anatolia. *Anatolian Studies* 36: 165-171.

Payne, S. 1995. The equids from Tumulus KY. In *The Gordion Excavations (1950–1973). Final Reports,* vol. II. *The Lesser Phrygian Tumuli*, vol. I. *The Inhumations*, edited by E.L. Kohler, 237–244. University Museum Monograph 88. University Museum, University of Pennsylvania, Philadelphia.

Penner, S. 1998. *Schliemanns Schachtgräberrund und der europaïsche Nordosten. Studien zur Herkunft der frühmykenischen Streitwagenausstattung.* Saarbrücker Beiträge zur Altertumskunde 60. Habelt, Bonn.

Piggott, S. 1983. *The Earliest Wheeled Transport from the Atlantic Coast to the Caspian Sea.* Thames & Hudson, London.

Piggott, S. 1992. *Wagon, Chariot and Carriage. Symbol and Status in the History of Transport.* Thames & Hudson, London.

Postgate, N. 1986. The equids of Sumer, again. In *Equids in the Ancient World*, vol. I, edited by R. H. Meadow and H. P. Uerpmann, 196–204. Tübinger Atlas des Vorderen Orients, 19/1 Reichert, Wiesbaden.

Postgate, N. 1992. *Early Mesopotamia. Society and Economy at the Dawn of History.* Routlegde, London [etc.].

Potratz, J. A. H. 1966. *Die Pferdetrensen des Alten Orient.* Analecta Orientalia 41. Pontificium Institutum Biblicum, Roma.

Potts, D. T. 2001. Ceremonial threshing or mobile sherpherds' huts? In Meyer, Novák and Pruss 2001, 410–411.

Raulwing, P. 1993. Pferd und Wagen im Alten Ägypten Forschungsstand. Beziehungen zu Vorderasien und methodenkritische Aspekte. *Göttinger Miszellen* 136, 71–83.

Raulwing, P. 2000. *Horses, Chariots and Indo–Europeans. Foundations and Methods of Chariotry Research from the Viewpoint of Comparative Indo–European Linguistics.* Archaeolingua. Series minor 13. Archaeolingua Alapítvány, Budapest.

Raulwing, P. 2004. Indogermanen, Indoarier und Maryannu in der Streitwagenforschung. Eine rezeptions– und wissenschaftgeschichtliche Spurenlese. In *Rad und Wagen*, 515–531.

Raulwing, P. 2005. The Kikkuli–Text. Some interdisciplinary remarks on Hittite training texts for chariot horses in the second half of the 2nd millennium B.C. In *Les équidés dans le monde méditerranéen antique. Actes du colloque organisé par l'Ecole française d'Athènes, le Centre Camille Jullian, et l'UMR 5140 du CNRS, Athènes, 26–28 novembre 2003,* edited by A. Gardeisen, 61–75. Monographies d'archéologie méditerranéenne. Édition de l'Association pour le développement de l'archéologie en Languedoc–Roussillon Lattes.

Raulwing, P. and Clutton–Brock, J. 2009. The Buhen horse: fifty years after its discovery (1958–2008). *Journal of Egyptian History* 2, 1–106.

Raulwing, P. and H. Meyer. 2004. Hippologische und methodenkritische Überlegungen zum Trainung von Streitwagenpferden im Alten Orient. In *Rad und Wagen*, 491–506.

Richter, T. 2004. Der Streitwagen im Alten Orient im 2. Jahrtausend v.Chr.—eine Betrachtung anhand der keilschriftlichen Quellen. In *Rad und Wagen*, 507–514.

Salonen, A. 1951. *Die Landfahrzeuge des alten Mesopotamien* (mit besonderer Berücksichtigung der 5. Tafel der Serie ḪAR–ra = *ḫubullu*). Eine lexikalische und kulturgeschichtliche Untersuchung. Suomalaisen Tiedeakatemian toimituksia, Sarja B 72,3. Suomalainen Tiedeakatemia, Helsinki.

Sandor, B. I. 2004a. The rise and decline of the Tutankhamun–class chariot. *Oxford Journal of Archaeology* 23, 153–175.

Sandor, B. I. 2004b. Tutankhamun's chariots. Secret treasures of engineering mechanics. *Fatigue and Friction of Engineering Materials and Structures* 27, 637–646.

Sarianidi, V. I. 2006. Goňurdepe: Şalaryň we hudaýlaryň şäheri [*Gonurdepe Türkmenistan. City of Kings and Gods*; in Russian, Turkish and English]. Miras, Ašgabad.

Schlichtherle, H. 2004. Wagenfunde aus den Seeufersiedlungen im zirkumalpinen Raum. In *Rad und Wagen*, 295–314.

Schulman, A. 1979. Chariots, chariotry, and the Hyksos. *Journal of the Society for the Study of Egyptian Antiquities* 10/2, 105–1533.

Shaw, I. 2001. Egyptians, Hyksos and military technology: causes, effects or catalysts? In *The Social Context of Technological Change. Egypt and the Near East, 1650–1550 B.C.*, edited by A. J. Shortland, 59–71. Oxbow Books, Oxford.

Sherratt, A. 2003. The horse and the wheel: the directions of change in the circum–Pontic region and adjacent areas, 4500–1500 B.C. In *Prehistoric Steppe Adaption and the Horse*, edited by M. Levine, C. Renfrew and K. Boyle, 233–252. McDonald Institute Monographs. McDonald Institute for Archaeological Research, Cambridge.

Sherratt, A. 2004. Wagen, Pflug, Rind: ihre Ausbreitung und Nutzung—Probleme der Quelleninterpretation. In *Rad und Wagen*, 409–428.

Spruytte, J. 1983. *Early Harness Systems. Experimental Studies. A Contribution to the History of the Horse.* [Translated by M. A. Littauer]. Allen, London.

Spruytte, J. 1993. Le trige de guerre assyrien au IXe siècle avant J.C. *Préhistoire anthropologie méditerranéenne* 2, 193–200.

Steinkeller, P. 1990. Ceremonial threshing in ancient Near East [Part] II. Threshing implements in ancient Mesopotamia: cuneiform sources. *Iraq* 52: 19–23.

Strommenger, E. 1990. Planwagen aus dem mittleren Euphrattal. In *Resurrecting the Past. A Joint Tribute to Adnan Bounni*, edited by P. Matthiae, M. van Loon and H. Weiss, 297–306. Uitgaven van het Nederlands Historisch–Archaeologisch Instituut te Istanbul 67, Nederlands Historisch–Archaeologisch Instituut, Istanbul.

Strommenger, E. and J. Bollweg. 1996. Onager und Esel im alten Zentralvorderasien. In *Collectanea Orientalia. Histoire, arts de l'espace et industrie de la terre. Études offertes en hommage à Agnès Spycket*, edited by H. Gasche and B. Hrouda, 349–366. Civilisation du Proche–Orient Serie 1, Archeologie et environnement 3. Recherches et publication, Neuchâtel and Paris.

Teufer, M. 1999. Ein Scheibenknebel aus Džarkutan (Süduzbekistan). *Archäologische Mitteilungen aus Iran* 31, 69–143.

Teufer, M. 2012. Der Streitwagen: eine "indo–iranische" Erfindung? Zum Problem der Verbindung von Sprachwissenschaft und Archäologie. *Archäologische Mitteilungen aus Iran und Turan* 44, 271–312.

Vila, E. 2006. Data on equids from late fourth and third millennium sites in northern Syria. In *Equids in Time and Space. Papers in Honour of Véra Eisenmann.* Ninth International Council of Archaeozoology, Durham 2002, edited by M. Mashkour, 101–123. Oxbow, Oxford.

Vita, J. P. 2008. Le char de guerre en Syrie et Palestine au Bronze récent. In *Les armées du Proche–Orient ancien (IIIe–Ier mill. av. J.–C.)*, edited by P. Abrahami and L. Battini, 51–70. BAR International Series 1855. BAR Publishing, Oxford.

von den Driesch, A. 1993. "Hausesel contra Hausonager". Eine kritische Bemerkung zu einer Untersuchung von J. Bollweg und W. Nagel über die Equiden Vorderasiens. *Zeitschrift für Assyriologie* 83, 258–267.

Yadin, Y. 1963. *The Art of Warfare in Biblical Lands in the Light of Archaeological Discovery*. Weidenfeld & Nicolson, London.

Zarins, J. 1976. *The Domestication of Equidae in Third Millennium B.C. Mesopotamia.* Vols. I–II. PhD Dissertation The University of Chicago.

Zarins, J. 1986. Equids Associated with Human Burials in Third Millennium B.C. Mesopotamia: Two Complementary Facets. In *Equids in the Ancient World*, edited by R. H. Meadow and H.–P. Uerpmann, 164–193. Beihefte zum Tübinger Atlas des Vorderen Orients, Reihe A—Naturwissenschaften 19/1. Wiesbaden: Reichert.

Zarins, J. 2014. [with the assistance of R. Hauser]. *The Domestication of Equidae in Third Millennium B.C. Mesopotamia.* Cornell University Studies in Assyriology and Sumerology 24. CDL Press, Bethesda, Maryland [update of *The Domestication of Equidae in Third Millennium B.C. Mesopotamia*, PhD Thesis University of Chicago 1976].

4

Some Notes on Pictograms Interpreted as Sledges and Wheeled Vehicles in the Archaic Texts from Uruk

Stefan Burmeister, Theo J. H. Krispijn and Peter Raulwing

Pictographic signs preserved on the so–called "Archaic Texts" from Uruk have drawn the attention of scholars since their first publication by Adam Falkenstein (1936). One set of signs is showing a sledge–like vehicle with various house–like constructions, another set is almost identical with the exception of depicting two round impressions underneath those sledge–like vehicles. This interpretation has been accepted by all scholars in the field. The listing and classification of these two sets of pictographic signs in connection with the number $2N_{14}$ (● ●), used in Uruk over 5000 years ago, however, could be misunderstood in the way that the two signs depicting various sledge–like vehicles with houses, but no wheeled–vehicles as the interpretation of wheels must be ruled out due to use of the number $2N_{14}$. As our study shows, the two round impressions for the number $2N_{14}$ must be distinguished from the two round impressions right underneath the sledge signs as they depict two different types of markings. In addition, this article analyses written evidence for wheeled vehicles and their parts from the "Archaic Texts" from Uruk. We come to the conclusion that the set of sledge–like vehicles depict, as Falkenstein proposed, a type of wheeled vehicle depicting the wheels exactly at the place underneath the runners where they would be expected in a pictographic sign for Wagon. In the context of this edition in memory of Mary Littauer, it is our goal to review the Mesopotamian pictograms and their contribution to the discussion of early wheeled transport.

Introduction

For students of ancient wheeled transport[1] the pictographic signs preserved on the so–called "Archaic Texts"[2] from Uruk in southern Mesopotamia have drawn much attention since their first publication in 1936[3] by Adam Falkenstein, one of the most renowned sumerologists of his time. Before the calibration of radiocarbon analysis overturned Old World chronology, these tablets had been regarded as the oldest evidence for the use of wagons and were seen as crowning witnesses for the origin of this innovation in the early city states of Mesopotamia.[4] They still provide evidence for the earliest use of wagons, but now have counterparts in other regions of the ancient world of similar early date.[5] The focus of this chapter is on a group of six tablets (Figures 1–7) showing vehicles interpreted as wheeled vehicles; five of the tablets were found in Uruk. Three of the Uruk tablets are kept in the Vorderasiatisches Museum in Berlin, two tablets in the Iraq Museum in Baghdad. The sixth tablet surfaced in the art trade and is thought to be from Tell Uqair.

The origin of the seminal innovation of wheeled vehicles is still in debate. Between the Atlantic (North Sea) and the Persian Gulf, the Alps and Central Asia wagons suddenly seem to emerge without any distinctive origin and spread. It is therefore necessary to scrutinise the evidence again. In the context of this edition in memory of Mary Littauer, it is our aim to review the Mesopotamian pictograms and evaluate how they contribute to the discussion of early wheeled transport (Figure 8).

1 For a concise overview see the contribution of Joost Crouwel on wheeled vehicles in this edition. See also Littauer and Crouwel 1979; Fansa and Burmeister 2004.
2 See Introduction in ATU 5: 11–14; Englund 1998: 32 ff.
3 Falkenstein, *Archaische Texte aus Uruk* (1936); see ATU 1 in our bibliography.
4 See e. g. Childe 1951: 178; Sherratt 1981: 263.
5 Burmeister 2011.

Catalogue of sledge–like vehicles with two round impressions marking wheels

1. Uruk W 9656,ag

CLAY TABLET:	38 × 29 × 16 mm (museum catalogue) / 39 × 31 × 18 mm (cdli)
SIGN:	ATU 743; ZATU 248
INSCRIPTION:	Obverse only; administrative text
TRANSLITERATION:	1N$_{14}$; GURUŠ$_{a}$+2N$_{14}$ UTUL^{d} ŠEG$_{9}$ DAa
PROVENIENCE:	Uruk; excavation no. W 9656,ag; findspot Qa XVI,2
URUK EXCAVATION DIARY:	"unter der Abgleichung der Schicht III, 1,5 m über der Höhe des Kalksteinsockels// t.a.q. [terminus ante quem] IIIc"
DATE:	Uruk Period IV
MUSEUM:	Vorderasiatisches Museum Berlin, VAT 14672
CDLI:	P001449
PUBLICATION:	ATU 1: 58; ATU 5: pl. 090

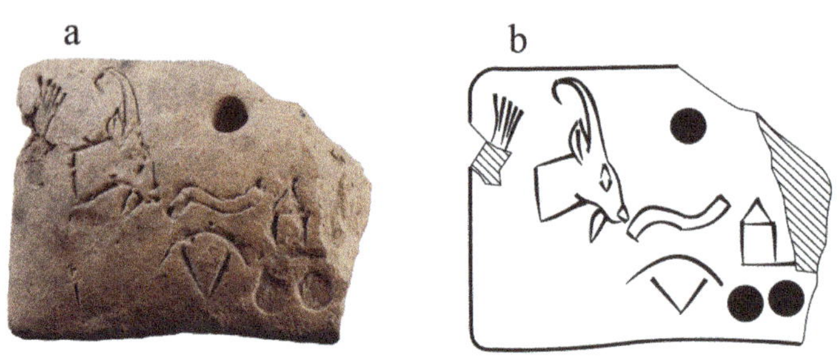

Figure 1. Uruk Tablet W 9656,ag; scale 1:1 and 4:1 (© Staatliche Museen zu Berlin—Vorderasiatisches Museum, Photos: Olaf M. Teßmer).

2. Uruk W 9656,h1

CLAY TABLET:	89 × 61 × 19 mm
SIGN:	ATU 744; ZATU 248
INSCRIPTION:	Obverse and reverse; lexical text, Archaic Lú A (witness)
TRANSLITERATION:	1(N$_{01}$), | GURUŠ$_{a}$+ 2N$_{14}$ (obverse; in this view rotated 90° clockwise 3rd line, 2nd column)
PROVENIENCE:	Uruk; excavation no. W 9656,h1; findspot Qa XVI,2
URUK EXCAVATION DIARY:	"unter der Abgleichung der Schicht III, 1,5 m über der Höhe des Kalksteinsockels// t.a.q. [terminus ante quem] IIIc"
DATE:	Uruk Period IV
MUSEUM:	Vorderasiatisches Museum Berlin, VAT 15003
CDLI:	P000006
PUBLICATION:	ATU 1: 340; ATU 3: pl. 023; Vaiman 1974: 25, fig. 9 (vs. line 3) [which is a schematic representation of the pictogram]

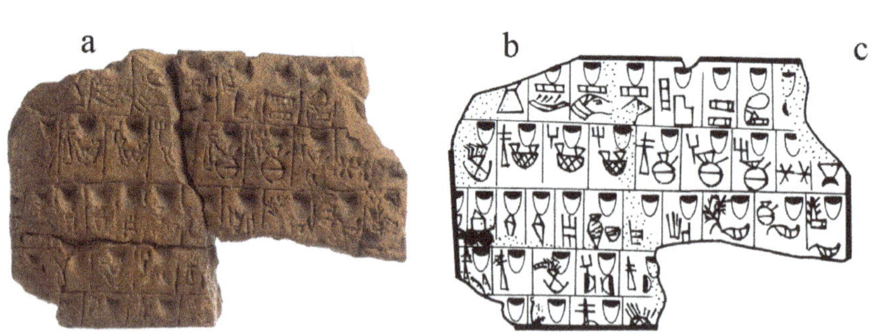

Figure 2. Uruk Tablet W 9656,h1; scale 1:1 and 4:1 (© Staatliche Museen zu Berlin—Vorderasiatisches Museum, Foto: Olaf M. Teßmer).

3. Uruk W 9579,bk

CLAY TABLET:	41 × 30 × 14 mm (museum catalogue) / 43 × 31 × 16 mm (cdli)
SIGN:	ATU 744; ZATU 248
INSCRIPTION:	Obverse and reverse; administrative text
TRANSLITERATION:	1N_{7a}; DU BA GURUŠ$_b$+ 2N_{14} MUŠ$_{3a}$ (obverse)
PROVENIENCE:	Uruk; excavation no. W 9579,bk; findspot Qa XVI,2
URUK EXCAVATION DIARY:	"unter der Abgleichung der Schicht III, 1,5 m über der Höhe des Kalksteinsockels// t.a.q. IIIc"
DATE:	Uruk Period IV
MUSEUM:	Vorderasiatisches Museum Berlin, VAT 14821
CDLI:	P001308
PUBLICATION:	ATU 1: 218; ATU 5: pl. 069

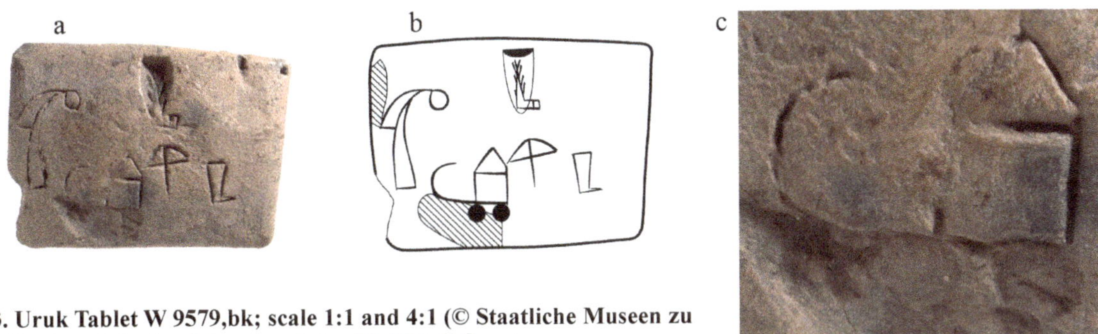

Figure 3. Uruk Tablet W 9579,bk; scale 1:1 and 4:1 (© Staatliche Museen zu Berlin—Vorderasiatisches Museum, Foto: Olaf M. Teßmer).

4. Uruk W 19568,a

CLAY TABLET:	62 × 45 × 28 mm
SIGN:	ATU 744; ZATU 248
INSCRIPTION:	Obverse and reverse; administrative text
TRANSLITERATION:	6N_{34} 2N_{14} X GURUŠ$_b$ X 2N_{14} X
PROVENIENCE:	Uruk; excavation no. W 19568,a; findspot Nc XVI,2
URUK EXCAVATION DIARY:	"1 m nördlich der S-Grenze in der Mitte, Füllschutt in Höhe des neubabylonischen Hauses"
DATE:	Uruk Period IV
MUSEUM:	National Museum of Iraq, IM — [no inventory number tracable]
CDLI:	P003216
PUBLICATION:	unpublished

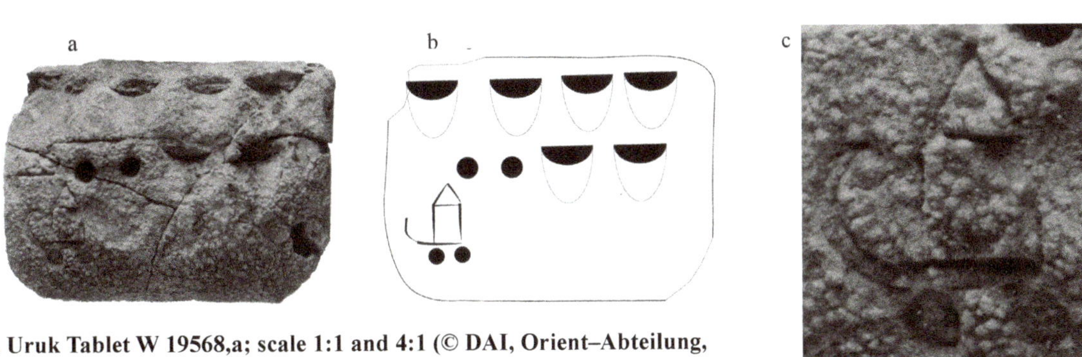

Figure 4. Uruk Tablet W 19568,a; scale 1:1 and 4:1 (© DAI, Orient–Abteilung, Uruk–Archiv).

5. Uruk W 19584,d

CLAY TABLET:	48 × 36 × 23 mm
SIGN:	ATU 743; ZATU 248
INSCRIPTION:	Obverse and reverse; administrative text
TRANSLITERATION:	2N$_1$ TUG$_{2a}$ GURUŠ$_a$ + 2N$_{14}$ ERIM$_a$ (obverse)
PROVENIENCE:	Uruk; excavation no. W 19584,d; findspot Nc XVI,2
URUK EXCAVATION DIARY:	"Südseite Mitte, im archaischen Füllschutt, etwa 4 m vor der Mauer aus gebrannten Ziegeln"
DATE:	Uruk Period IV
MUSEUM:	National Museum of Iraq, IM 064852
CDLI:	P003252
PUBLICATION:	unpublished

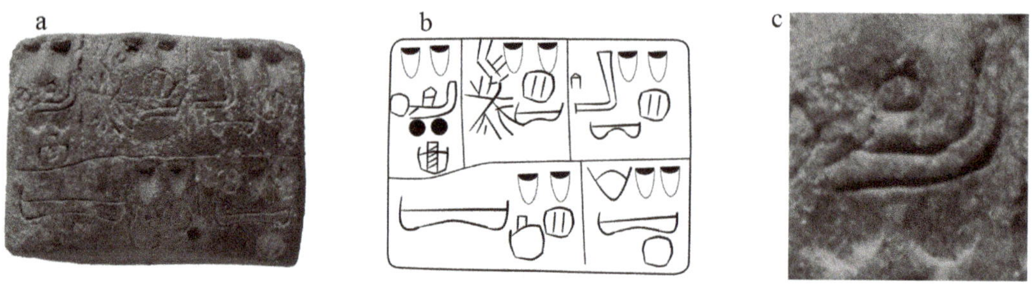

Figure 5. Uruk Tablet W 19584,d scale 1:1 and 4:1 (© DAI, Orient–Abteilung, Uruk–Archiv).

6. Tell Uqair

CLAY TABLET:	52 × 62 × 19 mm (museum catalogue) / 57 × 64 × 20 (cdli)
SIGN:	ATU 745; ZATU 248
INSCRIPTION:	obverse and reverse; administrative text
TRANSLITERATION:	GURUŠ$_a$ + 2N$_{14}$ (obverse; in this view rotated 90° clockwise 3rd line, 3rd column)
PROVENIENCE:	Tell Uqair; antiquity market, acquired 1903
URUK EXCAVATION DIARY:	"Südseite Mitte, im archaischen Füllschutt, etwa 4 m vor der Mauer aus gebrannten Ziegeln"
DATE:	Ǧemdet Nasr Period/Uruk III
MUSEUM:	Vorderasiatisches Museum Berlin, VAT 5308
CDLI:	P005431
PUBLICATION:	ATU 1: 639; MSVO 4: 23, pl. 16; Burmeister 2011: 212, fig. 2

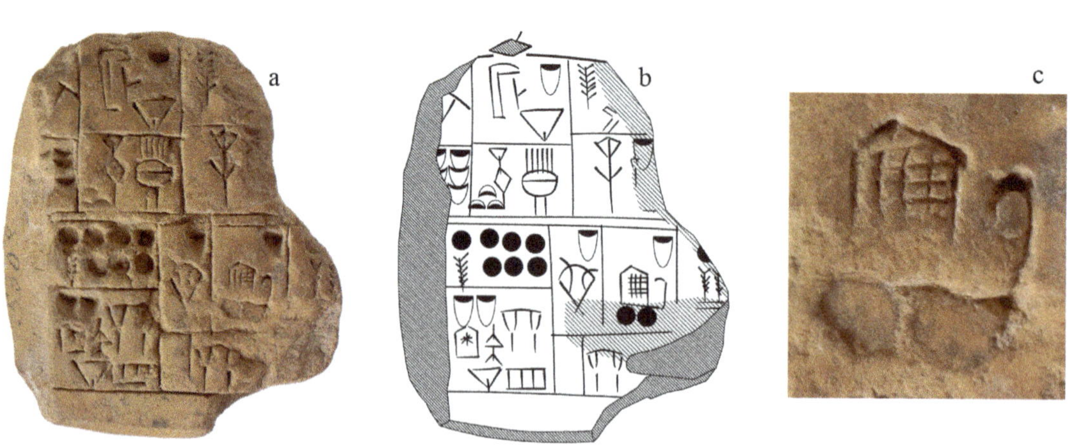

Figure 6. Tell Uqair Tablet; left: scale 1:1 and 4:1 (© Staatliche Museen zu Berlin—Vorderasiatisches Museum, Foto: Olaf M. Teßmer).

Figure 7. Photographic details all of sledge–like vehicles shown in Figures 1–6 with two round impressions marking wheels. (Figures 1c, 2c, 3, 4c, 5c, 6c).

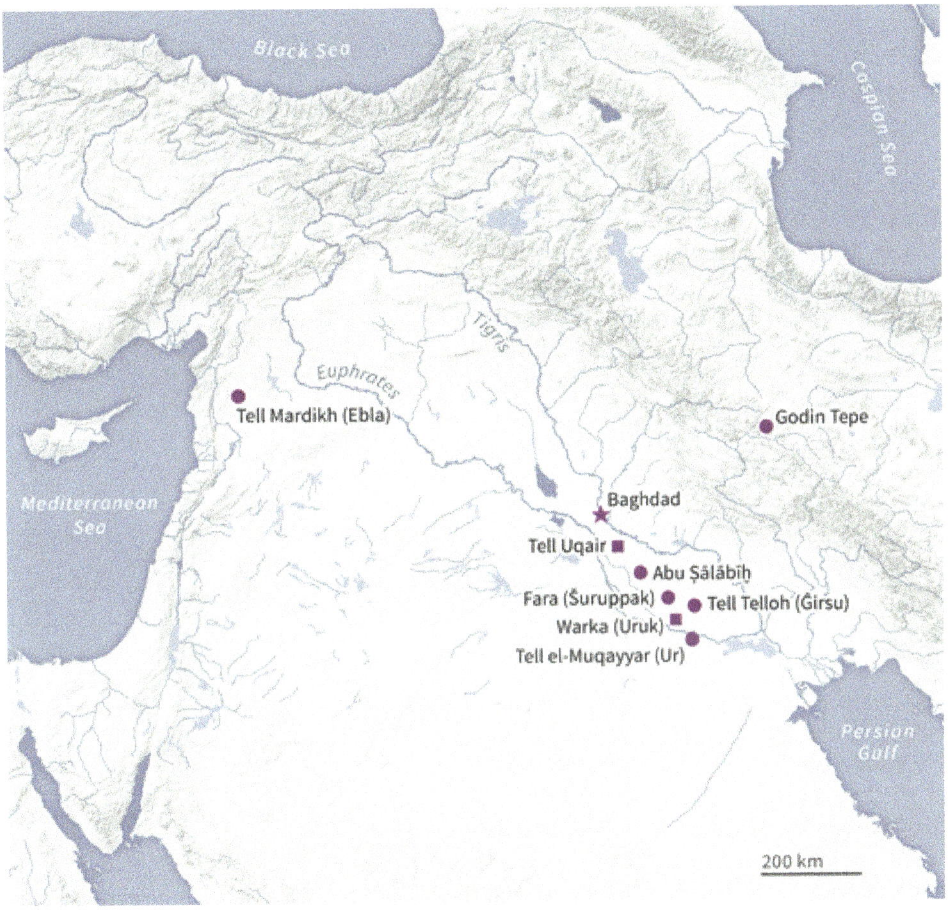

Figure 8. Map of Archaic settlements in Mesopotamia.

The Archaic Texts[6] can be divided into so–called "Administrative Texts"[7] and "Lexical Lists".[8] They derive from the so–called "Late Uruk" Period and date to the second half of the 4th millennium BCE.[9] Two phases of the Archaic Texts can be distinguished: Uruk IVa and Uruk III, named after the archaeological strata established during the excavation of Uruk, the Erech mentioned in the Bible.[10] However, the stratigraphically established layers

[6] For a general overview see Radner and Robson 2011.

[7] Published by R. K. Englund as *Archaic Administrative Texts from Uruk. The Early Campaigns* (1994); see ATU 5 in our bibliography.

[8] Published by R. K. Englund and H. J. Nissen, *Die lexikalischen Listen der archaischen Texte aus Uruk* (1993); see ATU 3 in our bibliography and the latest study by N. Veldhuis, *History of the Cuneiform Lexical Tradition* (2014).

[9] Hans Nissen has given a framework for the relative chronology of the earliest script (ZATU: 21–51). Dietrich Sürenhagen (1999) has refined this chronology, distinguishing the archaeological strata Uruk V, IVc, IVb, IVa, IIIc, IIIb, IIIa. The *Cuneiform Digital Library Initiative* (CDLI): http://cdli.ucla.edu distinguishes only Uruk V, IV, and III.

[10] The archaeological reports are published in the series *Vorläufiger Bericht über die von dem Deutschen Archäologischen Institut aus Mitteln der Deutschen Forschungsgemeinschaft unternommenen Ausgrabungen in Uruk-Warka*: 1 (1930)–5 (1934), *Vorläufiger Bericht über die von der Deutschen Forschungsgemeinschaft in Uruk-Warka unternommenen Ausgrabungen*: 6 (1935)–11 (1940)—single volumes are part of *Preußische Akademie der Wissenschaften / Philosophisch-Historische Klasse: Abhandlungen der Preußischen Akademie der Wissenschaften, Philosophisch-Historische Klasse* [Deutsche Akademie der Wissenschaften zu Berlin. Philosophisch-Historische Klasse. Abhandlungen; 1929, no. 7; 1930, no. 4; 1932, no. 2, 6; 1933, no. 5; 1935, no. 2, 4; 1936, no. 13; 1937, no. 11; 1939, no. 2; 1940, no. 3]—and

do not automatically date the texts, because the tablets were found as part of the debris used to level the ground before construction of a new building phase started. In addition to the much–discussed absolute and relative chronological dates, scholars are still debating the name and linguistic classification of the language in which the Archaic Texts are written.[11]

A particular group of signs has been interpreted as depicting sledges with runners curving upwards at one end.[12] On the six tablets illustrated and catalogued above underneath the runners two round impressions are placed which may either represent two (captive) rollers or four solid disk wheels.[13] In case of the latter, we would have to speak of wagons. These pictograms have been interpreted as GURUŠ$_a$ (*sledge*) and as GURUŠ$_b$ (*wagon*). However, this reading of the pictograms has been questioned based on philological[14] and archaeological[15] grounds.

It is our purpose to re–examine the "wagon" pictograms and the lines of argument brought forward since Falkenstein's publication in 1936. In the second part of this chapter Theo Krispijn presents a philological and linguistic analysis of the signs PU$_2$/GIGIR (**ZATU 430**), GURUŠ (**ZATU 247, 248**) and MAR (**ZATU 252**). Although it is not possible to solve all problems here, we offer a conclusion, hoping that it will serve as a starting point for a new debate of this most fascinating material, in which Mary Littauer would have certainly engaged.

History of Research

A. Falkenstein: Archaische Texte aus Uruk (1936): The Signs ATU 741–745

In his publication *Archaische Texte aus Uruk* (1936)—abbreviated in scholarly studies as ATU[16]—Falkenstein analyzed all cuneiform signs available to him[17] and provided a list of signs as well as their variations identifiable by individual numbers (Figure 9). The pictograms discussed in this article follow Falkenstein's numbers **ATU 741, 742, 743, 744,** and **745**.

According to Falkenstein, the signs **ATU 741** and **742** (Table 1) represent two groups based on five types.[18]

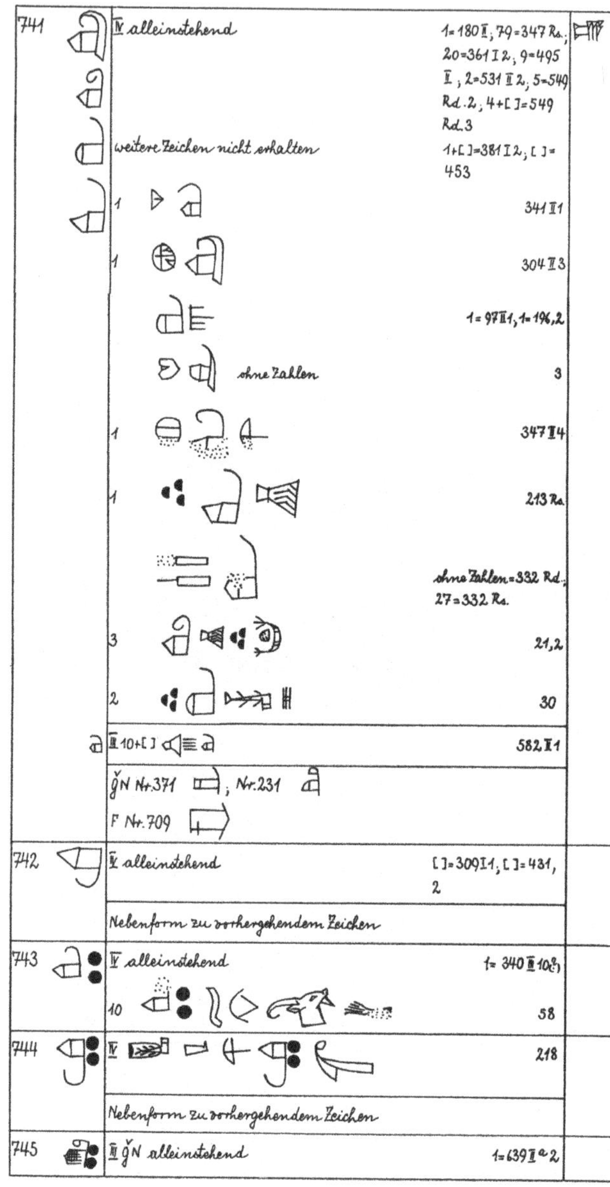

Figure 9. Signs ATU 741–745. From A. Falkenstein, *Archaische Texte aus Uruk* (1936), p. 167 f. [Listed under ATU in our bibliography].

Table 1. Summary of group ATU 741–742

ATU	Direction	House-like Construction	2 Runners (no wheels)
741 	facing right	pointed roof	double line
741 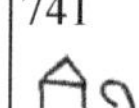	facing right	pointed roof	single line
741 	facing right	round roof	single line
741 	facing right	pointed roof	single line
742 	facing left	pointed roof	single line

Vorläufiger Bericht über die von dem Deutschen Archäologischen Institut und der Deutschen Orient-Gesellschaft aus Mitteln der Deutschen Forschungsgemeinschaft unternommenen Ausgrabungen in Uruk-Warka: 12/13 (1953/55 [1956])–31/32 (1973/74 [1983]) and as *Ausgrabungen in Uruk-Warka. Endberichte* (Deutsches Archäologisches Institut, Orient-Abteilung (1987 ff.). For the topic of this study, see Eichmann 1989; 2007 and Sürenhagen 1993; 1999.

[11] Discussed by Rubio (1999; 2005) and Whittaker (1998 ff.). See also Englund 1998: 73 ff.

[12] See next chapter for more details.

[13] Burmeister 2011: 213 f.

[14] ZATU 247; 248; Englund 2009.

[15] Nagel/Strommenger 1994: 207.

[16] See abbreviations at the end of this appendix.

[17] ATU 5: 11–14.

[18] The ideograms illustrated are flipped 90 degrees clockwise, respectively counterclockwise.

Table 2. Summary of group ATU 743–745

ATU	Direction	House-like Construction	Wheels	Runners
 743	facing right	pointed roof	four (2 round markings)	single line
 744	facing left	pointed roof	four (2 round markings)	single line
 745	facing right	pointed roof with a hatched house or baldachin-like construction	four (2 round markings)	single line[1]

[1] A close look at the photographs shows a slightly different line drawing than depicted by Falkenstein. The runners are not visible. Here, Falkenstein's overview is too schematic and not accurate.

These signs depict, as most scholars since Falkenstein's publication (1936) agree, sledges with schematic house– or baldachin–like constructions shown in side view resting on runners.[19]

Falkenstein's group "**ATU 743–745**" (Table 2) shows house or baldachin–like constructions with pointed roofs. In contrast to group **ATU 741–742**, beneath the runners are two round impressions. Since their first publication in 1936 during the peak of the Uruk campaigns in southern Mesopotamia these marks have been interpreted as depictions of wheels, as they are in positions where wheels would be expected at real wagons. For Falkenstein there was no doubt in his comments in ATU (1936: 56) that these signs represent sledges and sledges on wheels, which he called "Wagen"[20] (*wagon, four–wheeler*), as did others.[21]

A photograph of **ATU 745**, illustrated first in Burmeister 2011: 212, Fig. 1(2), shows that the hatches in Falkenstein's sign list should be modified showing two vertical lines on the left side of the house–like construction.[22]

M. W. Green and H. J. Nissen: Zeichenliste der Archaischen Texte aus Uruk (1987): The Signs ZATU 247 and 248

In their sign list *Zeichenliste der Archaischen Texte aus Uruk* (ZATU) H. J. Nissen and M. W. Green have listed Falkenstein's group **ATU 741–742** (sledges) and **ATU 743–745** (wheeled vehicles) under **ZATU 247** and **ZATU 248** (Figure 10). **ZATU 248** is here transliterated as GURUŠ + TAR, a reading that was rejected by p. Steinkeller.[23]

R. Englund, editor of ATU vol. 5 (*Archaic Administrative Texts from Uruk. The Early Campaigns*; 1994: 131), transliterated the two round impressions underneath the runners in **ATU 743–745 / ZATU 248** as (GURUŠ$_a$+) $2N_{14}$ and (GURUŠ$_b$+) $2N_{14}$. The transliteration of the two round impressions underneath the sledge as $2N_{14}$ is, however, misleading. $2N_{14}$ stands for the numerical sign N_{14} occurring twice next to each other. The classification as numerical sign deviates from a reading as wheel representations (Figures 11–14). Englund argues for a reading of GURUŠ as sledge, though leaves open how to interpret the two impressions transliterated as $2N_{14}$.[24]

The Numerical Sign $2N_{14}$ in connection with ZATU 248: GURUŠ$_a$+$2N_{14}$ and GURUŠ$_b$+$2N_{14}$

The Archaic Texts consist of ca. 1200 different signs, of which 60 have been identified as numerical signs[25] (Figure 15). N_{14} represents the fourteenth sign (N = number) in this list established by sumerologists.

[19] Littauer and Crouwel 1979: 13 ("[…] simple pictographic signs on inscribed clay tablets found in Uruk level IVa […] which may be called 'sledge cars'").

[20] Falkenstein, ATU: 56: "Auch den Wagen finden wir schon in Schicht IV, und zwar den vierrädrigen (Nr. 743–45). Durch seine enge Verwandtschaft mit dem 'Schlitten'' (Nr. 741–42) wird es sehr wahrscheinlich, daß der Wagen in Babylonien daraus entstanden ist[5]. Dieses letztere Zeichen hat sich vielleicht bis in die späte Schrift gehalten (Zeichen KAL), die ursprüngliche Bedeutung ist aber verloren gegangen. Auch das Zeichen für den Wagen ist aufgegeben worden, dafür tritt das Bild eines Wagenrades und zwar eines Vollrades ein[6]. Eine genauere Vorstellung von der Form des Wagens zur Zeit der Schicht IV würde uns eine Siegelabrollung auf einer Tontafel geben können; leider ist sie unvollständig erhalten. Man erkennt nur den Wagenlenker, der vier Zügel hält, und den Kopf der Zugtiere[7]." Fn. 5: "Zur Deutung des Zeichens Nr. 741 als Schlitten s. schon E. Herzfeld, AMI 6 (1934), 203; dort (S. 223) die Abbildung eines Steatittäfelchens, das genau die Zeichenform des Schlittens auf den archaischen Tafeln wiedergibt [Littauer and Crouwel 1979: Fig. 2; 1990: Fig. 3; Burmeister 2011: Fig. 2]. Herzfelds Deutung des Aufbaus über den Schlittenkufen als Thron mit Sonnendach ist unsicher. Ich möchte darin den gewöhnlichen Wagenkasten erkennen. Zu vergleichen ist noch der Schlitten der Šubad (L. Woolley, Royal Cemetery, Taf. 122–25)." Fn. 6: "Zur Bedeutung von mar=*marru* s. B. Landsberger, ZA NF 3, 93 und F. Thureau-Dangin, RA 24, 147 f." Fn. 7: "S. 4. Vorbericht, S. 29 und Taf. 14c–f; 5. Vorbericht, Taf. 28c. Die Tafel stammt aus dem großen Depot im Planquadrat Qa XVI 2."

[21] Kramer 1963: 102; Piggott 1978: 33 f.; 1983: 38, Fig. 8 ("A, sledge symbols; B, sledge-on-wheels vehicle symbols"); Sherratt 1981: 264, Fig. 10.1 ("[…] pictographic script including the sign for 'cart' […]"); 1983: 99; 1997: 160, Fig. 6.1; Littauer and Crouwel 1979: 13, Fig. 1; Drenth and Bakker 2005–06: 11 f. (with illustration); Schmandt-Besserat 1992: 5 ("Adam Falkenstein, the German scholar who studied the texts, noted that when writing began in Mesopotamia, truly pictorial signs were rare exceptions. The signs for 'plow,' 'chariot,' 'sledge,' or 'wild boar' were not only few but of uncommon use, represented by a single in one tablet alone"); Parpola 2008: 10, Fig. 2.

[22] The two vertical lines are also missing in MSVO 4: 23 (Pl. 16, Fig. 29).

[23] Steinkeller 1990: 22 fn. 33: "Against Green [ZATU] p. 220, no. 248 the 'wheels' show no relation to the sign TAR." See also Krebernik 1994: col. 383 ("Die Analyse als AB$_2$ + TAR stimmt nicht zu den unter TAR angeführten Formen; in LATU, S. 180 richtig als AB$_2$+ $2N_{14}$ wiedergegeben. Auch unter 248 und 320 wird $2N_{14}$ als TAR transliteriert").

[24] See also Englund 2009: 8 f. fn. 19.

[25] *Archaic Bookkeeping*: 25–27.

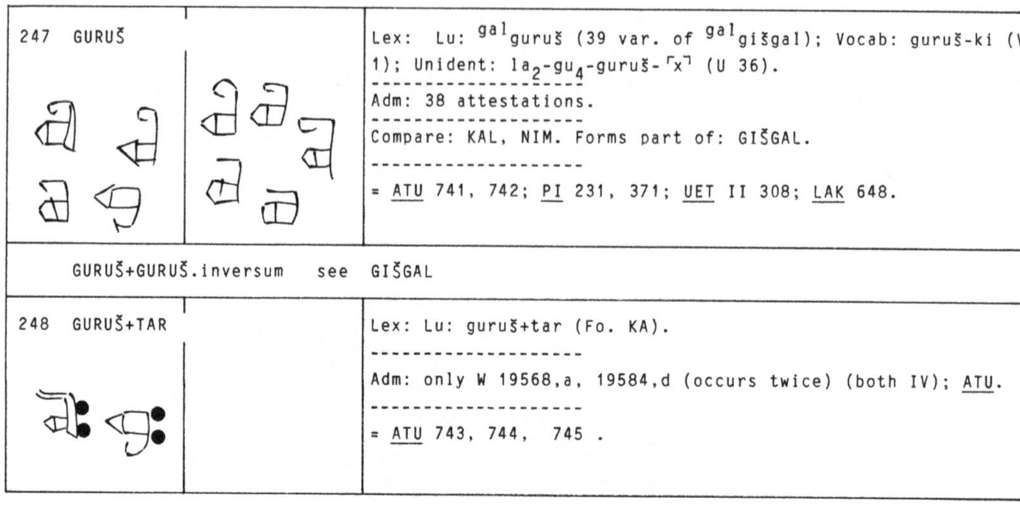

247 GURUŠ	Lex: Lu: ^{gal}guruš (39 var. of ^{gal}gišgal); Vocab: guruš-ki (V 1); Unident: la$_2$-gu$_4$-guruš-⸢x⸣ (U 36).
	Adm: 38 attestations.
	Compare: KAL, NIM. Forms part of: GIŠGAL.
	= ATU 741, 742; PI 231, 371; UET II 308; LAK 648.
GURUŠ+GURUŠ.inversum see GIŠGAL	
248 GURUŠ+TAR	Lex: Lu: guruš+tar (Fo. KA).
	Adm: only W 19568,a, 19584,d (occurs twice) (both IV); ATU.
	= ATU 743, 744, 745 .

Figure 10. Signs ZATU 247–248. From M. W. Green and H. J. Nissen, *Zeichenliste der Archaischen Texte aus Uruk* (1987) p. 220. [Listed under ZATU in our bibliography as well as the abbreviations].

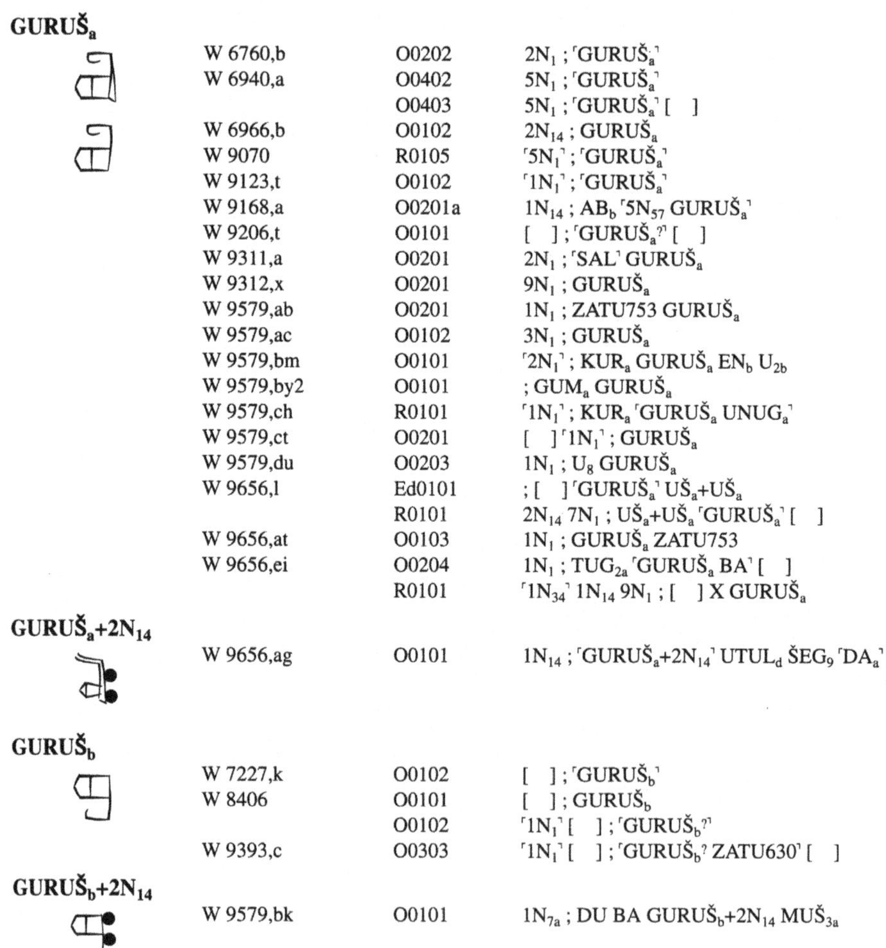

GURUŠ$_a$

W 6760,b	O0202	2N$_1$; ⸢GURUŠ$_a$⸣
W 6940,a	O0402	5N$_1$; ⸢GURUŠ$_a$⸣
	O0403	5N$_1$; ⸢GURUŠ$_a$⸣ [ ]
W 6966,b	O0102	2N$_{14}$; GURUŠ$_a$
W 9070	R0105	⸢5N$_1$⸣ ; ⸢GURUŠ$_a$⸣
W 9123,t	O0102	⸢1N$_1$⸣ ; ⸢GURUŠ$_a$⸣
W 9168,a	O0201a	1N$_{14}$; AB$_b$ ⸢5N$_{57}$ GURUŠ$_a$⸣
W 9206,t	O0101	[ ] ; ⸢GURUŠ$_a$?⸣ [ ]
W 9311,a	O0201	2N$_1$; ⸢SAL⸣ GURUŠ$_a$
W 9312,x	O0201	9N$_1$; GURUŠ$_a$
W 9579,ab	O0201	1N$_1$; ZATU753 GURUŠ$_a$
W 9579,ac	O0102	3N$_1$; GURUŠ$_a$
W 9579,bm	O0101	⸢2N$_1$⸣ ; KUR$_a$ GURUŠ$_a$ EN$_b$ U$_{2b}$
W 9579,by2	O0101	; GUM$_a$ GURUŠ$_a$
W 9579,ch	R0101	⸢1N$_1$⸣ ; KUR$_a$ ⸢GURUŠ$_a$ UNUG$_a$⸣
W 9579,ct	O0201	[ ] ⸢1N$_1$⸣ ; GURUŠ$_a$
W 9579,du	O0203	1N$_1$; U$_8$ GURUŠ$_a$
W 9656,l	Ed0101	; [ ] ⸢GURUŠ$_a$⸣ UŠ$_a$+UŠ$_a$
	R0101	2N$_{14}$ 7N$_1$; UŠ$_a$+UŠ$_a$ ⸢GURUŠ$_a$⸣ [ ]
W 9656,at	O0103	1N$_1$; GURUŠ$_a$ ZATU753
W 9656,ei	O0204	1N$_1$; TUG$_{2a}$ ⸢GURUŠ$_a$ BA⸣ [ ]
	R0101	⸢1N$_{34}$⸣ 1N$_{14}$ 9N$_1$; [ ] X GURUŠ$_a$

GURUŠ$_a$+2N$_{14}$

W 9656,ag	O0101	1N$_{14}$; ⸢GURUŠ$_a$+2N$_{14}$⸣ UTUL$_d$ ŠEG$_9$ ⸢DA$_a$⸣

GURUŠ$_b$

W 7227,k	O0102	[ ] ; ⸢GURUŠ$_b$⸣
W 8406	O0101	[ ] ; GURUŠ$_b$
	O0102	⸢1N$_1$⸣ [ ] ; ⸢GURUŠ$_b$?⸣
W 9393,c	O0303	⸢1N$_1$⸣ [ ] ; ⸢GURUŠ$_b$? ZATU630⸣ [ ]

GURUŠ$_b$+2N$_{14}$

W 9579,bk	O0101	1N$_{7a}$; DU BA GURUŠ$_b$+2N$_{14}$ MUŠ$_{3a}$

Figure 11. ATU 5: 131.

As seen above in the edition of the Archaic and Lexical Texts from Uruk (ATU 3 and ZATU 5), as well as in the Sign List (ZATU), the authors did not interpret the sign **ZATU 248** as "wagon", but rendered it as GURUŠ$_a$+2N$_{14}$ and GURUŠ$_b$+2N$_{14}$. There are other numerical signs documented in the context of the semantic field of "wagon". Since Early Dynastic Period III UMBIN is also used to designate "wheel" (ca. 2600–2450/2340 BCE); there is an occurrence for 2N$_{14}$ [ ] UMBIN and for 2N$_{58}$ [ ] UMBIN.[26]

[26] Salonen 1951; http://psd.museum.upenn.edu/epsd1/nepsd-frame.html.

2N$_{14}$	Compare AB$_2$+2N$_{14}$, GURUŠ$_a$+2N$_{14}$, GURUŠ$_b$+2N$_{14}$, LAGAB$_a$+2N$_{14}$

Figure 12. ATU 5: 220.

GURUŠ$_a$

Lú 39	W 21418,1	O0308	⸢GAL$_a$⸣ GURUŠ$_a$
Wood 65	W 24008,10	O0305	⸢GURUŠ$_a$ KU$_{3a}$?⸣ [ ]
Vocabulary 1	W 13948	O0102	GURUŠ$_a$ KI$_a$
Unid. 36, 24	W 19771,g+	O0204	GURUŠ$_a$ LA$_2$ ⸢GU$_4$⸣ X

GURUŠ$_a$+2N$_{14}$

Lú Vorläufer	W 9656,h	O0310	GURUŠ$_a$+2N$_{14}$

Figure 13. ATU 3: 237.

GURUŠ$_a$

MSVO 4, 2	O0101	⸢2N$_{45}$⸣ 8N$_{14}$; GAN$_2$ U$_4$+1N$_{57}$ A ⸢GURUŠ$_a$? NAM$_2$⸣ RAD$_a$ [ ]
MSVO 4, 26	R0104	⸢5N$_1$⸣ ; ⸢ZATU759 GAL$_a$? GURUŠ$_a$⸣ [ ]
MSVO 4, 74	O0101	1N$_1$; SAG GURUŠ$_a$ UŠ$_a$
	R0201	3N$_1$; SAG GURUŠ$_a$

GURUŠ$_c$+2N$_{14}$

MSVO 4, 29	O0202a1	1N$_1$; GURUŠ$_c$+2N$_{14}$

Figure 14. MSVO 4: 58.

Analyzing the impressions themselves, the two round circles pressed into the wet clay underneath the runners are clearly distinct from the numerical signs used elsewhere. The impressions of the numerical signs on the other hand are definitely deeper. The tablet from Berlin VAT 14672 [our No. 1 above] and from Baghdad IM (no invenory number = our No. 4) in fact show both kinds of impression: Two shallow impressions beneath the runners and clearly deeper impressions above the vehicle. Both kinds of impressions are similar in diameter and only differ in their depth. In addition, the impressions related to GURUŠ$_a$ and GURUŠ$_b$ are directly connected to the house or baldachin–like construction and positioned exactly where wheels would be expected when trying to depict a sledge with a wheel–like under–carriage. It could, therefore, be assumed that these two impressions are semantically related to the signs (and its variations) and are, in fact, marking wheeled vehicles. Furthermore, the numerical signs usually are noted on the cuneiform tablets considerably distant from the object they denote; a common practise by the scribes of these tablets. However, this would require a verification of all numerical signs in question. To sustain this observation, another tablet from Uruk, now kept in the Vorderasiatisches Museum Berlin (VAT 14908) shows two round impressions right above the sledge construction (Figure 16): both impressions are clearly deeper than those impressions beneath the runners at our six tablets under discussion.[27] It seems, therefore, less likely that the two impressions underneath the runners in ATU 743–745 can be interpreted as numerical sign 2N$_{14}$, though, obviously, they have been impressed with the same stylus. In personal communication Robert Englund remarked that the sign designation was a simple graphic description with no semantic background.[28] The non–numerical use of "numerical" signs has to be taken into account (Damerow and Englund 1987: 144).

Wheels versus Rollers (Rolling Beams)

Although A. Falkenstein without a doubt identified the tablets ATU 743–745 as simplified representations of wagons[29]—which formed a *communis opinio* over the next decades—an interpretation of the cuneiform tablets depicting wheeled vehicles has been questioned in archaeological studies.[30] The variations of house–like construction of the signs representing wagons match with the exception of **ATU 745** the tablets depicting sledges. The main differences, though, are two round impressions underneath the runners made by a stylus with a round end used by the scribes. It has been assumed that the two round impressions might represent rollers (rolling beams).[31] From the technological point of view, it is possible, albeit inefficient and quite labor–intensive, to move a sledge on

[27] http://www.cdli.ucla.edu/dl/photo/P000924.jpg.

[28] Email from July 29th, 2015.

[29] Falkenstein (ATU 1: 56).

[30] Nagel and Strommenger 1994: 207.

[31] See discussion in Littauer and Crouwel 1979: 13; Englund 1998: 81; Burmeister 2011: 213.

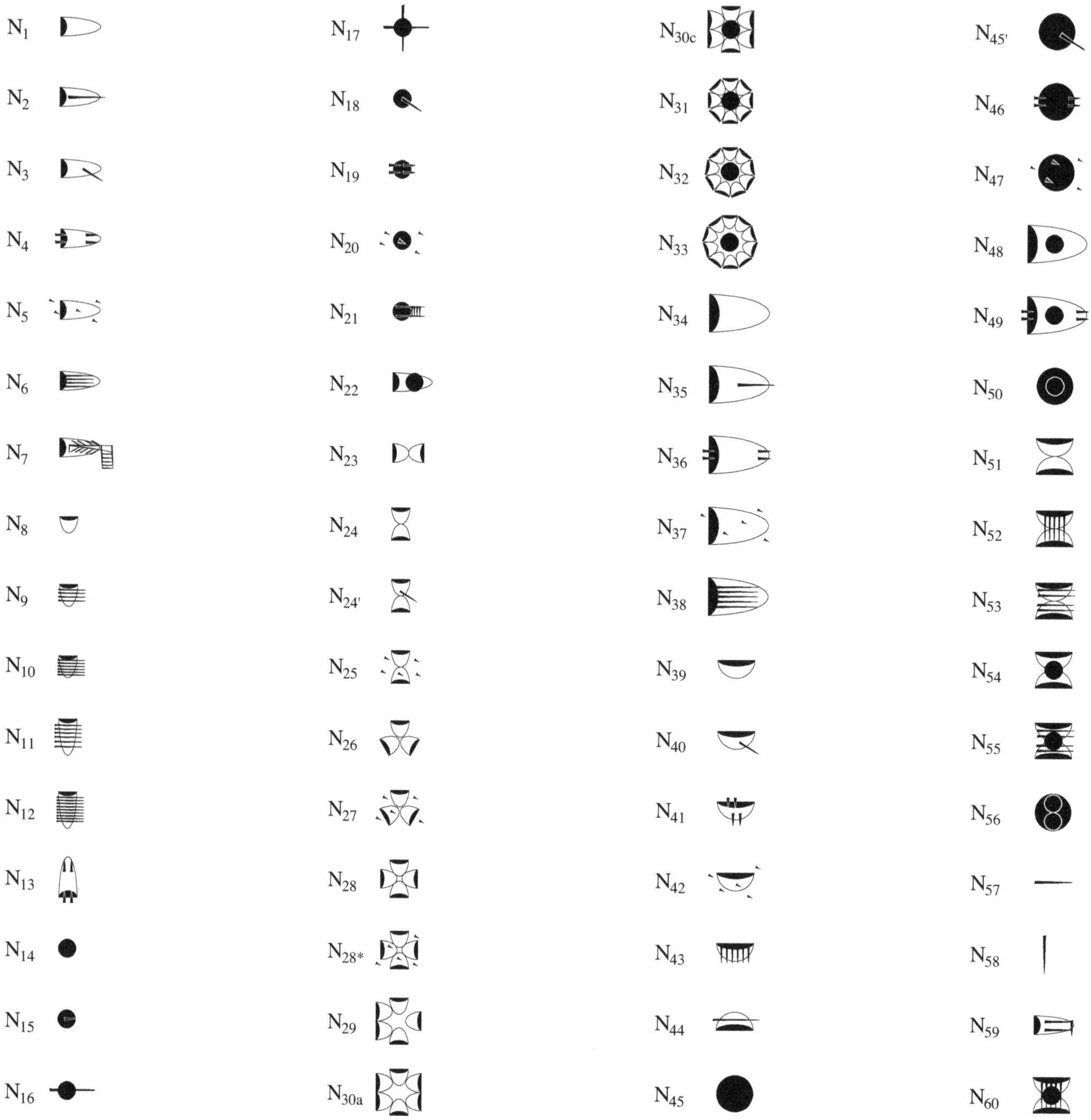

Figure 15. The numerical signs of the proto–Sumerian ("archaic") texts from Uruk together with their corresponding transcriptions. N_{14} is the third sign from the bottom in the left row. (*Archaic Bookkeeping*: Fig. 27, p. 26 = ATU 5: Fig. 11, p. 20).

rollers. However, rollers would, practically, offer no real advantage. At least three rollers are needed, which would have to be put in front of the sledge, and then picked up from behind after the sledge fully rolled over it. At least two persons would then have to carry a roller back to the front of the sledge so the sledge could roll over the next beam. Each new roller would have to be placed in position between the draught team and the front of the runners. This would have to be done by skilled persons who could handle the pole in the rear in reach of the animals' hoofs and pass the roller through underneath the pole shaft while the vehicle was moving (Figure 17).

Significant labor and skills had to be invested to move a sledge forward. A divine procession or parade, as the sledge ride often has been viewed, would have probably lost its solemnity with all of these side–activities which might have been more disturbing than practical as we can doubt that rollers would provide any technological advantage. For these reasons it is very likely that the two impressions underneath the runners represent solid disk wheels rather than rollers.

The Pictograms as Images of Vehicles

A close look at the pictograms themselves clearly show that Falkenstein's pictographic drawings are very schematic and do not represent the actual signs on the tablets. First of all, it can be said that all the wagon pictograms are unique in their formal composition. Solely the two round

Figure 16. Tablet from Uruk showing a sledge pictogram and two round impressions above (© Staatliche Museen zu Berlin—Vorderasiatisches Museum, Foto: Olaf M. Teßmer).

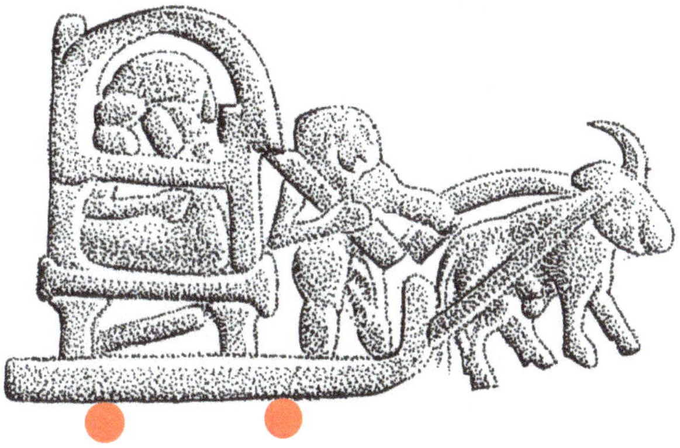

Figure 17. Stone plaque, Iraq. Unknown provenience. Late Uruk period. London, British Museum 12885. The red dots indicate rollers to illustrate challenges and obstacles when using rollers instead of wheels (after Burmeister 2011: Fig. 2 modified after Littauer and Crouwel 1979: Fig. 2 = Littauer and Crouwel 1990:Fig. 3. Drawing by Jaap Morel).

impressions are of identical make; all have a diameter of about 3 mm. The impressions are very shallow, only on tablet in Figure 4 (IM – [our No. 4 above]) they seem to be slightly deeper. Would these impressions, in fact, depict wheels, the wheelbase (i.e., the distance between the two axles) would be different at each wagon. As the overall composition of the pictograms looks in most cases very inaccurate, this should not be interpreted as depicting a technological detail.

The superstructure has the form of a hut, which is a sign of its own (ATU 740). It is composed by four straight incisions forming a square and on top with two inclined incisions forming a pointed roof. Only on one tablet on Figure 6 (VAT 5308), the "hut" shows an interior structure with vertical and horizontal lines. These might represent constructive features of the superstructure. On tablets in Figure 2 (VAT 15003) and Figure 5 (IM 064852) the "hut" is in its details not clearly recognizable. As the composition cannot be deduced, we are not able to assign it to a specific form. These divergent shapes can be due to damages on the tablets or to swift and inaccurate, non conventional writing. A very swift making can also be seen on tablet in Figure 1 (VAT 14672); this in contrast to tablet in Figure 3 (VAT 14821) and tablet in Figure 5, where the incisions and lines are remarkably accurate.

The common interpretation of these signs as sledges on wheels is certainly influenced by the curved line underneath the superstructure looking like runners in side view. A "hut" and a curved line below define the signs for "sledges" ATU 741 and ATU 742, alternatively ZATU 247. Examining Falkenstein's sign list, it is only the pictogram on tablet in Figure 3 that correlates with his table—and ATU 743 on tablet in Figure 6, which is in several aspects outstanding. The two pieces now kept in the National Museum of Iraq (Baghdad) do not fit in this scheme either, as the "runners" here are composed of a double line. The pictograms on tablet in Figure 1 and tablet in Figure 2 again show no resemblance. Below the superstructure on tablet in Figure 1 there is the rudiment of a straight line pointing to the right. The tablet is damaged here, so that the course and shape of this line cannot be identified. The same holds true for the pictogram on tablet in Figure 2. Here the sign is inserted in such a narrow box that anything protruding to the sides is omitted. This sign is composed only of the "hut" and the two "wheels". As already mentioned, the pictogram on tablet 6 is exceptional in so far as it is the only tablet using this sign and not originating from Uruk. The "hut" shows a divergent composition and the line between the superstructure and round impressions does not resemble anything like runners. The "hut" is hovering above this line which has a horizontal course and behind the "wheels" bends sharply in an almost 90° angle upright. In technological terms this is a front shield common in many wagon models and pictures of the 3rd millennium BCE in Mesopotamia. The upright bend of the "runner" on tablet in Figure 5 also evoke doubts on the interpretation as sledge runners. The double line might depict the platform with a front shield.

Overall it can be resumed that there was no canonical concept for this particular sign. It can be reduced to a "hut" and two "wheels". The analogy to the sledge signs is obvious, though problematic as it cannot be confirmed in all cases and is contradicted in the tablet from Tell Uqair. When thinking of wagons the "runners" prefigure our interpretation as sledges on wheels. This is a technological concept which is problematic as it imposes the idea that the wagon is a technical derivate of the sledge and not a vehicle in its own right. In summary, it can be concluded with much plausibility that the pictograms ATU 743–745/ ZATU 248 depict wagons with four wheels.

Dating of the Archaic Text Tablets in Uruk

The chronological position of these wagon pictograms is crucial for discussing the contribution of the Mesopotamian early city–states to the invention and spread of wheeled vehicles. To estimate how these tablets as early evidence of wagons are chronologically related to other early evidence in Europe and Asia, we need more secure and established grounds for their dating.

As an aggravating starting point, which is often overlooked, we have to realise that none of the thousands of cuneiform tablets and fragments from the Late Uruk Period (ca. 3500–3100 BCE) were found *in situ*. The tablets and fragments were part of debris used to level the ground for new building constructions.[32] Therefore, all tablets were displaced from their original locations after falling out of use (for whatever reason).[33] Three out of four cuneiform tablets from Uruk were found in Archaic levels of the northeast corner of the so–called "Red Temple"[34] (Figure 18).The Uruk excavation reports and excavation diaries (kept in Berlin) do not provide sufficient information on a finer stratification of the tablets within their find context,[35] which would sustain in their chronological assessment. A relative–chronological allocation of the tablets earlier than Uruk III Period (ca. 3100–2900 BCE)[36] would—at best—lead us to the Late Uruk period; strictly speaking toward the end of Uruk IV, because the tablets were incorporated in the leveled construction phase at the end of the Uruk IV period.

In his study on the relative chronology of Uruk Dietrich Sürenhagen (1999) came to the conclusion that it is indeed possible to differentiate specific stratigraphic layers for certain findings with cuneiform tablets. For the area of the "Red Temple", as Sürenhagen concluded, 87 cuneiform tablets can be identified with certainty, 73 tablets with a very high probability, and 634 tablets under certain caveats as part of the debris of the Uruk IVc phase—which marks the earliest building activities in the Uruk IV period.[37] However, the three wagon pictograms (VAT 14672 [our No. 1 above], VAT 14821 [No. 3], VAT 15003 [No. 2]) were found in the north–east corner of the "Red Temple" for which debris of the Uruk IVc phase could not be confirmed.[38]

From another point of view Sürenhagen argues for a chronological setting of these tablets already in early Uruk IV phases.[39] As the semi–pictographical script developed before Uruk III, the first evidence of pictographical signs like the wagon have to be sought in earlier layers. He stated that the pictographic script system was already fully developed at the beginning of Uruk IVb and has to be traced back to Uruk IVc.[40]

However, the development of the script does not provide any further chronological information about wagon pictograms due to the lack of specific stratigraphic data regarding their find context. Still, it is striking that the published tablets with sledges and wagons (GURUŠ$_a$ and GURUŠ$_b$ = **ZATU 247** and **248**; ATU 5: 131) were found mainly in the north–east corner of the "Red Temple". The mentioned three wagon pictograms (VAT 14672, VAT 14821, VAT 15003) from this area belong to a group of tablets that also contained the majority of sledge pictograms. The combination of pictograms with sledges and wheeled vehicles as a group in the debris could indicate that they might have originated from the same location. One tablet showing a sledge pictogram[41] belongs to a group which can be dated before the Uruk IVb phase; a second tablet[42] to the Uruk IVc phase. If sledge and wagon pictograms can be connected semantically—which is indicated by the similar concept and variations of the signs—it might be possible to date the wagon pictograms into the Uruk IVc phase as well, although the stratigraphic evidence does not support this conclusion with certainty.

The architectural compound called the "Red Temple" itself does not offer further information on the character of the finds as its function is far from being understood. The name implying religious or spiritual activities is misleading since the function has yet to be determined. As Ricardo Eichmann pointed out, the Uruk excavators Julius Jordan and Ernst Heinrich, have coined the name of this building because of a "prune–red" wall plaster and its location near a ziqqurrat.[43] Although the term "temple" is misleading, as the excavators were fully aware, the name was kept in the Uruk excavation reports and diaries for reasons of consistency and practicality as part of the Uruk architectural terminology in the preliminary excavation reports.

In summary, pictograms with wagons are documented in the context of the Uruk IVa phase, but could already have been developed in Uruk IVb or even in Uruk IVc.

Trying to assess absolute–chronological data is as problematic as reconstructing relative chronological information. Absolute dates for the tablets derive from a small number of ^{14}C samples from Uruk IV layers. Three samples taken from charred beams of the ceiling

[32] Eichmann 1989; 2007; Sürenhagen 1999.
[33] Sürenhagen 1999: 107 ff.; 115.
[34] Eichmann 2007: 54–57.
[35] ATU 5: 16.
[36] ATU 2: 33.
[37] Sürenhagen 1999: 108 ("Insgesamt können also 87 Tafeln und Tafelfragmente mit Gewißheit, 73 mit an Sicherheit grenzender Wahrscheinlichkeit und 634 unter gewissem Vorbehalt als Teil des IVc-zeitlichen Füllschuttes im Bereich des Roten Tempels bezeichnet werden").
[38] Sürenhagen 1999: 277–280: "Liste 4. Uruk. Numerische, numerisch-piktographische und piktographische Tontafeln (Duktus 'IV') mit den termini ante quos IVb und IIIc aus dem Bereich südöstlich der Umammu-Ziqqurrat"; here p. 279 no. 50.
[39] Sürenhagen 1999: 117 sub "**d** Schriftzeugnisse".
[40] Sürenhagen 1999.
[41] VAT 15124; ATU: Text no. 431; Sürenhagen 1999: "Liste 4. Uruk", p. 277 no. 13.
[42] VAT 15114; ATU: Text no. 531; Sürenhagen 1999: "Liste 4. Uruk", p. 278 no. 20.
[43] Eichmann 2007: 54 with fn.s 221 and 222.

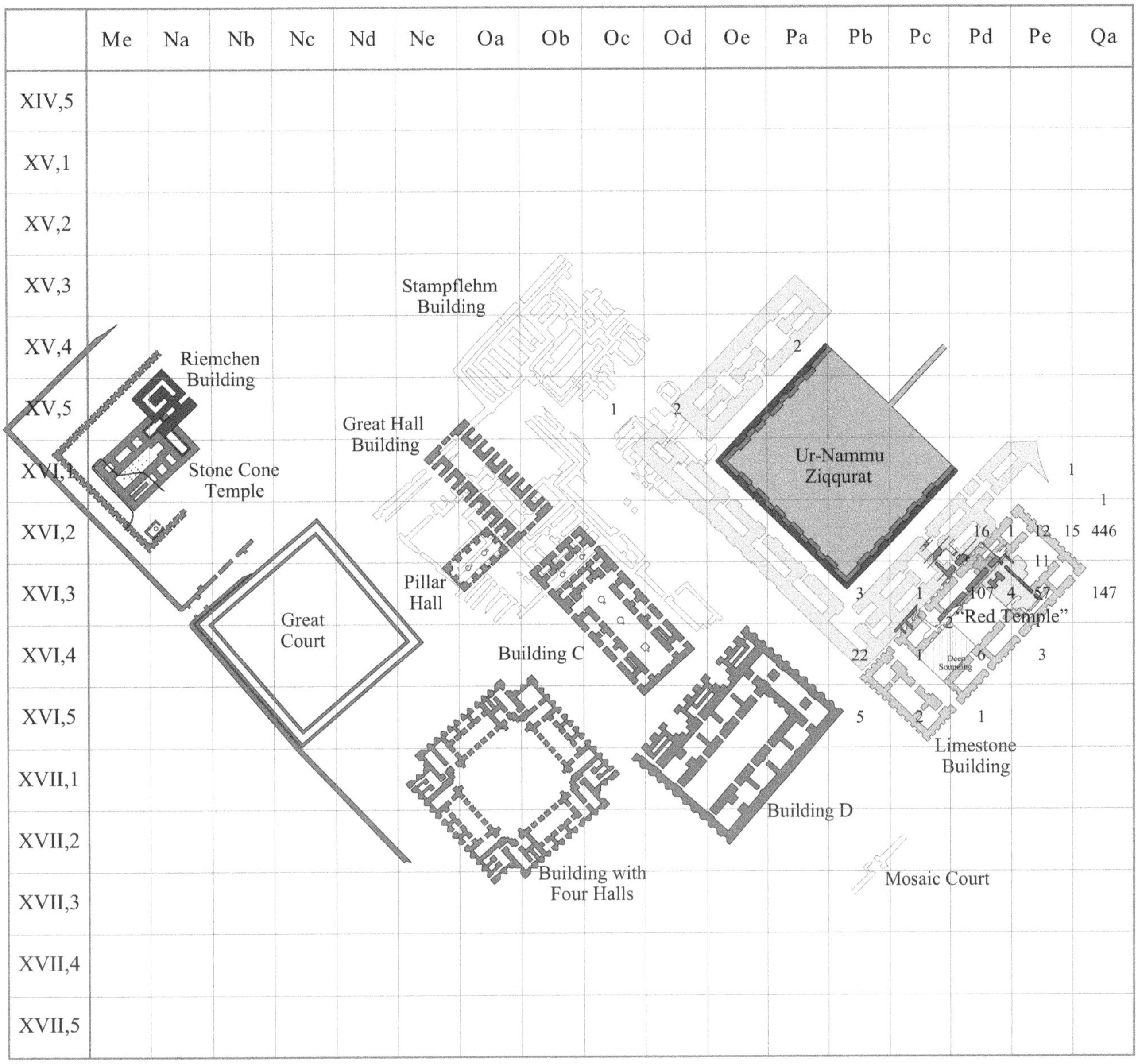

Figure 18. Plan of the Eanna excavations, levels Uruk V–III. The excavation squares shown here—for example, Pe XVI,3—measured 20 x 20 m. Numbers in the plan represent archaic Tablets found in the early Uruk campaigns. After ATU 5: Fig. 2, p.13.

construction of "Building C" range with their median between 3565 and 3428 cal BCE.[44] With refined methods new dates were recently achieved from the same samples.[45] These dates are slightly younger: 3370–3310 cal BCE (2σ). Building C was located ca. 160 m west of the "Red Temple".[46] The building was erected following larger leveling activities in the temple area[47] and marks the beginning of the Uruk IVa phase. Due to the leveling of all buildings in stratum IVa in the Eanna area at the end of this phase "Building C" was also demolished. The stratigraphic position of "Building C" in the architectural context of Uruk IVa and the construction of five—or six partial—stratigraphic layers document continuous usage throughout phase Uruk IVa.[48] In his analysis of Uruk architecture R. Eichmann has meticulously documented and evaluated the stratigraphy and its problems. An overall interpretation of relative chronological layers is restricted by the fact that the architectural phases of "Building C" cannot be synchronised with other buildings of that phase, making it impossible to clearly distinguish between the phases Uruk IVb and IVc.[49] In addition, the postulated separation of layers between the Uruk IV and Uruk III periods is based on challenging assumptions, so that an allocation of "Building C" into the Uruk III period cannot be excluded.[50]

[44] Boehmer *et al.* 1993: 67; Wright and Rupley 2001: 92.
[45] Van Ess 2013.
[46] Eichmann 2007: chapter 6.6.6.1, pp. 236–49.
[47] Eichmann 1989: 79 ff.
[48] Sürenhagen 1999: 79 ff.
[49] Eichmann 1989: 115.
[50] Eichmann 1989: 175.

The ^{14}C samples of building C derive from pinewood (*Pinus* spec.), which was probably imported from the upper area of the Euphrates river; different species of wood are already mentioned in cuneiform tablets in the lexical lists.[51] The timber was taken from trees not older than 80 years.[52] The absolute chronological date only provides a *terminus post quem*, meaning that the "dated" "Building C" was erected sometime after the calibrated ^{14}C date of 3370–3310 cal. BCE.

Therefore, the calibrated ^{14}C dates are deceptive since we do not have any reliable evidence regarding the chronological relationship of the ^{14}C data and the cuneiform tablets with wagon pictograms. This also holds true for several ^{14}C dates of cedar and juniper branches from the Anu Ziggurat, level C. They range between 3517 and 3358 cal. BCE (2σ) and give a rough date for Uruk IV.[53]

In summary, because only a few ^{14}C dates for a larger area of the city of Uruk are available, the absolute chronology of the Uruk period, which could furnish further detailed information, has not been fully established yet. Wright and Rupley (2001) date the beginning of the Late Uruk phase IVc in Mesopotamia not before 3400 BCE, and—based on the only data available at that time (from Godin Tepe)—its end in south–west Iran around 3100 BCE.[54] However, doubts have been voiced regarding this absolute chronological interpretation, because comparison among the artifact inventories indicate that the dates for Godin Tepe should be 200–300 years older.[55] Several new ^{14}C dates place the transition between Uruk IV and III around 3300 BCE.[56] However, since the number of reliable ^{14}C–dates is still low, the statistical foundation of any absolute chronological scheme for Late Uruk period is weak. Still, we gain some evidence for dating the wagon pictograms: On the basis of the relative and absolute–chronological data, we can assume that the pictograms with wheeled vehicles mark the beginning of the script in the ancient Near East and that depicted wagons already belonged to the cultural inventory of the Late Uruk phase IVc, in absolute terms by the beginning of the second half of the 4th millennium BCE.

This conclusion might find support in a small number of objects that have been mostly neglected so far in the synopses of evidence of wheeled vehicles in Mesopotamia.[57] The excavator of Ur Leonard Woolley noted in a short message several finds of wheel models in a Uruk period stratum. One of these wheels is painted in "Ubaid III"–technique.[58] Ubaid III in Woolley's terminology has to be paralleled with Uruk period.[59] Though it is far from clear that these objects were model wheels, the simultaneous occurrence of wagon depictions and these small wheel–like discs might strengthen this assumption.

The Lexical Evidence

One problem regarding the logographic signs indicating wheeled vehicles occurs in the so–called "Lexical Lists" among the documents of the Uruk IV and Uruk III periods in southern Mesopotamia. These are polysemic, which indicates they can have multiple meanings. The sign AN, for example, can signify god, divine being, or a special type of bronze, and it can be used phonetically at the beginning of the sign GA_{2b}×AN (=am_6), which is pronounced */ama/*.

The documents from the Late Uruk period mostly comprise administrative texts and are still difficult to read. The lexical texts, however, are mostly understandable, because the signs and combinations of signs in these texts often occur in a clear context of semantic groups within the 'lexical series'. Lexical texts are already found among the earliest cuneiform documents. These lexical texts contain mostly thematically grouped words and signs. Together with the exercise texts, they form the earliest traces of school activities in Uruk. From the many lexical tablets and fragments lexical series can be reconstructed. R. Englund and H. J. Nissen have reconstructed the early lexical series of the Uruk IV, III and Ur I periods (published as *Die lexikalischen Listen der archaischen Texte aus Uruk* (1993) [= ATU 3]). The editors named most series after their main subject, e. g. 'Wood' for a lexical list of wooden objects.[60] The order of entries of most of these series became more or less fixed in the Uruk III period after a process of standardization. Lexical texts from the Uruk IV period with more or less the same content as the standardised series are called 'forerunners' (precursors) of those series. Versions of the series from later periods with better–known shapes of signs or alternative entries help us considerably in reading and understanding the signs from the Uruk period. The latest versions of this lexical repertoire come from the early Old Babylonian period (ca. 1800 BCE). A new lexical repertoire was developed mainly in the city of Nippur after that period.[61]

There are three cuneiform signs suggesting the idea of wheeled vehicles: PU_2/GIGIR, GURUŠ and MAR. The signs from this period are often cited according to the mentioned sign list compiled by M. W. Green and H. J. Nissen *Zeichenliste der Archaischen Texte aus Uruk* (1987

[51] See next section.
[52] Van Ess 2013.
[53] Van Ess 2013.
[54] Wright and Rupley 2001: 121.
[55] Petrie 2014
[56] Van Ess 2013: 363; Görsdorf and Kohl 2014; Deckers et al. 2015: 420, fig. 14. Deckers et al. let their phase Early Middle Euphrates (EME) 1 start around 3270. EME 1 is a transition phase after the decline of the "Uruk world system" at the end of late Uruk period; see Sconzo 2015: 90 ff.; Finkbeiner et al. 2015: 431 f.
[57] Piggott 1979; Bakker et al. 1999; Crouwel 2004; Burmeister 2011.
[58] Woolley 1955, 28.
[59] Oates 1960: 41.
[60] The indication Professions Lú A originate from another type. The indication Professions is clear, but Lú A or Professions Lú A derived from the later list of professions, that start with the Sumerian word *lú* "man" (incipit). Since there are several Early Dynastic lists of professions, they are distinguished by the letters A, B, and C etc.
[61] For a recent introduction to the lexical lists see Veldhuis 2014. See also Englund 1998 for the texts from Uruk.

Table 3. PU_2/GIGIR "vehicle" in the series 'Wood'

Lexical Series	Uruk Versions	Early Dynastic Versions	Old-Babylonian Versions	Translation
Wood 62	GIŠ-tenû E_2 W 20044,45 00306 GIŠ-tenû E_{2b} W 20327,2 00510 GIŠ-tenû E_{2a} W 24008,10 00302 NIMGIR	—	—	**"wooden cabin (?)"** [1]
Wood 63	GIŠ PU_2/GIGIR W 20044,45 00307 GIŠ PU_2/GIGIR W 20327,2 00511 GIŠ PU_2/GIGIR W24008,1O 00303 GIŠ PU_2 X_2	—	—	**"wooden wagon"**
Wood 64	GIŠ GIGIR GUB_{3b} W 20044,45 00308 GUB_{3b} (?)GIŠ PU_2/GIGIR W 24008,10 00304 GIŠ PU_2/GIGIR [ ]	—	—	**"wooden wagon with side boards (?)"**
Wood 65	[GIŠ?] $GURUŠ_a$ KU_{3a} W 24008,10 00305 $GURUŠ_a$ KU_{3a}? [ ]	—	—	**"sacred/silver (wooden) sledge" (?)**
Wood 66	X KI GIŠ	—	—	"wooden ..."
Wood 67	GIŠ $APIN_b$ KI_a W20044,45 GIŠ $APIN_b$ KI_a	—	—	type of wooden seeding plough
	W20044,46+ UR_{4a} U_a W 20327,2 GIŠ $APIN_b$ KI_a			

[1] Terms related to chariots are highlighted in bold.
[2] W 20044,45. ATU 3: 23; 49; W 20327,2. ATU 3: 55. Published in Du Ry 1969: 37; ATU 2: Fig. 33; ATU 3: 23, Fig. 7 (line drawing); Damerow/Englund/Nissen 1988: 81; Nissen 1986: 327; W 20274,154. ATU 3: 55; W 24008,9. ATU 3: 63.

[= ZATU]). In the following section the occurrences of these signs in the lexical series, occasionally referring to the administrative texts, will be discussed.

PU_2/GIGIR (ZATU 430)

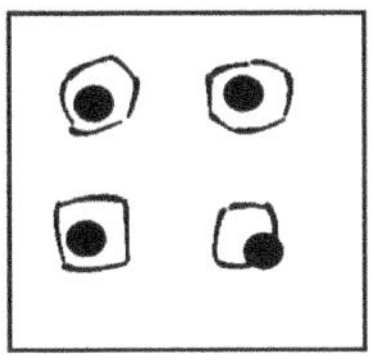

The signs PU_2 and GIGIR (**ZATU 430**) are not separable in the texts from the Uruk period. Only the determinative sign GIŠ "wooden"—usually inserted by the scribe in front of the following sign or group of signs to provide information to the reader of the tablet indicating the nature/classification of the following sign(s)—distinguishes PU_2 "well, pit" from GIŠ PU_2 = (GIŠ) GIGIR "two-wheeler", Akkadian *narkabtu* "chariot". The sign PU_2/GIGIR is either a pictogram of a solid wheel with an axle—from the viewpoint of the observer seen from the side and, perhaps, emphasizing its rotating effect—or a pictogram of a hole in the ground; seen from the top down. The earliest attestations of PU_2/GIGIR are from the Uruk III period (3100–2900 BCE).

P. Steinkeller has extensively discussed the paleography of PU_2 and GIGIR in connection with the signs TIL, BAD and similar signs and comes to the following conclusion: In texts from Fara (the ancient Šuruppak) and Abu Ṣalābīḫ[62] (ca. 2500 BCE), the two signs are differentiated into GIGIR (LAK 770) "wagon, cart", a pictogram of a bolt through the axle, and PU_2/TUL_2 (LAK 771) a pictogram of a hole in the ground.

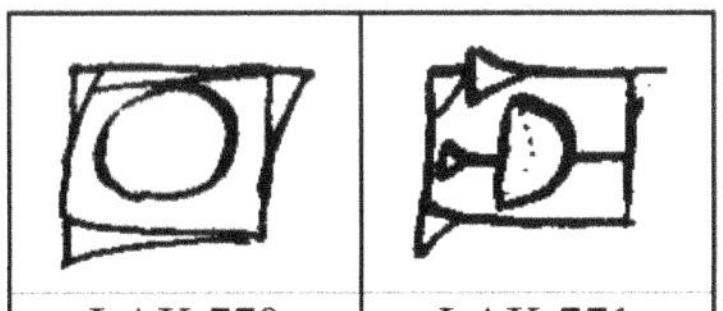

LAK 770 LAK 771

[62] The ancient name is uncertain. Presumably it was called GIŠ NAM_2 after its first entry. For the texts from Fara and Abu Ṣalābīḫ, see the instructive overview of Krebernik 1998.

Table 4. PU$_2$/GIGIR "two-wheeler" in the Lexical Series 'Metal'

Lexical Series	Uruk Versions	Early Dynastic Versions	Old-Babylonian Versions	Translation
Metal 36	NAGAR$_b$ TAK$_{4a}$	ED: bulug$_4$.šu.gur	Kiš: [ ]	"hand chisel" (NAGAR = bulug$_4$)[1]
Metal 36	NAGAR$_b$ TAK$_{4a}$N$_{57}$ = 3 horizontal wedges (?)	—	—	another type of "hand chisel"
Metal 37	AN NAGAR$_b$ TAK$_{4a}$	ED: bulug$_4$.šu.gur.an	Kiš: [ ]	"bronze (?) hand chisel"
Metal 37	AN NAGAR$_b$ TAK$_{4a}$N$_{57}$ (?)	—	—	another type of "bronze (?) hand chisel"
Metal 38	NAGAR$_b$ BUR$_a$	ED: bulug$_4$.bur (H: bulug$_4$.HAR)	Kiš: bulug$_4$.bara$_2$^{uruda}	"chisel for stone vessels"
Metal 39	AN NAGAR$_b$ BUR$_a$	ED: bulug$_4$.bur.an	Kiš: bulug$_4$.bara$_2$.an ^{uruda}	"bronze (?) chisel for stone vessels"
Metal 41	AN NAGAR$_b$ PU$_2$	ED: bulug$_4$.gigir.an	Kiš: bulug$_4$.gigir.an ^{uruda}	**"bronze (?) chisel for wheels or carts"**
	W 15895,be 00101 [ ]			
	W 20266,12 00203 AN NAGAR$_b$ [ ]			
	W 20266,41 00203 NAGAR$_b$ X			
	W 21208,3+ 00303 AN NAGAR$_b$ PU$_2$			
	W23997,2 00208 [ ] PU$_2$ NAGAR$_a$			
	W 24018,5 00205 NAGAR$_b$ AN PU$_2$			
Metal 42	TAK$_{4a}$ MAR$_b$ / (X MAR$_a$)	ED: mar.šu	Kiš: uš.šu ^{uruda}	"spade"
		Ebla: mar.šu		
Metal 43	AN TAK$_{4a}$ MAR$_b$	ED: mar.šu.an	Kiš: uš.šu.an ^{uruda}	"bronze (?) spade"
		Ebla: mar.šu.an		
Metal 44	BAR UŠ$_a$	ED: bar.uš	Kiš: bar.uš ^{uruda}	"type of stick/goad"
Metal 45	AN BAR UŠ$_a$	ED: bar.uš.an	Kiš: bar.uš.an ^{uruda}	"type of bronze (?), stick/goad"

[1] NAGAR is differentiated into the sign NAGAR and the sign BULUG in the Ur III period (Civil 1965: 128).

Later in Presargonic Ĝirsu, the sign PU$_2$/TUL$_2$ was differentiated into two signs: PU$_2$ and TUL$_2$. These sign were registered together with GIGIR under Rosengarten 1967 (Ros.) No. 259.

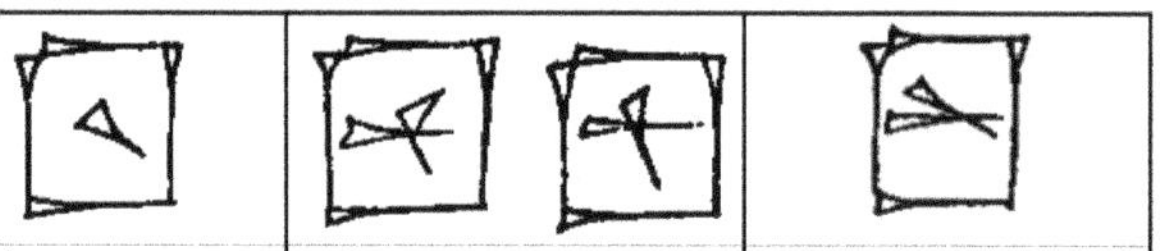

NIĜIN$_2$ × U = PU$_2$ Ros. No. 259A	NIĜIN$_2$ × BAD = GIGIR Ros. No. 259B/C	NIĜIN$_2$ × TIL = TUL$_2$ Ros. No. 259D

The inscribed sign U in PU$_2$ (Ros. No. 259A) is in fact the Sumerian logogram *buru$_3$* "hole", the BAD in GIGIR (Ros. No. 259B/C) is the pictogram for an axle with a bolt, and the sign TIL in TUL$_2$ (Ros. No. 259D) is the logogram *idim* "spring, underground water".[63] TUL$_2$ is also used for *hab$_2$* "stinking" in Presargonic Sumerian.[64]

The Uruk sign PU$_2$/GIGIR occurs twice in a context of vehicles in the series 'Wood' (Table 3).[65] Unfortunately the later tradition of the series 'Wood' is confused, so that we cannot correct the readings of the Uruk version from those later versions.

PU$_2$/GIGIR is attested in the chapter NAGAR (later *bulug$_4$*) "chisel" in the series 'Metal'. Among different types of chisels there is one "chisel for the wheel or cart" (Table 4).

[63] According to Krebernik (1998: 276 fn. 464) the sign IDIM must be distinguished from TIL in the ED IIIa period and identified with LAK 4. In the Ur III and Old-Babylonian period IDIMis distinguished as well (M. Civil in Steinkeller 1981: 21), but not in the intermediate periods.

[64] See Bauer 1989–1990: 78.

[65] Lú A: ATU 3: 69–86; 153–154; Officials: ATU 3: 86–89; Animals: ATU 3: 89–93; Fish: ATU 3: 93–98; Birds: ATU 3: 98–100; Swine [was: Dog]: ATU 3: 100–103; Wood: ATU 3: 103–112; 154–159; Tribute: ATU 3: 112–120; Plants: ATU 3: 120–122; Vessels: ATU 3: 123–34; 159–160; Metal: ATU: 3: 134–141; 159–160; Grain: ATU 3: 142–145; 160; Cities: ATU 3: 145–150; Geography: ATU 3– 150–151; 160–162; Vocabulary: ATU 3: 162–164; Practice: ATU 3: 164–165; Unidentified: ATU 3: 165–176.

GURUŠ (ATU 741–742 / ZATU 247 and ATU 743–745 / ZATU 248)

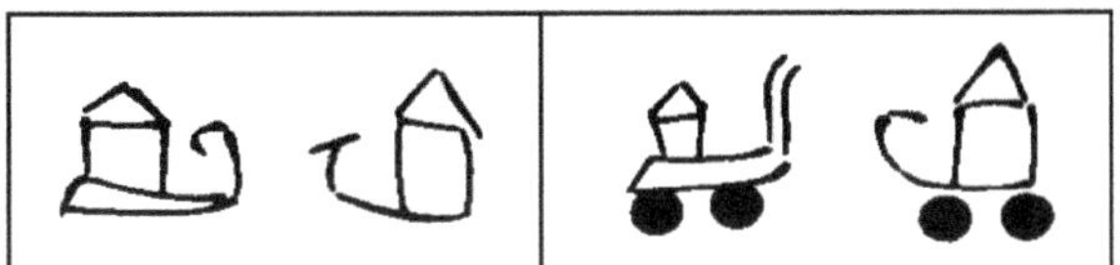

GURUŠ$_a$ (ZATU 247) | GURUŠ$_b$ (ZATU 248)

The pictogram GURUŠ clearly depicts a type of sledge or upper part of a wheeled vehicle, sometimes supplied with two circular impressions, resembling the number sign N$_{14}$, indicating wheels.[66] We must therefore take into consideration that "sledge" and "sledge on wheels" could be meanings for these signs.

GURUŠ occurs in a clear context of vehicles (see the series 'Wood' 65 in Table 3). The earliest attestations of ATU 741–742 / ZATU 247, as well as ATU 743–745 / ZATU 248, are from the Uruk IV–period. ZATU 248 is found only in Uruk IV.

Recently, R. Englund has given his thoughts about GURUŠ$_a$ in a footnote to his article "The Smell of the Cage".[67] On the basis of his thoughts, the attempt will be made to try to reconstruct the history of the signs GURUŠ and KAL in the Uruk and ED IIIa period.

The development of GURUŠ$_a$ (ZATU 247) is difficult to trace, since later parallels for Uruk period lexical entries with GURUŠ$_a$ are lacking. However, the opposite is also the case, precursors for the sign GURUŠ (LAK 709A) are not found in Uruk lexical texts. The only entry that is completely preserved (Lú A 39, Table 5), only contains gal.ĝišgal and not the later version of the Uruk variant GAL$_a$.GURUŠ$_a$. The two other lexical entries: GURUŠ$_a$ KI "place of the sledge (?)" (Vocabulary 1 I 2) and GURUŠ$_a$ LA$_2$ GU$_4$ X "… oxen yoked to the sledge" (Unidentified 36,24 II 4), have no later parallels.[68] The last entry confirms that GURUŠ$_a$ is a vehicle drawn by animals. Therefore, we must conclude that there is no direct proof that GURUŠ$_a$ is the precursor of ED IIIa GURUŠ from the lexical lists.

M. Krebernik has demonstrated that the sign LAK 709A GURUŠ/LAMA/KAL has to be distinguished from LAK 709B KALAG/RIB. LAK 709A GURUŠ is the equivalent of later ĝuruš, "male adult worker, manpower".[69] In the Sargonic period LAK 709A and LAK 709A converge into one sign.[70] Because we are not sure what the Uruk precursor of LAK 709A is (Tables 6–7), Englund (2009) has made it plausible that KUR$_a$ and SAL (e. g. MS 3035 = P006268; W 20274,2 and W 23999,1) are the equivalents of later Urdu "slave" and *geme*$_2$ "slave–girl" and ĜURUŠ$_a$

Table 5. Parallel entries of archaic lexical series with GURUŠ$_a$ in the Uruk and ED IIIa period

	Uruk (ATU 3)	Early Dynastic IIIa/b (Pettinato 1982)
Lú A 39	GAL$_a$ GIŠGAL	gal.ĝišgal
	(var W21418,1: GAL$_a$.GURUŠ$_a$) GIŠGAL	
	GURUŠ$_a$	GIŠGAL
Lú A 118	GAL$_a$ [ ]	gal.ĝuruš
		ĝuruš
Wood 65	GURUŠ$_a$ KU$_{3a}$.[GIŠ]	*no parallels available*

Table 6. Parallel entries of archaic lexical series with ĝuruš (LAK 709A) in the Uruk and ED IIIa period

	Uruk (Englund 1993)	Early Dynastic IIIa/b (Pettinato 1982)
Lú A 118	GAL$_a$ [ ]	gal.ĝuruš
		ĝuruš
Officials / Word List E 91	*no parallels available*	ĝuruš
Officials / Word List E 92	*idem*	aš.ĝuruš
Officials / Word List E 93	*idem*	gur$_4$.gur$_4$.ĝuruš
Officials / Word List E 94	*idem*	ĝuruš.an.dul$_3$.ak

and SAL (e. g. MSVO 1,1; ATU 5: Pl. 66 W 9579,ac) the equivalents for ĝuruš "male adult worker" and *geme*$_2$ "female adult worker" in the administrative texts of the Uruk period.[71] Therefore it is probable, that GURUŠ$_a$ was already pronounced /ĝuruš/.

The question of how the pictogram of a sledge, likely pronounced /ĝuruš/, can be used for "male adult worker" is intriguing. Some think that GURUŠ "sledge" is a Pre–Sumerian word[72], but in our opinion a Sumerian etymology is not excluded: GURUŠ = */gurus/* = **gur/gir* "wheel"[73] + *us*$_2$ "to lean on" = "something that rests on wheels" = sledge on wheels. The problem with this last etymology is that later Sumerian distinguishes /ĝ/ from /g/, but we do not know whether this distinction applies in the earlier periods.[74] The later Sumerian word for "wheel" is

[66] See also Burmeister 2011, 211–214; Steinkeller 1990.
[67] Englund 2009, 8 f. fn. 19.
[68] Vocabulary 1 (ATU 3: 162; Pl. 81); Unidentified 36,24 (ATU 3: 168; Pl. 89).
[69] Cf. Krebernik 1986: 162 and fn. 4; 1998: 277 and fn. 471.
[70] Gelb 1961: 229 No. 173.

[71] Cf. Englund 2009: 8–9 fn. 19; 12–15.
[72] Cf. Englund 1998: 81.
[73] *gigir* "two-wheeler" might be a reduplicated form of the (Pre-) Sumerian word **gir/gur* "hoop, wheel"; cf. Sumerian *gur*$_2$ "hoop" and Akkadian *kakkaru* < *karkaru* "disk, round loaf of bread".
[74] The Old-Babylonian Syllabary ProtoEa (ca. ± 1750 BCE) with very few exceptions uses the sign GA$_2$ for /ĝa/ :: GA for */ga/*, MI for /ĝe/ and /ĝi/ :: GI/GI$_4$ for /ge/ and */gi/*, and MU for /ĝu/ :: GU/GU$_2$ for / *gu/*. For the rendering of the Sumerian /ĝ/ sound in Presargonic Lagaš and in Ebla, see Krispijn 2012: 199; 213. The situation before Ebla is difficult to establish, since phonetic indicators are rare. The only phonetic indicator from earlier periods containing the phoneme /ĝ/ which we know is GIŠ KUŠU KI = [giš]KUŠU[ki] = /ĝišša/ a part of the city of Umma, that is rendered phonetically ĝiš-ša3 in Pre-Sargonic PN's from Umma. See

Table 7. Parallel entries of archaic lexical series with KAL (LAK 709B) in the Uruk and ED IIIa period

	Uruk[1]	**Fara/Abu Ṣalābiḫ[2]**	**Ebla[3]**	**Ur III[4]/Old-Babylonian**
	ZATU 281A	LAK 709B	ELLes 348 = kal	
	= KAL$_a$	= kal		
Tribute 12	1 (N$_1$) KAL$_a$ AB$_2$	1 ab$_2$-KAL		ab$_2$-ga-ri-a$_5$
Tribute 40	1 (N$_1$) KAL$_a$ AB$_2$	1 ab$_2$-KAL		ab$_2$-ga-g[u$_7$]-eš
				ab$_2$-ga-ri (?)-a
Fish 36	PAD$_b$ ku$_{6a}$	pad.kal^{ku6}	Ebla: pad^{ku6}	
	KAL$_{b1}$			
	KAL$_{b2}$			
Swine 41	KAL$_{b1}$ ŠUBUR	*no parallels available*		
Vocabulary 8-2 I 3	NIMGIR KAL$_{b2}$	*no parallels available*		
Vocabulary 8-2 I 4	ŠE$_a$+NAM$_2$ KAL$_{b2}$	*no parallels available*		
Unidentified 62 II 1	[ ] KAL$_{b2}$	*no parallels available*		

[1] Cf. ATU 3; Steinkeller 1990: 22 fn. 32.
[2] Cf. Pettinato 1981; Biggs 1974.
[3] Cf. Pettinato 1981.
[4] Metal: cf. Iraq 31, 3–7. Tribute List score edition: Gelb 2010: 215–228; Civil 2013.
[5] In contrast with Englund (2009: 8 fn. 12) Civil (2013: 31) has another explanation of Tribute List 12// 40 AB$_2$ KAL$_a$ (ZATU 281A/LAK 709B): not "cow of the sledge" (KAL = ĝuruš), but "suckling cow " (KAL = /sub/), which is corroborated by the explanatory phonetic spelling of these lines in the Old-Babylonian versions (Civil 2013: 26–27) containing the word ga "milk", gu$_7$ "to eat, to feed" and ri "to bring into".

umbin "claw, wheel". The supposed precursor of UMBIN in the Uruk period, UMBIN$_{(a)}$ with question mark under ZATU 581 in ATU 2, is only attested to in administrative texts in Uruk (Englund 1994) and therefore identified with later UMBIN (LAK 289; Ros. no. 62). The context of the occurrences of UMBIN$_{(a)}$ does not indicate the meaning "wheel" in this period.[75]

Since GURUŠ$_a$ (ZATU 248) is the earliest confirmation (Uruk IV period) of a wheeled vehicle, the occurrence needs special attention. Piotr Steinkeller (1995) states that the 2N$_{14}$ is part of the sign; two circular impressions with the back of the stylus is not the sign TAR (ZATU), but the pictogram of rollers or wheels.[76]

The only attestation in lexical texts, in a 'forerunner' of the series Professions Lú A, W 9656,h III 10 is listed in a context of the professions "gardener" and "prayer priest".[77] It must indicate a profession linked with the depicted vehicle. A possible parallel from 'Professions Lú A' might be GAL$_a$ GIŠGAL "man of the pedestal" with the variant GAL$_a$ GURUŠ$_a$ "man of the sledge" (Lú A 39). GIŠGAL usually indicates a pedestal for divine statues. Since GURUŠ$_a$ "sledge" is a variant of GIŠGAL in one lexical fragment, it can be interpreted as movable pedestal for a divine statue. This might be supported by the entry [GIŠ?] GURUŠ$_a$ KU$_{3a}$ "wooden sacred sledge" (Series 'Wood' 65), although an interpretation KU$_{3a}$ "silver" i.e., "with silver or golden fittings" cannot not be excluded.[78]

MAR$_a$ and MAR$_b$ (ZATU 252)

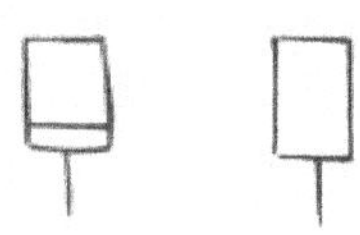

The last sign that might indicate a wheeled vehicle is ^{ĝiš}MAR.[79] ^{ĝiš}MAR is a pictogram of a shovel, which is the primary meaning of the sign. The earliest attestations for a type of vehicle, indicating a four–wheeled transport vehicle, come from Early Dynastic (IIIb) Ĝirsu (ca. 2450 BCE). Since the Ur III period (2100–2000 BCE) the longer form ^{ĝiš}*mar–gid$_2$–da* "drawn *mar*" comes up. Since MAR$_b$ occurs in the context of metal objects, it seems likely that it indicates a metal shovel, presumably made of copper,[80] so we could think of a shovel with a predominant metallic part (the shovel with a wooden stick). MAR$_a$ is combined with APIN$_a$ "plough, leader of the plough team" in the lexical series Lú A 74 and elsewhere with ŠE$_{a,}$ "grain", which could indicate a "shovel" or "winnow".[81] As far as we

Frayne 2008: 357–359. Steinkeller (1995: 694–695) lists GAR in the sign combination GA$_2$+GAR = ĝalga among phonetic indicators of the Uruk Period. But it is not clear in which part of /ĝalga/ GAR functions phonetically: GAR could equally well be a gloss on the first part = */ĝar/* or /ĝal/ as on the second part = */ga/.*

[75] There are some occurrences of UMBIN in a context of animals, e.g. ATU 5: Pl. 055, W 9335,s and ATU 5: Pl. 066, W 9579,ai and ATU 5: Pl. 109, W 9656,fv. This sign might be the precursor of the sausage-shaped container of fat of sheep and pigs as UMBIN in the late presargonic 'early' Isin text IM 11053/156 XVI 4 (Steinkeller and Postgate 1992: 29).

[76] The same circular impression indicates the number 10 and is coded as Number Sign 14 = N$_{14}$ (ZATU 337).

[77] "gardener" W 9656,h III 9 NIMa // GAL$_a$ NIM$_a$/NI$_b$ Professions Lú A 23 Uruk III and "prayer priest" W 9656,h III 11 ŠITA$_{a1}$ // GAL$_a$ GURUŠ$_a$ Professions Lú A 14 Uruk III.

[78] See Krispijn *forthcoming*: 263 f.

[79] ATU: 592–593; 56 fn. 5: "In dem Zeichen MAR (Nr. 592–593) sehe ich das Abbild eines Spatens, nicht das eines Wagens [so A. Deimel, ŠL 307, I und E. Herzfeld, AMI 6 (1934), 202]".

[80] TAK$_{4a}$ MAR "hand shovel" 'Metal' 42 and AN TAK$_{4a}$ MAR "bronze hand shovel" 'Metal' 43.

[81] 'Wood' (?) W 24012,2 III 2, Geography 3 II 6, Unid. 68 r I 3 (ATU 3).

know there are no clear attestations of ĝišMAR as "vehicle" either in the Uruk lexical texts or in the administrative texts. The later form ĝiš*mar–gid*$_2$*–da* "drawn *mar*"[82], can help us to understand the development of the term ĝišMAR as vehicle. It is very likely that the four–wheeled vehicle ĝišMAR originated from a transporting device drawn by oxen resembling a shovel, and it cannot be excluded that this device was already used in the Uruk period.

Conclusion

The German sumerologist Adam Falkenstein distinguished two sets of pictograms on Archaic Texts—mainly from Uruk in southern Mesopotamia—which, on one hand, depict sledges with variations of house–like constructions and, on the other hand, sledges with house–like constructions showing two round impressions beneath the runners (ATU 1: 56). These two circular impressions beneath the runners have been identified by Falkenstein as representations of solid–disc wheels. Over the following decades most scholars have followed Falkenstein's interpretation. In the 1980s members of the Berlin Uruk project, such as R. K. Englund, H. J. Nissen and M. W. Green, transliterated the two round impressions beneath a sledge as the numeric sign $2N_{14}$ (● ●). Furthermore, it was concluded that the meaning of the sign GURUŠ was reduced exclusively to "worker, slave, etc.". Following this reading the depiction of wheeled vehicles on the Archaic Texts must be excluded. However, an analysis of all available material raises questions about that interpretation, first summarised by S. Burmeister (2011: 211–215), provides the following evidence: The two circular impressions beneath a sledge are exactly where we would expect them to be, if they indicate wheels as part of pictographic representations of wheeled vehicles. The two round impressions beneath the runners seem to be flatter when compared to the numerical sign $2N_{14}$. It has also been observed that the numerical signs are usually not directly in close vicinity to the object they count. p. Steinkeller (1995) concluded that $2N_{14}$ is not the number, but part of the ideogram, and that the two circular impressions are not the sign TAR (as listed in ZATU under sign 248), but have to be interpreted as an ideogram of rollers or wheels. GURUŠ occurs in a clearly defined context of vehicles (see the series 'Wood' 65 in Table 1). GURUŠ "sledge" has been interpreted as a Pre–Sumerian word. However, a Sumerian etymology is not excluded: GURUŠ = */gurus/* = **gur/gir* "wheel" + *us*2 "to lean on" = "something that rests on wheels" = sledge on wheels. The problem with this last etymology is that later Sumerian distinguishes /ĝ/ from /g/, but we do not know whether this distinction applies to the earlier periods. In summary, it is most likely that the two impressions beneath a sledge pictogram represent wheels; it is not very likely that they have to be read as the numerical sign $2N_{14}$. Falkenstein's original observation still has the most plausibility. The small number of early cuneiform tablets with wagon depictions provide evidence that the Mesopotamian city states in the 4th millennium BCE belonged to those societies that used wheeled vehicles from the earliest time.

Acknowledgements

The first six sections have been written by Stefan Burmeister and Peter Raulwing; section "The Lexical Evidence" by Theo Krispijn, the conclusions and the references cited by all three authors. We would like to thank Joost Crouwel for his valuable comments and discussion, Barbara Hellwing (Sydney), Paola Sconzo and Katleen Deckers (both Tübingen) for kindly discussing the radiocarbon chronology and providing us with helpful literature. Further, we would like to thank Markus Hilgert, Olaf Teßmer, and Alrun Gutow from the *Vorderasiatisches Museum* Berlin, and Margarete van Ess and Irmgard Wagner from the German Archaeological Institute, Orient Department, for kindly providing us with photos of the tablets shown here as illustration and with information on the objects. Last but not least, we thank Robert Englund for discussing the topic of GURUŠ with us and kindly granting us permission to use his data from The *Cuneiform Digital Library Initiative* (http://cdli.ucla.edu) for our article.

Abbreviations[83]

Archaic Bookkeeping H. J. Nissen, p. Damerow and R. Englund, *Archaic Bookkeeping. Early Writing and Techniques of the Economic Administration in the Ancient Near East.* University of Chicago Press, Chicago. [Translation of *Frühe Schrift und Techniken der Wirtschaftsverwaltung im alten Vorderen Orient. Informationsspeicherung und –verarbeitung vor 5000 Jahren.* Franzbecker, Bad Salzdetfurth, 1990 by Paul Larsen; exhibition catalogue].

ATU [1] A. Falkenstein, *Archaische Texte aus Uruk.* Ausgrabungen der Deutschen Forschungsgemeinschaft in Uruk–Warka 2. Deutsche Forschungsgemeinschaft, Berlin and Harrassowitz, Leipzig, 1936.[84]

ATU 2 See ZATU

ATU 3 R. K. Englund and H. J. Nissen, *Die lexikalischen Listen der archaischen Texte aus Uruk.* Archaische Texte aus Uruk 3 = Ausgrabungen der Deutschen Forschungsgemeinschaft in Uruk–Warka 13. Gebr. Mann, Berlin, 1993.

[82] The form *gid2-da* is a participle form of the *gid2* "to draw". Less likely is, that *gid2-da* comes from *gid2* "long", because a form *mar-gid2* is more likely then.

[83] See http://cdli.ox.ac.uk/wiki/abbreviations_for_assyriology.

[84] H. J. Nissen reflecting on the series ATU in ZATU, preface: "Als Zeichen der Wertschätzung dieser früheren Arbeit und um die Kontinuität sichtbar werden zu lassen, habe ich mich entschlossen, in die neue Reihe 'Archaische Texte aus Uruk' den genannten Band Falkensteins mit demselben Titel als Band 1 einzubeziehen. Die Reihe der Veröffentlichungen der Ergebnisse unserer Arbeit beginnt daher mit Band 2. Band 3 wird die Lexikalischen Listen behandeln, Band 4 wird einen vollständigen, kommentierten Katalog aller Archaischen Texte aus Uruk enthalten, und von Band 5 an werden die Texte in Kopie veröffentlicht". In this appendix we use ATU and ZATU + a number exclusively to refer to cuneiform signs. When the authors are quoted, they are referenced as Falkenstein, ATU 1: followed by a page number or other references other than cuneiform signs. Steinkeller 1990: 19 fn.* uses ZAT for ZATU and ZATU (1995–1996: 211). ATU 3 is also abbreviated as LATU (Krebernik 1994: 381).

ATU 5 — R. K. Englund, *Archaic Administrative Texts from Uruk. The Early Campaigns*. Archaische Texte aus Uruk 5 = Ausgrabungen der Deutschen Forschungsgemeinschaft in Uruk–Warka 15. Gebr. Mann, Berlin, 1994.

ED — Early Dynastic.

LAK — A. Deimel, *Die Inschriften von Fara I. Liste der archaischen Keilschriftzeichen*. Ausgrabungen der Deutschen Orient–Gesellschaft in Fara und Abu Hatab 1 = Wissenschaftliche Veröffentlichungen der Deutschen Orientgesellschaft 40. Hinrichs, Leipzig, 1922. [Reprint Zeller, Osnabrück, 1970].

MSVO 4 — R. K. Englund, *Proto–Cuneiform Texts from Diverse Collections*. Materialien zu den frühen Schriftzeugnissen des Vorderen Orients 4. Gebr. Mann, Berlin, 1996.

PI — S. Langdon, *Pictographic Inscriptions from Jemdet Nasr*. The Herbert Weld Collection in the Ashmolean Museum. Oxford University Press et al., London et al., 1928.

UET — *Ur Excavations. Texts. Joint Expedition of the British Museum and of the Museum of the University of Pennsylvania to Mesopotamia*. London: Printed by Order of the Trustees of the two Museums, 1928 ff. [Vol. II. *Archaic Texts*, by E. Burrows, 1935].

ZATU — M. W. Green and H. J. Nissen, *Zeichenliste der Archaischen Texte aus Uruk*. Archaische Texte aus Uruk 2 = Ausgrabungen der Deutschen Forschungsgemeinschaft in Uruk–Warka 11. Gebr. Mann, Berlin, 1987.

Bibliography

Bakker, J. A., J. Kruk, A. E. Lanting and S. Milisauskas. 1999.The earliest evidence of wheeled vehicles in Europe and the Near East. *Antiquity* 73: 778–790.

Bauer, J. 1989–1990. Altsumerische Wirtschaftsurkunden in Leningrad. *Archiv für Orientforschung* 36/37: 76–91.

Boehmer, R. M., G. Dreyer and B. Kromer. 1993. Einige frühzeitliche ^{14}C–Datierungen aus Abydos und Uruk. *Mitteilungen des Deutschen Archäologischen Instituts, Abteilung Kairo* 49: 63–68.

Burmeister, S. 2011. Innovationswege—Wege der Kommunikation. Erkenntnisprobleme am Beispiel des Wagens im 4. Jt. v. Chr. In *Sozialarchäologische Perspektiven: Gesellschaftlicher Wandel 5000–1500 v. Chr. zwischen Atlantik und Kaukasus*. [International conference (Kiel 2007)], edited by S. Hansen and J. Müller, 211–240. von Zabern, Darmstadt.

Childe, G. 1951. The first wagons and carts—from the Tigris to the Severn. *Proceedings of the Prehistoric Society* 17: 177–194.

Civil, M. 1965. *Le débat sumérien entre la houe et l'araire*. Thèse de 3e cycle: Lettres, Paris. [unpublished].

Crouwel, J. 2004. Der Alte Orient und seine Rolle in der Entwicklung von Fahrzeugen. In: *Rad und Wagen—Der Ursprung einer Innovation. Wagen im Vorderen Orient und Europa,* edited by M. Fansa and S. Burmeister 2004: 69–86.

Damerow, p. and R. K. Englund. 1987. Die Zahlzeichensysteme der archaischen Texte aus Uruk. In: ZATU: 117–166.

Damerow, P., R. K. Englund, and H. J. Nissen. 1988. Die Entstehung der Schrift, *Spektrum der Wissenschaft*. Februar 1988, 74–85 [reprinted in: J. Dittami, ed., *Signale und Kommunikation. Mechanismen des Informationsaustauschs in lebenden Systemen*. Spektrum Akademie Verlag Heidelberg etc. 1993, 150–161 and B. Riese, ed., *Sprache und Schrift. Ein Lesebuch*. Spektrum Akademie Verlag, Heidelberg etc. 1994] 90–101).

Deckers, K., p. Drechsler and p. Sconzo. 2015. Radiocarbon Chronology. In: *Middle Euphrates*, edited by U. Finkbeiner, M. Novák, F. Sakal, and p. Sconzo, 401–421. ARCANE. Associated Regional Chronologies for the Ancient Near East and the Eastern Mediterranean 4. Brepols, Turnhout.

Drenth, E. and J.A. Bakker. 2005–06. Im memoriam Albert Lanting (10 April 1941–13 December 2004). *Palaeohistoria* 47–48: 3–37.

Du Ry, C. 1969. *Völker des Alten Orient*. Holle, Baden–Baden.

Eichmann, R. 1989. *Uruk. Die Stratigraphie. Grabungen 1912–1977 in den Bereichen "Eanna" und "Anu–Ziqqurat"*. Ausgrabungen in Uruk–Warka. Endberichte 3. Von Zabern, Mainz.

Eichmann, R. 2007. *Uruk. Architektur I. Von den Anfängen bis zur frühdynastischen Zeit Ausgrabungen*. 2 vols. Uruk–Warka. Endberichte 14. Leidorf, Rahden/Westf.

Engel, T. and H. Kürschner. 1993. Weitere Holzkohlen und Holzkohlenreste aus Uruk. *Baghdader Mitteilungen* 24: 127–135.

Englund, R. K. 1998. Texts from the Late Uruk period. In *Mesopotamien. Späturuk–Zeit und Frühdynastische Zeit. Annäherungen 1*, edited by p. Attinger and M. Wäfler, 15–233. Orbis Biblicus et Orientalis 160/1. Universitätsverlag, Freiburg (Switzerland) and Vandenhoeck & Ruprecht, Göttingen.

Englund, R. K. 2009. The smell of the cage. Cuneiform Digital Library Journal 2009:4 Version: 21 August 2009. http://www.cdli.ucla.edu/pubs/cdlj/2009/cdlj2009_004.html. [accessed January 24, 2015; printed in: T. M. Nikolaeva (ed.), *Issledovanija po lingvistike i semiotike. Sbornik statej k jubileju Vjačeslava Vs. Ivanova* Jazyki Slavjanskich Kul'tur, Moskva, 2010].

van Ess, M. 2013. Neue Radiokarbondatierungen aus Uruk. In: *Uruk 5000 Jahre Megacity*, edited by N. Crüsemann, M.van Ess, M. Hilgert, and B. Salje, 362–363. Vorderasiatisches Museum—Staatliche Museen zu Berlin, Reiss–Engelhorn–Museen, Deutschen Archäologischen Institut, und Deutschen Orient–Gesellschaft, Berlin.

van Ess, M. and R. Neef. 2013. Bauholz für die Tempel. In: *Uruk 5000 Jahre Megacity*, edited by N. Crüsemann, M.van Ess, M. Hilgert, and B. Salje, 58–59. Vorderasiatisches Museum—Staatliche Museen zu Berlin, Reiss–Engelhorn–Museen, Deutschen Archäologischen Institut, und Deutschen Orient–Gesellschaft, Berlin.

Fansa, M. and S. Burmeister (eds.). 2004. *Rad und Wagen—Der Ursprung einer Innovation. Wagen im Vorderen Orient und Europa.* Archäologische Mitteilungen aus Nordwestdeutschland, Beiheft 40. Von Zabern, Mainz.

Finkbeiner, U., M. Novák, F. Sakal and p. Sconzo. 2015. Conclusion. In: *Middle Euphrates*, edited by U. Finkbeiner, M. Novák, F. Sakal, and p. Sconzo, 431–438. ARCANE. Associated Regional Chronologies for the Ancient Near East and the Eastern Mediterranean 4. Brepols, Turnhout.

Frayne, D. 2008. *Presargonic Period (2700–2350 B.C.)*. Royal Inscriptions of Mesopotamia. Early Periods 1. University of Toronto Press, Toronto.

Gelb, I. 1961. *Sargonic Texts from the Diyala Region.* 2nd edition. Materials for the Assyrian Dictionary, 1. University of Chicago Press, Chicago.

Görsdorf, J. and G. Kohl. 2014. Radiokohlenstoffdatierungen. In: *Die Kleinfunde von Habuba Kabira–Süd 1, Text*, edited by E. Strommenger and K. Kohlmeyer, 437. Ausgrabungen in Habuba Kabira. Harrassowitz, Wiesbaden.

Kramer, S. N. 1963. *The Sumerians. Their History, Culture, and Character.* Chicago University Press, Chicago.

Krebernik, M. 1986. Die Götterlisten aus Fara. *Zeitschrift für Assyriologie* 76: 161– 204.

Krebernik, M. 1994. Review of ZATU. In: *Orientalistische Literaturzeitung* 89: 380–385.

Krebernik, M. 1998. Die Texte aus Fāra und Tell Abū Ṣalābīḫ. In *Mesopotamien. Späturuk–Zeit und Frühdynastische Zeit. Annäherungen 1*, edited by p. Attinger and M. Wäfler, 235–427. Orbis Biblicus et Orientalis 160/1. Universitätsverlag, Freiburg (Switzerland) and Vandenhoeck & Ruprecht, Göttingen.

Krispijn, T. J. H. 2012. Writing Semitic with cuneiform script. The interaction of Sumerian and Akkadian orthography in the second half of the third millennium BC. In: *The Idea of Writing. Writing Across Borders*, edited by A. de Voogt and J. F. Quack, 181–218. Brill, Leiden and Boston.

Krispijn, T. J. H. *Forthcoming.* Early silver: thoughts about the sign *KU*3 in the earliest documents of Uruk. In: *Studies in Silver, Money and Credit. A Tribute to Robartus J. van der Spek on Occasion of his 65th Birthday on 18th September 2014*, edited by K. Kleber & R. Pirngruber. Nederlands Instituut voor the Nabije Oosten, Leiden.

Littauer, M. A. and J. Crouwel. 1979. *Wheeled Vehicles and Ridden Animals in the Ancient Near East.* Handbuch der Orientalistik, Abt. 7: Kunst und Archäologie 1: Der alte Vordere Orient, Abschn. 2: Die Denkmäler, B: Vorderasien, Lfg. 1 Brill, Leiden and Köln.

Littauer, M. A. and J. Crouwel. 1990. Ceremonial threshing in the ancient Near East [Part I]. Archaeological evidence. *Iraq* 52: 15–19.

Nagel, W. and E. Strommenger. 1994. Der frühsumerische Kultschlitten—ein Vorläufer des Wagens? In: *Beiträge zur Altorientalischen Archäologie und Altertumskunde. Festschrift für Barthel Hrouda zum 65. Geburtstag*, edited by p. Calmeyer, 202–209. Harrassowitz, Wiesbaden.

Nissen, H. J. 1986. The Archaic Texts from Uruk. *World Archaeology* 17, 3 (Early Writing Systems): 317–334.

Oates, J. 1960. Ur and Eridu, the Prehistory. *Iraq* 22: 32–50.

Parpola, A. 2008. Proto–Indo–European speakers of the Late Tripolye Culture as the inventors of wheeled vehicles: linguistic and archaeological considerations of the PIE homeland problem. In: *Proceedings of the 19th Annual UCLA Indo–European Conference: Los Angeles, November 2–3, 2006* [recte 2007]*: Selected Papers*: 1–59, edited by K. Jones–Bley. Journal of Indo–European studies. Monograph 54. Institute for the Study of Man, Washington/DC.

Petrie, C. A. 2014. Iran and Uruk Mesopotamia: chronologies and connections in the 4th millennium BC. In: *Preludes to Urbanism in the Ancient Near East*, edited by A. MacMahon, H. Crawford and J. N. Postgate. A Conference in Honour of Joan Oates, 137–155.

Piggott, S. 1978. Chinese chariotry: an outsider's view. In: *Arts of the Eurasian Steppelands*, edited by p. Denwood, 32–51. Colloquies on Art and Archaeology in Asia 7. Percival David Foundation of Chinese Art, London.

Piggott, S. 1979. The First Wagons and Carts: twenty–five years later. *Bulletin of the Institute of Archaeology London* 16: 3–17.

Piggott, S. 1983. *The Earliest Wheeled Transport: From the Atlantic Coast to the Caspian Sea.* Thames and Hudson, London.

Radner, K. and E. Robson (eds.). 2011. The Oxford Handbook of Cuneiform Culture. Oxford University Press, Oxford and New York.

Rubio, G. 1999. On the alleged "pre–Sumerian substratum." *Journal of Cuneiform Studies* 51: 1–16.

Rubio, G. 2005. On the linguistic landscape of early Mesopotamia. In: *Ethnicity in Ancient Mesopotamia. Ethnicity in Ancient Mesopotamia. Papers Read at the 48th Rencontre Assyriologique Internationale Leiden, 1–4 July 2002*, edited by W. H. van Soldt, 316–332. Uitgaven van het Nederlands Instituut voor het Nabije Oosten te Leiden 102.

Salonen, A. 1951. *Die Landfahrzeuge des alten Mesopotamien* (mit besonderer Berücksichtigung der 5. Tafel der Serie ḪAR–ra= ḫubullu). Eine lexikalische und kulturgeschichtliche Untersuchung. Suomalaisen Tiedeakatemian toimituksia, Sarja B 72,3. Suomalainen Tiedeakatemia, Helsinki.

Schmandt–Besserat, D. 1992. *Before Writing 1: From Counting to Cuneiform*. University of Texas Press, Austin.

Sconzo, p. 2015. Ceramics. In: *Middle Euphrates*, edited by U. Finkbeiner, M. Novák, F. Sakal, and p. Sconzo, 85–202. ARCANE. Associated Regional Chronologies for the Ancient Near East and the Eastern Mediterranean 4. Brepols, Turnhout.

Sherratt, A. 1981. Plough and pastoralism: aspects of the Secondary Products Revolution. In: *Pattern of the Past: Studies in Honour of David Clarke*, edited by I. Hodder, G. Isaac and N. Hammond, 261–305. Cambridge University Press, Cambridge.

Sherratt, A. 1983. The secondary exploitation of animals in the Old World. *World Archaeology* 15/1 [Transhumance and Pastoralism]: 90–104.

Sherratt, A. 1997. Economy and Society in Prehistoric Europe. Changing Perspectives. Edinburgh University Press, Edinburgh.

van Soldt, W. H. (ed.). 2005. *Ethnicity in Ancient Mesopotamia. Ethnicity in Ancient Mesopotamia. Papers Read at the 48th Rencontre Assyriologique Internationale Leiden, 1–4 July 2002*. Uitgaven van het Nederlands Instituut voor het Nabije Oosten te Leiden 102.

Steinkeller, p. 1981. Studies in third millennium palaeography, 1: Signs TIL and BAD. *Zeitschrift für Assyriologie* 71: 19–28.

Steinkeller, p. 1990. Ceremonial threshing in ancient Near East [Part] II. Threshing implements in ancient Mesopotamia: cuneiform sources. *Iraq* 52: 19–23.

Steinkeller, p. 1995. Review of ZATU. *Bibliotheca Orientalis* 52: 689–713.

Steinkeller, p. 1995–1996. Review of ATU 3. In: *Archiv für Orientforschung* 42/43: 211–214.

Steinkeller, p. and J. N. Postgate. 1992. *Third–Millennium Legal and Administrative Texts in the Iraq Museum, Baghdad*. Eisenbrauns, Winona Lake.

Sürenhagen, D. 1993. Relative Chronology of the Uruk Period. *Bulletin of the Canadian Society for Mesopotamian Studies* 25 (May): 57–70.

Sürenhagen, D. 1999. *Untersuchungen zur relativen Chronologie Babyloniens und angrenzender Gebiete von der ausgehenden ʿUbaidzeit bis zum Beginn der Frühdynastisch–II–Zeit*. Heidelberger Studien zu Alten Orient 8. Heidelberger Orientverlag, Heidelberg.

Vaiman, A. A. 1974. Über die protosumerische Schrift. *Acta antiqua Academiae Scientiarum Hungaricae* 22: 15–27.

Veldhuis, N. 2014. *History of the Cuneiform Lexical Tradition*. Guides to the Mesopotamian Textual Record 6, Ugarit–Verlag, Münster.

Whittaker, G. 1998. Traces of an early Indo–European language in southern Mesopotamia. *Göttinger Beiträge zur Sprachwissenschaft* 1: 111–147.

Whittaker, G. 2001. The dawn of writing and phoneticism. In: *Hieroglyphen, Alphabete, Schriftreformen. Studien zu Multiliteralismus, Schriftwechsel und Orthographienregelungen*, edited by D. Borchers, F. Kammerzell and S. Weninger, 11–50. Lingua Aegyptia. Studia Monographica 3. Seminar für Ägyptologie und Koptologie, Göttingen.

Whittaker, G. 2004. Word formation in Euphratic. In: *Indo–European Word Formation. Proceedings of the Conference Held at the University of Copenhagen October 20th–22nd 2000*, edited by J. Clackson and B. A. Olsen, 381–423. Museum Tusculanum Press, Copenhagen.

Whittaker, G. 2004–2005. Some Euphratic adjectives. *Göttinger Beiträge zur Sprachwissenschaft* 10/11: 119–139.

Whittaker, G. 2005. The Sumerian Question: Reviewing the issues. In: *Ethnicity in Ancient Mesopotamia. Ethnicity in Ancient Mesopotamia. Papers Read at the 48th Rencontre Assyriologique Internationale Leiden, 1–4 July 2002*, edited by W. H. van Soldt, 409–429. Uitgaven van het Nederlands Instituut voor het Nabije Oosten te Leiden 102.

Whittaker, G. 2008. The case for Euphratic. *Bulletin of the Georgian National Academy of Sciences* 2/3: 156–168.

Woolley, L. 1955. *Ur Excavations, vol. IV. The Early Periods*. British Museum Press, London and University of Pennsylvania Museum, Philadelphia.

Wright, H. T. and E. S. A. Rupley 2001. Calibrated radiocarbon age determinations of Uruk–related assemblies. In: *Uruk Mesopotamia & its Neighbors. Cross–Cultural Interactions in the Era of State Formation*, edited by M. S. Rothmann, 85–122. School of American Research Advanced Seminar Series. School of American Research Press, Santa Fe, NM.

5

Equids in Mesopotamia—A Short Ride through Selected Textual Sources[1]

Joachim Marzahn

In this paper I am going to examine some of the cuneiform tablets in the ancient Near East—especially those from Mesopotamia and present some thoughts by re–examining the world of textual sources covering a period of three thousand years from the 4th to the 1st millennium BCE. My synopsis not only includes a view of the earliest written documents, but also different text genres. This review will, although condensed, provide an overview including all equids, not just horses.

When I received an invitation to give a talk at the Lexington Conference in Honor of Mary Littauer, initially I felt very honored, and this is the right opportunity to express my gratitude to the organizers, especially to Sandra Olsen. However, my second thought was: What is left to say about horses in the ancient Near East? Much has been written about the subject, and I do not want to repeat what is already widely known. To name only a few—Mary Littauer, Nicholas Postgate, Juris Zarins and others[2]—have contributed so much to our knowledge that it became a challenge to me to find new information; and, as the following overview hopefully shows, I hope I did.

I want to restrict myself to my field of expertise, the cuneiform tablets in the ancient Near East—especially those from Mesopotamia—and present some thoughts by re–examining the world of textual sources covering a period of three thousand years from the 4th to the 1st millennium BCE (Figure 1). My synopsis not only includes a view of the earliest written documents, but also different text genres. One the one hand, all this review will be somewhat condensed; on the other hand it will also include all equids, not just horses.

At the very beginning of writing in the ancient Near East we already find a variety of notations of equids. The above–mentioned variety concerns donkeys, the animals particularly used in the local economy. The so–called Archaic Texts from Uruk—written arguably in Sumerian, an isolated language that was spoken in southern Mesopotamia (in present–day Iraq)—show not only a broad use of related cuneiform signs denoting these animals, but we also find at least five different notations that can be identified with the later phonetic value of ANŠE "donkey" (Figure 2). The examples shown here have been connected to stratgraphic levels IV to II in Uruk excavated since 1928/29 by German archaeologists and

Figure 1. Top: Archaic cuneiform tablet with the image of an equid (obverse), Berlin, Vorderasiatisches Museum, VAT 17600. From Uruk, ca. 3350–3200 BCE. © Vorderasiatisches Museum. Bottom: Drawing of the tablet, from: http://cdli.ucla.edu, P001946. © CDLI.

[1] This contribution was given as a lecture at the the Symposium "Equids in the Ancient Near East, Egypt, and Arabia", June 1–3, 2010, Kentucky Horse Park, Lexington. It has been revised and supplemented with bibliography and references to be printed in this edition.

[2] Littauer/Crouwel 1979; Postgate 1986; Zarins 1976; 1986; 2014; von den Driesch/Raulwing 2004.

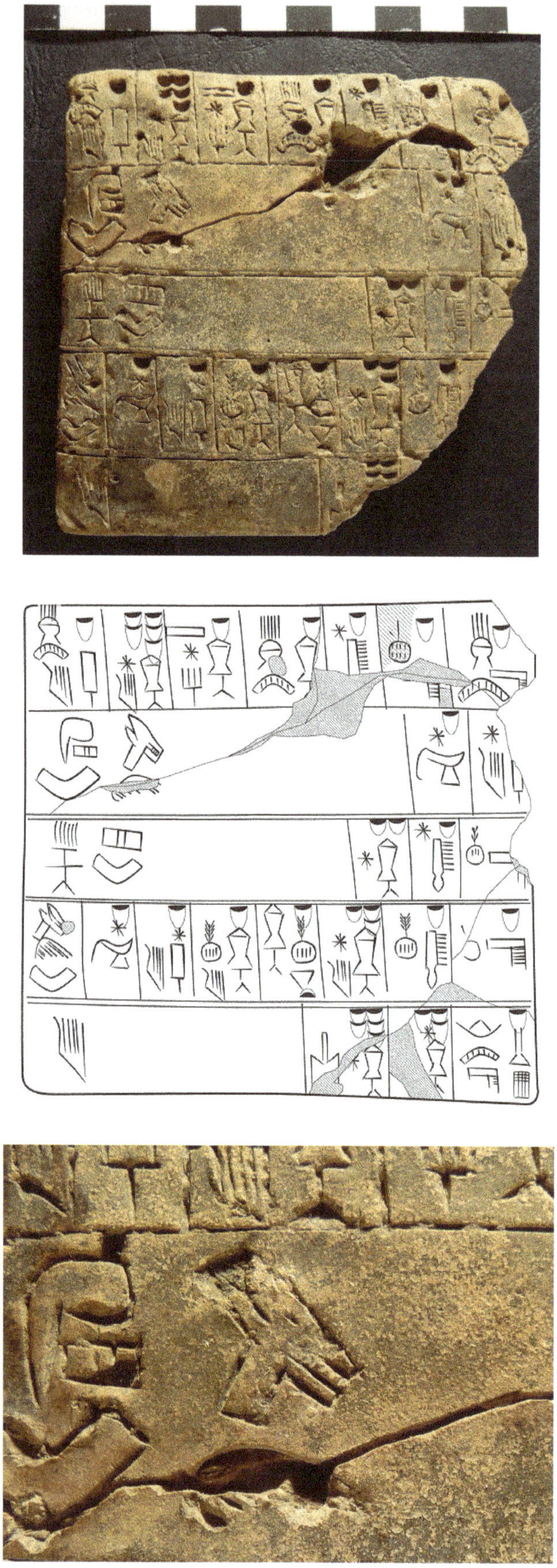

Figure 2. Top: Archaic cuneiform tablet with the image of an equid (obverse), Berlin, Vorderasiatisches Museum, VAT 16763. From Uruk, ca. 3350-3200 BCE, © Vorderasiatisches Museum. Middle: Drawing of the tablet, from: http://cdli.ucla.edu, P001993. © CDLI. Bottom: Detail image. © Vorderasiatisches Museum.

Figure 3. Document about the employment of donkeys as draft animals for ploughing (obverse), Berlin, Vorderasiatisches Museum, VAT 12433. From Shuruppag (Fara), ca. 2600 BCE. © Vorderasiatisches Museum.

Figure 4. Detail from reverse of VAT 12433. showing the ANŠE–sign (left). © Vorderasiatisches Museum.

architects (see the contribution of Burmeister, Krispijn and Raulwing in this volume).

In later texts—mainly administrative sources which were at that time undoubtedly written in Sumerian—equids became more frequently mentioned. Among the references in cuneiform texts, donkeys are distinguished as species and subspecies such as the well–known donkey, mule and the hybrid of donkey and onager. Most of these terms were also used to designate gender, age and other characteristics (Figures 3–5). The texts do not mention their usage for warfare, yet we know this was the case from inscriptions on the famous Standard of Ur (see Olsen/Culbertson 2010: 94).

Moreover, even among personal names the donkey appears as an element, but this evidence is rare and derives from the

Figure 5. Administrative notice about the feeding of donkeys (obverse). Berlin, Vorderasiatisches Museum, VAT 4693. From Girsu (Telloh), ca. 2440 BCE. © Vorderasiatisches Museum.

Figure 6. Terracotta plaque with the depiction of a scene from the poem "Gilgamesh and Huwawa". Berlin, Vorderasiatisches Museum, VA 7246. Purchase, ca. 18th–17th century BCE. © Olaf M. Teßmer, Vorderasiatisches Museum Berlin.

latest Sumerian period at the end of 3rd millennium called Ur III, named after the Third Dynasty of Ur in southern Mesopotamia. In one text we find:

> ur–^{d}anše–du$_{24}$–ur$_{3}$–re
>
> "Hero of the along going young donkey:"[3]

In this case ANŠE is marked with a so–called determinative sign "d" to be read as a god's name DINGIR.ANŠE = "God Donkey".

Of more interest might be the enormous number of notices in poems and songs in Sumerian literature. Here we find—for instance—a quote about donkeys in the narrative that reports a struggle between Gilgamesh the hero of Uruk and the ruler of Kish and a forerunner of the later Babylonian epic (Figure 6).

Regarding a dispute of the donkey with the elder's assembly we read:

> "In the convened assembly, his city's able–bodied men answered Gilgamesh: 'Standing on duty and sitting in attendance, escorting the king's son, and forever grasping the donkey's reins—who has that much breath?", as the saying goes. Gilgamesh answered: "You old men should not submit to the house of Kish! Should we young men not smite it with weapons?"[4]

In genres of Sumerian literature connected with real life activities about agricultural matters mention the equids of the land. They appear, for instance, in a so–called didactic poem, known as the Instructions of Šuruppag—a wise old man—in which he gives advices to his son regarding the right way to live as a farmer. The old man also includes remarks about onagers and donkeys:

> "[…] you should not buy an Onager: it lasts only until the end of the day. […]
>
> You should not have sex with your slave girl: she will chew you up.
>
> You should not buy a donkey at the time of harvest. A donkey which eats […] will […] with another donkey. […]
>
> A vicious donkey hangs its neck; however, a vicious man […], my son"[5]

In addition, the horse is known in Sumerian literature. In a self–praising poem of King Šulgi of Ur of the twentyfirst century BCE the king designates himself as:

> "I am a mule, most suitable for the road. I am a horse, whose tail waves on the highway. I am a donkey of Šakkan, who loves running."[6]

In texts providing information about domestic life, horses appear side by side with donkeys since at least the twentyfirst century BCE. Even so, we do not know to what extent they found a distinguished place in Sumerian literature. In cuneiform texts, horses are scarcely mentioned and even letters only deal with horses like the one of Aba–indasa, "overseer of the elite troops, the one

[3] Schneider 1931, pl. 36 = text 88, 4:16 = line 386

[4] Cited after the poem «Gilgamesh and Aga»: translation = Electronic Text Corpus of Sumerian Literature (ETCSL) = http://etcsl.orinst.ox.ac.uk/; translation, lines 9 14.

[5] See «Instructions of Shuruppag", ETCSL; http://orinst.ox.ac.uk/cgi-bin/etcsl.cgi?text=t.5.6.1# lines 48–49 and 216–219.

[6] In a praise poem of king Šulgi (Šulgi A), ETCSL; http://orinst.ox.ac.uk/cgi-bin/etcsl.cgi?text=t.2.4.2.01# lines 16–18. Šulgi A contains several equid terms: 1.) line 16: anše–gír–nun–na (which was translated as "donkey hybrid"), 2.) line 17: ANŠE.KUR.RA (all agree that the meaning is "horse"), 3.) line 18: dùr–^{d}šakan (which was traslated as "male donkey") and 4.) line 72: anše–edin–na (almost all agree to translate these terms as "Halbesel, Onager"). Also, Šulgi B and D contain terms for equids

who, thanks to the prayers for his king", to king Šulgi from the turn of the 2nd millennium:

"Say to my lord, and repeat to my kid of the mountains, with beautiful horns; to my horse of the mountains, with an eagle's claws; my date–palm, growing on untouched ground and with fresh dates hanging from it: this is what the captain of pledged troops, Aba–indasa—[...] your servant, says."[7]

In legal documents, so–called names for a year ("Jahresnamen") are often used to designate important events during a king's reign. These documents enable us to date cuneiform texts by further analyzing the 'year names' compared to historical events from other cuneiform sources. From the turn of the 3rd to the 2nd millennium 'year names' also deal with affairs alluding to draft animals. Although they are not directly mentioned in the texts, we can clearly deduce information about draft animals such as equids as follows: 'Year name' no. 15 of Ur–Namma of Ur: "The year that the chariot of Ninlil was made."[8]

But let us look at another interesting text genre, the proverbs of Sumer. Here predominantly donkeys are referenced, but in some cases also horses are mentioned. Compared to documents from other text genres, the proverbs of Sumer may be taken as a marker for the actual beginning of the expansion of the horse in Mesopotamia.

Donkeys in Sumerian Texts

"You should drive them like pack–asses into a death–stricken city."

"For a donkey there is no stench. For a donkey there is no washing with soap."

"One does not marry a three–year–old wife, as a donkey does."

"The donkey, after he had thrown off his packs, said: 'Now I can forget the burdens of former days!'"[9]

"He who rents a donkey for a whole year kindles a fire in the moonlight."

Horses in Sumerian Texts

"Like a horse you paw the ground where you drink."

"The horse, after throwing off his rider, said: 'Were my load to be like this forever, how weak I would become!'"[10]

We turn now to the Akkadian literature which incorporates all texts written in Akkadian—a Semitic language—focusing on texts from Babylonia as well as from Assyria. Overall, texts mentioning of horses become more common. Starting with older Babylonian evidence in early the second millennium BCE we should first examine the famous Codex Ḫammurapi, named after the 6th king of the Old Babylonian Dynasty.[11] The Codex Ḫammurapi deals—among wide range of various types of legal matters—also with donkeys in six paragraphs. Two of them are quoted here:

"**§ 8** If any one steal cattle or sheep, or a donkey, or a pig or a goat, if it belongs to a god or to the court, the thief shall pay thirtyfold; if they belonged to a freed man of the king he shall pay tenfold; if the thief has nothing with which to pay he shall be put to death."

"**§ 224** If a veterinary surgeon performs a serious operation on a donkey or an ox, and cure it, the owner shall pay the surgeon one–sixth of a shekel as a fee."[12]

A donkey is also mentioned in an Akkadian literary composition and tells us about a wise man named *Atra–Ḫasis*, the builder of the first ark to survive the deluge. The mighty flood sent by the gods to devastate mankind—because of the noise they made. This comment compared the power and strength of a wild donkey:

"The sky turned dark, so no one could see anyone else. The flood roared toward them like a bull and, like a wild donkey, the winds screamed overhead. There was no sun and darkness hid the land."[13]

In the Old Babylonian period Sumerian literature was still cultivated, but Akkadian literature particularly flourished at that time. Even in private documents traces of horses can be found, whereas in official documents the donkey still takes the center stage. Again, in a letter we read (VAT 6088; Figure 7):

"To Aḫuni speak, as follows said Belanum. Šamaš and Marduk may save your life! Took 1 Gur barley as feed for the horses in order that they can scoff it. They shall not stay hungry."[14]

In the second half of the 2nd millennium BCE, horses gain attention. This is due to relations with New Kingdom Egypt and to the development of a new type of weapon—

[7] Letter from Aba–indasa to Šulgi about his neglect http://etcsl.orinst.ox.ac.uk/cgi-bin/etcsl.cgi?text=t.3.1.21# lines 1–8.

[8] Ur–Namma year name **o.** = mu ^{geš}gigir ^{d}nin–lil$_2$ ba–dim$_2$–ma "The year that the chariot of Ninlil was made", see http://cdli.ucla.edu/wiki/doku.php/ur-namma.

[9] Sumerian Proverbs collection 14 Segment B 1 No 15 (http://etcsl.orinst.ox.ac.uk/cgi-bin/etcsl.cgi?text=t.6.1.14#); collection 2 Segment A No 139, No 81(http://etcsl.orinst.ox.ac.uk/cgi-bin/etcsl.cgi?text=t.6.1.02#) collection 5 Segment B No. 39 (http://etcsl.orinst.ox.ac.uk/cgi-bin/etcsl.cgi?text=t.6.1.05#).

[10] Sumerian Proverbs collection 5 Segment B No. 37 and 38 (http://etcsl.orinst.ox.ac.uk/cgi-bin/etcsl.cgi?text=t.6.1.05#).

[11] Roth 1987; van de Mierop 2004.

[12] My translation.

[13] Translated from German by the author. See also Lambert/Millard/Civil 1999.

[14] Translated by the author.

Figure 7. Private letter from Old Babylonian times mentioning the feeding of horses (obverse), Berlin, Vorderasiatisches Museum, VAT 6088. Purchase, ca. 18th–17th century BCE. © Vorderasiatisches Museum.

Figure 8. Letter of the Babylonian king Burnaburiash to the Egyptian pharaoh negotiating a dynastic marriage and gifts, among them also horses. Berlin, Vorderasiatisches Museum, VAT 152. Tell el–Amarna, Egypt. 14th century BCE. © Olaf M. Teßmer, Vorderasiatisches Museum Berlin.

the chariot, a light, fast vehicle mainly used combat that had two spoked wheels and was drawn by a pair of horses harnessed with metal cheekpieces. Much has been written about horses and chariots, and I only refer to the famous correspondence from El–Amarna at the Nile River—about 300 kilometers south of Cairo—which contains hundreds of cuneiform letters exchanged between the kings of the predominant powers in the eastern Mediterranean. These texts partly deal with or at least mention horses (Figure 8).[15]

To the same period belong the Hittite and Assyrian equestrian texts of which examples were shown as part of the exhibition "A Gift from the Desert: The Art, History and Culture of the Arabian Horse" in Lexington (Kentucky) in 2010. Also, many studies about those horse training texts have published, so I will leave them aside here.[16] Instead let us have a look at the numerous political conflicts between Mesopotamia and their neighbors as well as the wars and combat between Assyria and Babylon.[17]

The Assyrian Tukulti–Ninurta Epic from end of thirteenth century[18] describes multiple attempts of this ruler to subdue the Kassite Babylonians. In his fight against Kaštiliaš of Babylon the Assyrian king Tukulti–Ninurta is accused of having "mounted his chariot and harangued the hordes" of Babylon (Figure 9). However, Tukulti–Ninurta prompts

Figure 9. Socle of the Assyrian king Tukulti–Ninurta I (1244–1207 BCE) from the temple of the goddess Ishtar in Ashur, Berlin, Vorderasiatisches Museum, VA 8146. © Olaf M. Teßmer, Vorderasiatisches Museum Berlin.

[15] Moran 1992.
[16] Raulwing 2006; Probst 2010, preface by P. Raulwing and J. Marzahn.
[17] Hamblin 2006.
[18] Machinist 1978.

Figure 10. Clay prism with dedicatory inscription of the Assyrian king Tiglath–Pileser I (1115–1076 BCE) with reports of his military campaigns mentioning chariotry, from Ashur, Berlin, Vorderasiatisches Museum, VA 8255. © Olaf M. Teßmer, Vorderasiatisches Museum Berlin.

Kaštiliaš to attack the Assyrians, because he thinks he is sufficiently armed in order to win the battle:

> "My army is camped not many leagues from you, and as for you, all your chariotry is in readiness and your army is massed. Attack me, then, like a brave man, fight the battle that you strive so hard to attain."[19]

One of the most famous examples of Assyrian royal text documents is the octagonal clay prism of Tiglath–Pileser I (1114–1076 BCE) in the Vorderasiatisches Museum, Berlin (Figure 10). This is the first known annalistic report that includes the ruler's military campaigns early in his reign. The text was composed to serve as a deposit under the foundations of a temple in Ashur, in northern Iraq at the left bank of the Tigris. The King's reports read:

> "I took my warriors trained for successful combat with thirty of my chariots escorting my aggressive shock troops. I marched to the land Išdiš [...]. Riding in my chariot when the way was smooth and going by foot when the way was rough, I passed through the rough terrain of mighty mountains. In mount Aruma, a difficult area which was impassable for my chariots, I abandoned my chariotry. Taking the lead of my warriors I slithered victoriously with the aggressiveness of a viper over the perilous mountain ledges."[20]

In these texts we witness of the mentioning of the chariot in full action as a weapon in Neo–Assyrian times. However, it is at this point where I quote a statement of Mary Littauer on the depictions of the later battle scenes on Assyrian palace reliefs. She concluded: "Unfortunately, neither the Assyrian reliefs nor textual accounts of military campaigns are sufficiently explicit for us to deduce the tactics use in actual battles."[21] Perhaps a little more light can be shed on this topic by evaluating some texts from Nuzi in northern Iraq near Kirkuk from the same period. One of the Nuzi archives allows us to meet a man who was the "head of 50 chariots, each 10 chariots followed by 20 riders and 200 infantrymen. Thus, the lord of this was a head of about 1000 warriors." And in another text, "One of the officers in these archive is mentioned as one of the fifteen officers of the right flank."[22] (Ninel Jankowska).

If this information reflects historical reality and we extrapolate, we would have two flanks of a military unit with one officer with 50 chariots and a crew of 100 men together with 100 cavalrymen and 200 infantrymen. Altogether on one flank alone: 18,000 men with 3000 horses in this example staggered in rows of chariotry, cavalry and infantry one after another. I am a not a mathematician, but the accounting appears to be erroneous. Whatever the historical reality, it seems that we have here a certain battle order in Nuzi and probably that same order can be conferred on the Assyrian army.[23]

Another fascinating element where archaeology and philology meet comes from Nimrud, Assyrian residence Kalḫu in northern Mesopotamia in ninth century BCE. Here, the excavator Austen Henry Layard (1817–1894) discovered extraordinarily wide doorways and suggested that they belonged to parts of the building where chariots were restored or repaired. Thus, this may be an example of a building that was specifically adjusted to the requirements of chariotry. A direct link to this architectural practice is documented in a later inscription of King Esarhaddon (681–661 BCE), a prism that contains the description of his arsenal in Niniveh (Prism Inscription Niniveh, Episode 21):

> "At that time the royal storehouse of Niniveh, [...] in order to store the material of the battle camps and to house the horses, mules, combat chariots, arms, equipments for the battle, booty of the enemies etc. which Ashur, the king of the gods, has let have a share my majesty, the place where horses were trained to gallop and chariots were driven, this place has become too tight."

All these pieces of information are enriched through knowledge of the so–called administrative Horse Texts from Fort Shalmaneser at Nimrud in northern Iraq during the reign of King Sargon II (722–705 BCE) where a number of reports were found. They deal mostly with ongoing and outgoing groups of horses of the so–called *ēkal māšarti* = the Review Palace of Nimrud. This palace was erected for "organizing the campaign, and checking the horses, mules, chariots, militäry equipment and enemy plunder,"[24] as King Esarhaddon pointed out. The time of the muster is described by him as follows (Instructions of Esarhaddon):

> "May I [...]. take stock there during the month of the New Years Festival, the first month, of all steeds ("Rösser"), mules, donkeys and camels, of the harness and battle gear of all of my troops and of the booty taken from the enemy."

Of special interest in these reports is the indication of colors of horses already going back to earlier times.[25] However, they become more frequent now. In these cases, we can determine black and red with certainty. Other expressions are not clear, but the coat of one horse seems to be decribed with distinct colored spots or markings.[26]

[19] After Foster 1993:211ff.
[20] Kirk Grayson 1991:16, lines II 64–77.
[21] Littauer/Crouwel 1979:137; see also Mayer 1995.
[22] Jankowska 1981:197–198.

[23] Kendall 1975:75 with a table.
[24] Ibid.
[25] See Balkan 1954:26–2; Kendall 1975:288 for Nuzi.
[26] Colored horses are definitively mentioned in a letter from the time of Sargon II, with a report about the arrival of a foreign prince who is bringing tribute in form of horses: "[...]16 red horses, 13 e/irginu–horses, 14 black horses, 1 haršāju–horse, 1 tuānu–horse, 6 female mules, 5 male mules, altogether 51 horses (sic!) of the crown–prince of Andia." For the full text see Lanfranchi/Parpola 1987:No. 171. These horses reappear abbreviated in a report of the so–called Calaḫ horse–list: "[...] ditto, black

Also of importance in this case are the scribes, because for the first time they are not only mentioned, but are distinguished as stable–scribes and chariotry–scribes.[27] Sometimes they dealt with no less than hundreds of horses or cavalrymen and charioteers. In addition, we learn much about the origin of military goods as, for example, ND 2672 and ND 2727 show (Figure 11):

> "1 horse 1 mare, total in Guzana, 1 mare 1–year old foal in Nisibin .... total 28 horses from Askelon......, yearly tribute and linen cloth, decorated harness clothes, fish, roasted emmer ..." (ND 2672)

> "50 riding horses in Ahsana, 184 horses across the river, 20 riding horses in the town of Halman, 100 horses at the depot?, Total 404 horses"

> "60 horses, 30 mules, iškaru tax(?) 20 horses namurtu payment (honorary gift)" (ND 2727)

Despite the fact that the sums in the last text were obviously wrong, all these notations provide an important overview about the income and administration of the equestrian sources of Assyria: tributes, gifts, taxes and duties of certain cities or provinces with one exception—amazingly enough—the actual breeding of horses.[28] Perhaps this was a fact that lay beyond the Assyrian accountancy, which was interested in results only (Figure 12).

Another group from Ashur are protocols for mustering of horses, datable to the time of Sargon II under which two documents are extraordinary, because they contain the names of Aramaean charioteers as well as names of Arrapchaeans and Chaldeans (Figure 13).[29]

With this evidence we can compare some Aramaic dockets dealing with loans of barley which contain the same names of charioteers.[30] From this textual evidence we can deduce that in addition to all cuneiform documents we would have to expect also texts in Aramaic. However, those records were mostly written on perishable materials and are therefore lost today.

Before returning to Royal texts we should have a look on legal documents about the purchase of land, houses or slaves. In these documents sometimes mention of a special penalty appears and suggests that horses could be used to pay a penalty. A fragment of a contract about the purchase of land of 14 ha (c. 150,000 square feet) of surface (!) by a recruitment officer of the army, the notes say:

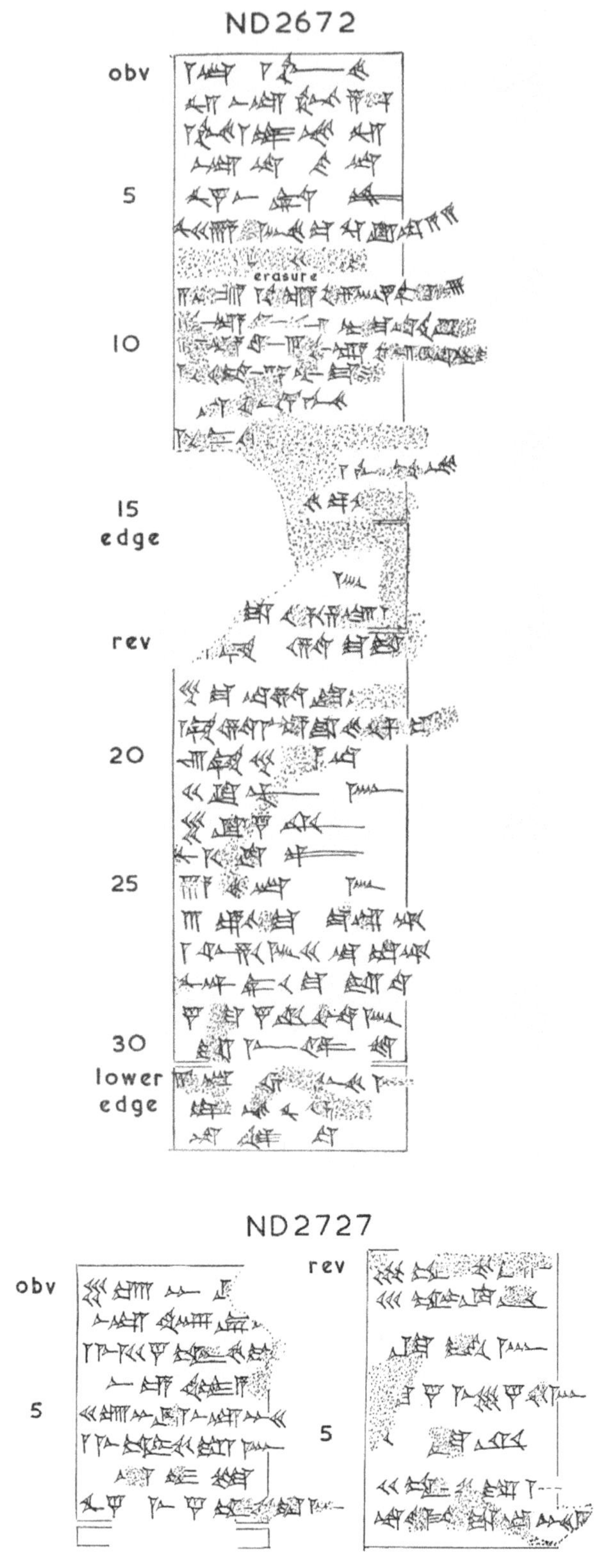

Figure 11. Autographs of the texts ND 2672 and ND 2727; from Iraq 23 (1961) Pls. XXII and XXIV.

> "Whoever in future at any time, breaks the contract and seeks a lawsuit or litigation against Šūmu–lēšir, shall [...] tie two white horses to the feet of Aššur and shall bring four *ḫarbakkannu*–horses to the feet of Nergal."[31]

(ones), 2 ditto, red (ones), 1 horse", and mentions ibid. before: e/irginu–coloured (sign IR), haršāju–coloured (sign ḪA), but also "speckled' (sign GUN$_3$)?, see e.g. ePSD (electronic Philadelphia Sumerian Dictionary) sub GUN$_3$. See for this text: Dalley/Postgate 1984:No. 116. See also Figure 11. See also Younger 2014.

[27] Parker 1961:23ff.

[28] Postgate 1974.

[29] Schroeder 1925:31–38; 131–132.

[30] Lidzbarski 1921:15–20; cf. Donner/Röllig 1962–64:No. 234–236.

[31] Kwasman/Parpola 1991:No 20, Rev 2.

Figure 12. Protocols of the muster of horses from Ashur, Berlin, Vorderasiatisches Museum, 8th century BCE. Left: VAT 10473. Right: VAT 8878. © Vorderasiatisches Museum Berlin.

Figure 13. Aramaic dockets from Ashur with names of charioteers; Berlin, Vorderasiatisches Museum, 8th cent. BCE, VAT 7496, 7497, 7499. © Olaf M. Teßmer, Vorderasiatisches Museum Berlin.

Political documents, royal inscriptions—mostly written on obelisks, reliefs and sculptures but also on numerous tablets, cones and cylinders—represent what we may call an official version of the hitherto quoted texts. These texts record limited mention of horses or numeration. In the annals of king Sargon II the Assyrian king describes year by year—among other things—his concern about horses and related matters. Thus, he notes for instance, for the 8th year of his reign from the campaign against Musasir in Urartu in eastern Anatolia:

> "Concerning Urzana of Musasir, who has broken his oath, Aššur, my lord, encouraged me, that I went against the hardly accessible mountains with not more than my single combat wagon and 1000 men of my cavalry full of bloodlust and of my foot–troops, experienced in combat, walking in difficult terrain and riding on horses when the way was smooth."

Compare the Tiglath–Pileser Prisma:

> "Scarcely Urzana of Musasir heard this he flew away like a bird and disappeared in unapproachable mountains."[32]

It seems obvious that we can take this as a clear evidence for the domination of the cavalry at that time.

[32] The prism inscription of Sargon II, 8th year, translated from German after: Fuchs 1994:321.

In the previous paragraphs we discussed a number of examples mentioning troops and their teams of horses. However, at the same time it should be pointed out that even the Assyrian king had to be educated and trained to exercise all of his military duties by being physically present and leading the campaigns against his enemies. Not all kings, but at least the last important Assyrian ruler, Ashurbanipal, reported in his so–called autobiography:

> "I, Aššurbanipal, laid hold upon the wisdom of Nabû, penetrated to all the intricacies of the art and writing, as practised by all kinds of craftsmen, learned how to shoot with the bow, to ride horses and chariots, to hold the reins".[33]

Several Assyrian literary sources mirror the hitherto described facts. A scribal text containing one of the earliest examples of philosophical poetry in the ancient Near East, called the "Dialogue of Pessimism"—a satirical composition—describes the conversation of a master and his servant (Figure 14). Whenever the master suggests a course of action, the servants supports it by providing further arguments for support. However, each time the master changes his opinion, his servant offers him equally good reasons for the second option. Some of the stanzas begin with "Servant, listen to me. Yes, master, yes." And at the end the story finds its surprising turn:

> "Quickly, get me the chariot and hitch it up for me so I can drive to the palace.
>
> Drive, master, drive, it will bring you where you want to go, the others will be outclassed, the prince will pay attention to you.
>
> No, servant, I will certainly not drive to the palace.
>
> Do not drive, master, do not drive. The prince will send you on a journey that you do not know."[34]

Along with the texts discussed above our attention may be turned to the last great period of Mesopotamian history, the time of the Chaldean kings such as Nebuchadnezzar II (634–562 BCE). However, since the rather few sources dealing with horses from this period are—more or less—quite similar to those we discussed in the context of the Assyrian textual evidence, I will leave this aside here. But I do not want to end without an attempt to demonstrate the sense of humor in the ancient Near East preserved in a scholarly text, called "The Jester", that unfolds the routine of this man by being satirical about professions. And herewith we will close the topic with the equid with which we started at the beginning of this paper: the donkey.

> "Jester, what can you do?"
>
> "I can sing laments (against a lament singer–priest)

Figure 14. Clay tablet with the "Dialogue of Pessimism", Ashur, 2nd half 1st millennium BCE, Berlin, Vorderasiatisches Museum, VAT 9933. © Olaf M. Teßmer, Vorderasiatisches Museum Berlin.

> I can squeeze out apple juice and brew beer."
>
> "Jester, what can you do?"
>
> "Of the whole exorcist's craft, nothing is beyond me."
>
> "Jester, how do you exorcise?"
>
> "Here is how: I take over the haunted house, I set up the holy water.
>
> [...] I skin a donkey and stuff it with straw.
>
> I tie a bundle of reeds, set it on fire, and toss it inside.
>
> I spare the boundaries of the house and its surroundings,
>
> but the haunt of the house, the serpent, the scorpion are not spared."[35]

Excursus: The Kikkuli Text

Among the cuneiform tablets found in the ancient Near East and adjacent regions, the Hittite instructions for the training of chariot horses form a special literary genre (Figure 15). From them as well as from Middle–Assyrian instructions on horse training from Ashur[36] we have

[33] Cited after the German version in Streck 1916:5, lines 31–34.
[34] After Foster 1993:815ff.
[35] My translation.
[36] Ebeling 1951.

Figure 15. Tablet IV of the so–called Hittite Kikkuli–Text. Ḫattuša, 13th century BCE, Berlin, Vorderasiatisches Museum, VAT 6693. © Vorderasiatisches Museum.

relatively rare first–hand information on certain aspects of training concepts for chariot horses in the second half of the 2nd millennium. It also includes how such training manuals were composed, structured and archived by the Hittite and Middle Assyrian scribes.[37]

The so–called *Kikkuli Text* is written on clay tablets in the Hittite language using Hittite cuneiform signs. Certain words are borrowed from other languages. The *Kikkuli Text* became famous mainly for two reasons: It is the oldest surviving horse training text of its kind, and it contains a handful of technical terms used by Kikkuli, such as his professional title, which can be linguistically classified as belonging to Old Indic.[38] The *Kikkuli Text* is named after its author who introduces himself in the first line of Tablet I of his training instructions as the "horse trainer from the land of Mittani", a powerful empire in northern Syria and southwest Anatolia for a certain period in the second half of the 2nd millennium BCE.

It was found in the first campaign in 1906–07 in Boğazköy in central Anatolia,[39] which is the Turkish name of the modern village where the ancient capital of the Hittite kings, named Ḫattuša, was discovered.

[37] Marzahn 1996; 2007; van den Hout 2002.
[38] Raulwing 2004, Table 1.
[39] Raulwing & Meyer 2004, Fig. 2; Raulwing 2006.

Bibliography

Balkan K. 1954. *Kassitenstudien.* 1. *Die Sprache der Kassiten.* [Translated by F.R. Kraus]. American Oriental Society 37, American Oriental Society, New Haven. [Reprint Kraus Reprint, Millwood (NY) 1978].

Borger, R. 1956. *Die Inschriften Asarhaddons von Assyrien.* Archiv für Orientforschung Beiheft 9, Im Selbstverlage des Herausgebers, Graz. [Reprint Biblio–Verlag, Osnabrück 1967].

Dalley, S., and J. N. Postgate 1984. *The Tablets from Fort Shalmaneser.* Cuneiform texts from Nimrud 3. British School of Archaeology in Iraq, London.

Donner, H. and W. Röllig. 1962–64. *Kanaanäische und aramäische Inschriften,* 2 vols. Harrassowitz, Wiesbaden [^{5}2002].

von den Driesch, A. and P. Raulwing. 2004. Pferd, (Esel*, Halbesel, Maulesel, Maultier*). D. Archäozoologisch. *Reallexikon der Assyriologie* vol. 10:493–503.

Foster, B. R. 1993. *Before the Muses. An Anthology of Akkadian Literature,* vol. 1. *Archaic, Classical, Mature,* CDL Press, Bethesda (MD).

Fuchs, A. 1994. *Die Inschriften Sargons II. aus Khorsabad.* Cuvillier, Göttingen.

Hamblin, W. F. 2006. *Warfare in the Ancient Near East to 1600 BC. Holy Warriors at the Dawn of History.* Routledge, London and New York.

van den Hout, T. P. J. 2002. Another view of Hittite literature. In *Anatolia Antica. Studi in memoria di Fiorella Imparati*, edited bz S. de Martino and F. Pecchioli Daddi. LoGisma, Firence, 857–879.

Jankowska, N. 1981. Life of the Military Élite in Arrapḫe. In *Studies on the Civilization and Culture of Nuzi and the Hurrians. In Honor of Ernest R. Lacheman*, edited by M.A. Morrison and D.I. Owen, 195–200. Studies on the Civilization and Culture of Nuzi and the Hurrians 1, Eisenbrauns, Winona Lake.

Kendall, T. B. 1975. *Warfare and Military Matters in the Nuzi Tablets.* PhD Thesis Brandeis University. Ann Arbor, University Microfilms International.

Kirk Grayson, A. 1991. *Assyrian Rulers of the Early First Millenium BC. I, 1114–859 BC.* The Royal Inscriptions of Mesopotamia, Assyrian Periods 2, University of Toronto Press, Toronto, Buffalo and London.

Klein, J. 1981a. *Three Shulgi Hymns: Sumerian Royal Hymns Glorifying King Shulgi of Ur.* Bar–Ilan Studies in the Near Eastern Languages and Culture, Publications of the Bar–Ilan University. Bar–Ilan University Press, Ramat Gan: Institute of Assyriology.

Klein, J. 1981b. *The Royal Hymns of Shulgi, King of Ur: Man's Quest for Immortal Fame.* Transactions of the American Philosophical Society 71/7. The American Philosophical Society, Philadelphia.

Kwasman, T., and S. Parpola 1991. *Legal Transactions of the Royal Court of Nineveh*, pt. 1. *Tiglath–Pileser III Through Esarhaddon.* State Archives of Assyria / Neo–Assyrian Text Corpus Project of the Academy of Finland in Cooperation with Deutsche Orient–Gesellschaft 6, Helsinki University Press, Helsinki.

Lambert, W. G., A. R. Millard, and M. Civil. 1999. *Atra–Ḫasis: The Babylonian Story of the Flood*, Eisenbrauns, Winona Lake, IN. [Reprint of the 1st ed. Published at Clarendon Press, Oxford 1969].

Lanfranchi, G. B., and S. Parpola. 1987. *The Correspondence of Sargon II*, pt. 2. *Letters from the Northern and Northeastern Provinces*, State Archives of Assyria / Neo–Assyrian Text Corpus Project of the Academy of Finland in Cooperation with Deutsche Orient–Gesellschaft 5, Helsinki University Press, Helsinki.

Lidzbarski, M. 1921. *Altaramäische Urkunden aus Assur* [...]. Ausgrabungen der deutschen Orient–Gesellschaft in Assur. E, Inschriften 5 / Wissenschaftliche Veröffentlichungen der Deutschen Orient–Gesellschaft 38, Hinrichs, Leipzig. [Reprint Zeller, Osnabrück 1970].

Littauer, M. A. and J. H. Crouwel. 1979. *Wheeled Vehicles and Ridden Animals in the Ancient Near East.* Handbuch der Orientalistik Abt. 7: Kunst und Archäologie, Bd. 1: Der alte Vordere Orient, Abschnitt 2: Die Denkmäler, B: Vorderasien, Lfg. 1. Brill, Leiden and Köln.

Machinist, P. 1978. *The Epic of Tukulti–Ninurta I: A Study in Middle Assyrian Literature.* PhD Thesis Yale. University Microfilms International, Ann Arbor.

Marzahn, J. 1996. Training instructions for horses from cuneiform texts: The Kikkuli Tablets and the related material from the Vorderasiatisches Museum in Berlin. In *Furusiyya. The Horse in the Art of the Near East.*, edited by D. Alexander. Vol. I, 22–25. King Abdulaziz Public Library, Riyadh.

Marzahn, J. 2007. Pferdetraining nach Keilschrifttexten. In *Pferdestärken. Das Pferd bewegt die Menschheit*, edited by A. Wieczorek et al., 45–50. Publikationen der Reiss–Engelhorn–Museen, 23. von Zabern, Mainz 2007.

Mayer, W. 1995. *Politik und Kriegskunst der Assyrer.* Abhandlungen zur Literatur Alt–Syrien–Palästinas und Mesopotamiens 9. Ugarit–Verlag, Münster.

van de Mieroop, M. 2004. *King Hammurabi of Babylon. A Biography.* Blackwell, Cambridge.

Moran, W.L. 1992. *The Amarna Letters.* The Johns Hopkins University Press, Baltimore.

Olsen, S. L. and C. Culbertson. 2010. *A Gift from the Desert. The Art, History and Culture of the Arabian Horse* [A Gift from the Desert Exhibition, International Museum of the Horse, Kentucky Horse Park]. Lexington (KY): International Museum of the Horse, Kentucky Horse Park.

Parker, B. 1961. Administrative tablets from the North–West Palace, Nimrud. *Iraq* 23:15–67.

Postgate, N. 1974. *Taxation and Conscription in the Assyrian Empire.* Studia Pohl. Series Maior 3. Biblical Institute Press. Roma.

Postgate, N. 1986. The equids of Sumer, again. In *Equids in the Ancient World*, vol. 1, edited by R.H. Meadow and H.P. Uerpmann, 196–204. Tübinger Atlas des Vorderen Orients, 19(1). Reichert, Wiesbaden.

Postgate, N. 1992. *Early Mesopotamia. Society and Economy at the Dawn of History.* Routlegde, London [etc.].

Probst, G. F. 2010. *The Kikkuli–Text* (2nd ed.). Translated from the German and with an Introduction by Gerhard Probst. The King Library Press, Lexington, KY.

Raulwing, P. 2004. Indogermanen, Indoarier und maryannu in der Streitwagenforschung.Eine rezeptions– und wissenschaftsgeschichtliche Spurenlese. In *Rad und Wagen. Der Ursprung einer Innovation. Wagen im Vorderen Orient und Europa*, edited by M. Fansa and S. Burmeister, 515–531. Beiheft der Archäologischen Mitteilungen aus Nordwestdeutschland, 40. Von Zabern, Mainz.

Raulwing, P. 2006. The Kikkuli Text (CTH 284). Some Interdisciplinary Remarks on Hittite Training Texts for Chariot Horses in the Second Half of the 2nd

Millennium B.C. In *Les Équidés dans le monde méditerranéen antique. Actes du colloque organisé par l'École française d'Athènes, le Centre Camille Jullian, et l'UMR 5140 du CNRS, Athènes, 26–28 Novembre 2003*, edited by A. Gardeisen, 61–75. Monographies d'archéologie méditerranéenne, 1er éd. de l'Association pour le développement de l'archéologie en Languedoc–Rousillon, Lattes.

Raulwing, P. and J. Marzahn. 2010. Foreword. In *The Kikkuli Text*, by G. F. Probst. The King Library Press, Lexington, KY.

Raulwing, P, and H. Meyer. 2004. Der Kikkuli–Text. Hippologische und methodenkritische Überlegungen zum Training von Streitwagenpferden im Alten Orient. In *Rad und Wagen. Der Ursprung einer Innovation. Wagen im Vorderen Orient und Europa*, edited by M. Fansa and S. Burmeister, 491–506. Beiheft der Archäologischen Mitteilungen aus Nordwestdeutschland, 40. Von Zabern, Mainz.

Roth, M. T. 1997. *Law Collections from Mesopotamia and Asia Minor*. Writings from the Ancient World 6.: W.L. Scholars Press, Atlanta.

Schneider, N. 1931. *Die Drehem– und Djoḫa–Urkunden der Strassburger Universitäts– und Landesbibliothek. In Autographie und mit systematischen Wörterverzeichnissen*, Analecta Orientalia 1, Pontificium Institutum Biblicum, Roma.

Schroeder, O. 1920. *Keilschrifttexte aus Assur verschiedenen Inhalts.* [...]. Ausgrabungen der deutschen Orient–Gesellschaft in Assur. E, Inschriften 3 / Wissenschaftliche Veröffentlichung der deutschen Orient–Gesellschaft 35. Hinrichs, Leipzig. [Reprint Zeller, Osnabrück 1970].

Streck, M. 1916. *Assurbanipal und die letzten assyrischen Könige bis zum Untergange Niniveh's*, Part 2 *Texte. Die Inschriften Assurbanipals und der letzten Assyrischen Könige.* Vorderasiatische Bibliothek 7(2), Hinrichs, Leipzig. [Reprint Zentralantiquariat, Leipzig 1975].

Younger, K. Lawson, Jr., An Assyrian Horse List (3.128). *Context of Scripture Online.* [Editor in Chief: W. Hallo]. Brill Online, 2014. Accessed 28 July 2014. (http://www.brillonline.nl/entries/context-of-scripture/an-assyrian-horse-list-3-128-aCOSB_3_128).

Zarins, J. 1976. *The Domestication of Equidae in Third Millennium B.C. Mesopotamia,* PhD Thesis University of Chicago.

Zarins, J. 1986. Equids associated with human burials in third millennium B.C. Mesopotamia: two complementary facts. In *Equids from the Ancient World*, vol. 1, edited by R.H. Meadow and H.–P. Uerpmann, 164–193. Tübinger Atlas des Vorderen Orients, 19(1). Reichert, Wiesbaden.

Zarins, J. 2014. [With the assistance of R. Hauser]. *The Domestication of Equidae in Third Millennium B.C. Mesopotamia.* Cornell University Studies in Assyriology and Sumerology 24. CDL Press, Bethesda, MD. [update of *The Domestication of Equidae in Third Millennium B.C. Mesopotamia*, PhD Thesis University of Chicago 1976].

6

Harnessing the Chariot Horse

Gail Brownrigg

Iconographic evidence and archaeological finds show how chariot harness in the Near East and the eastern Mediterranean was developed and embellished over two millennia. Experimental reconstructions have shown how it functioned and confirm that, far from being inefficient as has often been claimed, it was well adapted for its purpose and for changing needs.

Early Harnessing and Control

Early harnessing was by means of a yoke for cattle—still used in many parts of the world today—either attached to their horns or placed on their necks in front of the withers. At its centre point was secured the beam of an ard for making furrows to plant seed and perhaps for irrigation. Wheeled vehicles were drawn in the same way, by pairs of oxen under a yoke, the draught pole also performing the functions of steering and braking. Yokes continued to be used for all animal traction until the time of the Roman Empire in the West and the Han dynasty in the East.

The earliest images of harnessed equids date from the third millennium BCE in the Near East. Four donkeys or donkey/onager hybrids are yoked abreast, drawing two- or four-wheeled vehicles with disc wheels (see Crouwel, this volume, Figures 5, 8). They wear collars or fringed neck straps fastened to a yoke which lies on the necks of the central pair, and are controlled by lines to nose rings through the septum, like those still used for leading bulls (Figure 1). Tight nosebands or muzzles may be intended to stop them biting, as they are often stallions. Guiding them would have been difficult, but control by voice commands and a tug on the lines would have been adequate for prestige vehicles designed to carry kings, priests, warriors, or images of a deity in stately procession, accompanied by footmen and grooms. The lumbering "battle wagons" with their fixed axles—which would in any case preclude sharp turns—could not have been manoeuvred with precision.

Figure 1. A team of four equids controlled by nose rings. Inlay fragment from Nippur, Mesopotamia, 3rd millennium BCE. The Oriental Institute, University of Chicago, 6N–169. Author's drawing from Littauer and Crouwel 1979: Fig. 6, by permission of J. Crouwel.

Horses and Chariots

By the beginning of the second millennium BCE, horses make their appearance, harnessed in pairs to light, spoke-wheeled vehicles which the standing passengers entered from behind—the two-wheeled chariot. The four reins visible on cylinder seals indicate the use of bits (Figure 2).

It was the invention of the bit that enabled fast horses—more agile and more sensitive than donkeys—to be controlled from the vehicle. A bit is not needed for riding. A halter with a single rein, or a strap around the neck, can be used to guide or stop one's mount; the rider's weight, pressure from his heels and a short stick all give signals to turn or go faster. The Bedouin still ride their feisty Arab mares with just a braided rope on a tasselled halter. But for driving from behind, though reins can be attached either side of a noseband, steering is not easy because the pull is backwards with no lateral pressure, and the noseband tends to swivel. It is likely that the bit was invented for driving to control direction, especially at speed.[1] A pair of lively horses harnessed to a light, manoeuvrable vehicle needed both brakes and power steering.

This preoccupation with lateral control is reflected in the frequent use of spikes on the early cheekpieces. A bronze bit dated to about the sixteenth century BCE from Tel Haror in the northern Negev[2] is similar to one found at Tell el Ajjul, ancient Gaza (top of Figure 3). Both have rigid mouthpieces over 20 cm. long (as compared with modern bits ranging from 10 to 15 cm.), which run through circular cheekpieces studded on the inside. The long canon gave leverage, while the prongs would encourage the

[1] Littauer 1969: 489; Dietz 1992; Brownrigg 2004.
[2] Littauer and Crouwel 2001.

horse to turn his head: a special bit with a long "run–out" mouthpiece is still sometimes used on the trotting track for a stiff-necked horse that will not yield to its driver's signal to turn.[3]

Another type of bronze bit with jointed canons, known from both the Aegean and the Near East, also carried studs on the inner face of long, rectangular cheekpieces, which may also have been used in combination with a strap across the nose[4] (bottom of Figure 3).

Figure 2. A divine driver in a "platform car". Detail of a Syrian seal impression, 18th c. BCE. British Museum, London, 16815a. After Littauer and Crouwel 1979: Fig. 31.6, by permission of J. Crouwel.

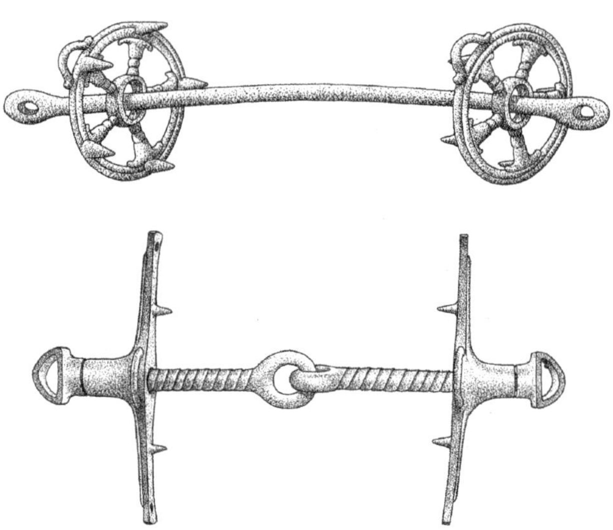

Figure 3. Top: Bronze bit with spiked cheekpieces from Tell el Ajjul (ancient Gaza), Levant. ca. 15th c. BCE. L. 25 cm. Rockefeller Museum, Jerusalem, 37.3271. After Littauer and Crouwel 1979: Fig. 48. Bottom: Bronze bit from Tell el Amarna, Egypt, 14th c. BCE. L. 20 cm. Ashmolean Museum, Oxford, 1933.1209. After Littauer and Crouwel 1979: Fig. 49, by permission of J. Crouwel.

Egypt

Both types of bit have been found in Egypt, where the beautifully executed reliefs and paintings show in detail the bridles and harnesses which were in use throughout the Near East during the second millennium BCE (Figure 4). Divided cheek straps support a bit or a low, reinforced noseband. Decorative discs or bosses and sometimes blinkers are depicted (Figure 5). A yoke with recurved ends rests on the necks of a pair of horses, held in place by a broad neck strap. Another long strap, which hangs under the belly, behind the elbows, would prevent the yoke from

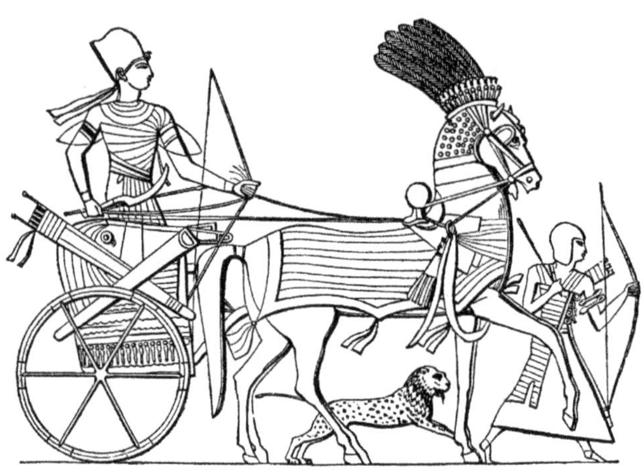

Figure 4. Rameses II's chariot at Abu Simbel, Egypt, 13th c. BCE. After Bonomi 1853: Fig.108. Work published over 75 years ago.

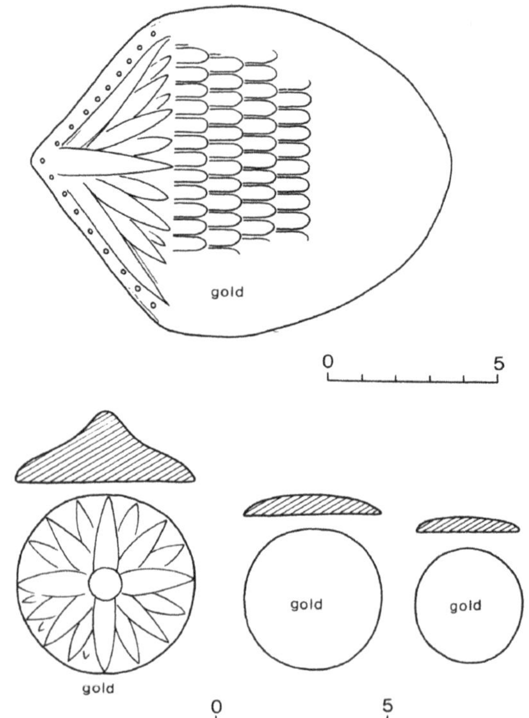

Figure 5. Top: Blinker decorated with gold on gesso. Bottom: bridle bosses. All from the tomb of Tutʿankhamūn, Thebes, Egypt, 14th c. BCE. Egyptian Museum, Cairo. After Littauer and Crouwel 1985: Pl. LX, by permission of J. Crouwel. Scale in cm.

[3] Littauer 1969: pls. 203, 206.
[4] Littauer 1969.

sliding forward when slowing down and would act as a backing element. The teams drawing royal chariots were gorgeously caparisoned. Howard Carter described how splendid Tut'ankhamūn's chariots must have looked:

> "From various paintings of the king in his chariot, we know that the horses were decked with sumptuous housings, neck coverings, and that a crest of ostrich feathers was fastened to the headstall and bridle.... Their effect when in motion under Egyptian skies must have been one of dazzling splendour, with their jewelled trappings flashing back the light, the horses' plumes waving, in a pageant of brightness, colour, gleam and richness".[5]

Finds from the tomb of Tut'ankhamūn testify to the use of sheet gold, silver, faience, obsidian, and coloured glass on various parts of the harness, as well as on the chariots themselves. The yoke saddles, shaped to fit over the horses' necks, bore calcite finials, as did the yokes. One pair was carved with the head of Bes, covered with gesso, gilded, and inlaid with glass, faience, and calcite (Figure 9). The ends of the harness straps passing through the hole that formed his mouth would have resembled a tongue. Matching bridle bosses repeated the Bes-head motif on the chariot to which they must have belonged.[6]

Blinkers decorated with the *wedjat*–eye motif or gilded scales on a wooden base were found in pairs, which could perhaps indicate that they were only used on the outside of each team (Figure 5). Bridle plumes have not survived, but the remains of an embroidered linen housing[7] confirm that the horses' blankets, like those on the painted lid of a wooden chest now in the Cairo Museum,[8] were as colourful as Joseph's coat.

As well as for parade and for personal transport by royalty and the nobility, during the first half of the second millennium BCE the chariot was used as a mobile firing platform for archers in hunting and warfare. Each would be accompanied by a charioteer, leaving both hands free to manipulate his weapon and allowing him to turn and take aim. As a symbol of his might, the pharaoh is sometimes depicted with the reins tied around his waist as if he alone is responsible for conquering his enemies.

The Function of the Harness

For over half a century, historians and archaeologists believed that early harness tended to choke the horses—and this widely–held view can still be found in modern publications.

Commandant Richard Lefebvre des Noëttes (1856–1936) gathered a large body of iconographical evidence for two well-illustrated books.[9] He made a reproduction of what he called "antique" harness (Figure 6) and concluded that because this system put pressure on the windpipe, efficient equine traction only became possible after the invention of the horse collar in the early Middle Ages. His theory had a wide impact and was accepted by scholars and writers in the fields of archaeology, ancient history, and medieval studies alike.

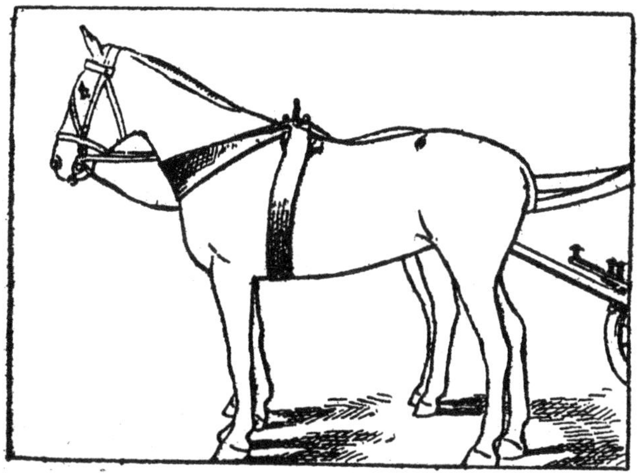

Figure 6. "Ancient harnessing" experiment in 1910. After Lefebvre des Noëttes 1931: p.163 (detail). Work published over 75 years ago.

It was not until 1977 that Jean Spruytte, a retired French army officer and riding teacher, and an eminently practical man, showed by his extensive experiments with reconstructed chariots and harness that Lefebvre des Noëttes had confused two different harness systems: the neck yoke documented in the Near East, Central Asia, and China, and the dorsal yoke developed in Iron Age Greece[10] (Figures 7, 17).

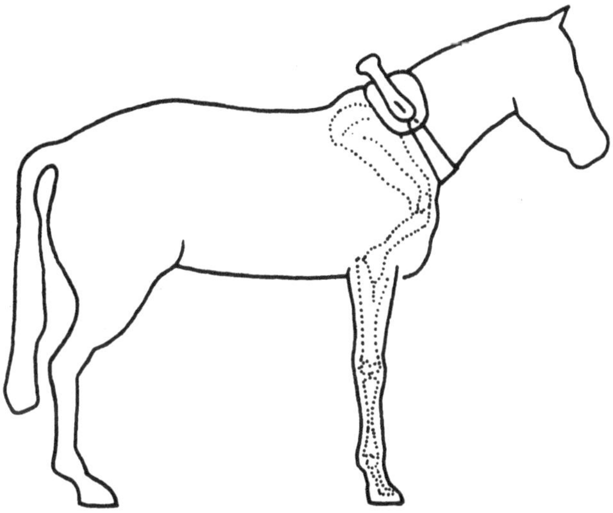

Figure 7. Harnessing system with neck yoke and yoke saddle. After Spruytte 1983: p.14, by permission of A. Spruytte.

[5] Carter 1972: 116 f.
[6] Littauer and Crouwel 1985: 28, 33.
[7] Littauer and Crouwel 1985: 88, pl. LXII.
[8] Rommelaere 1991: pl. 8.
[9] Lefebvre des Noëttes 1924; 1931.
[10] Spruytte 1977; 1983.

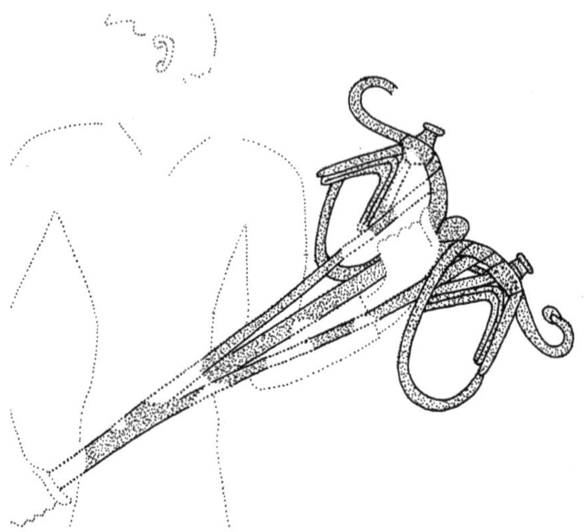

Figure 8. Yoke and harness on a chariot brought as tribute by Levantines. Detail from a wall painting in the tomb of Rekhmire, Thebes, Egypt, 15th c. BCE. After Littauer and Crouwel 1979: Fig. 43 (detail), by permission of J. Crouwel.

Figure 9. Yoke saddle with calcite finial from one of Tut'ankhamūn's chariots, Thebes, Egypt, 14th c. BCE. Height 30 cm. Egyptian Museum, Cairo, One of a pair, object nos. 158 and 167. After Littauer 1968: Fig. 1, by permission of J. Crouwel.

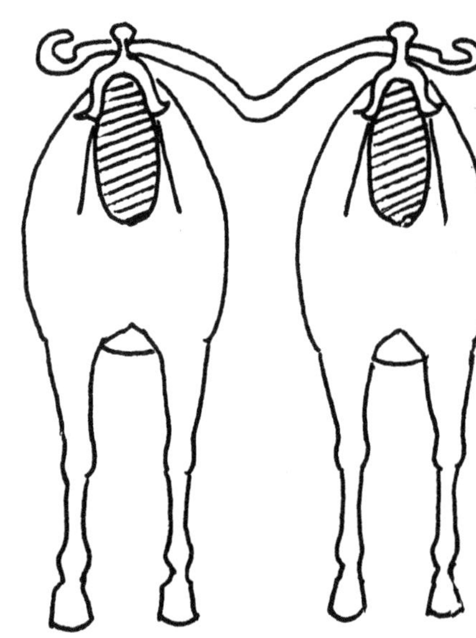

Figure 10. Position of yoke and yoke saddles on the horses' necks. After Brownrigg 2004: Abb. 6 (author's drawing from Spruytte 1982).

In the ancient Near East, traction was by means of "yoke saddles" in the shape of an inverted "Y," which enabled a harnessing system originally designed for cattle to be adapted to equine anatomy (Figures 8, 9). Lashed to the yoke, which rested on the stallions' strong necks, the yoke saddle sat firmly in the slight natural dip just ahead of the horse's withers. Its legs lay in front of the scapula on either side, transferring the draught to the upper part of the shoulders (Figures 7, 10). They were held in position by neck straps joining the lower ends of the fork across the muscular base of the horses' necks—not the throat—and which also served to prevent the team from moving out from under the yoke. The remains of two such bands made of coarse, folded linen from the tomb of Tut'ankhamūn show that these not only widened towards the front, as is usually depicted, but were of crescent shape matched by appliqués of gold foil found in the Antechamber. The curved design would fit the neck's contour and allow them to lie somewhat lower than straight ones.[11] Saddle pads with a rectangular or rounded profile protected the neck from bruising and chafing.

The position of the axle at the rear meant that some of the weight of the occupants was carried by the draught pole, holding the yoke down and thus helping to keep it in place. Spruytte found that this amounted to some 9.5 kg per animal with two people in the vehicle. A long strap from the front of the pole, passing under the animals' bellies and attached to the outside leg of each yoke saddle, served for braking and reversing.

It was at the same time as Jean Spruytte was working on his experiments with Egyptian, Greek, and Saharan chariots and harnesses that Mary Littauer published a paper in *Antiquity* explaining the use of yoke saddles like those found in Tut'ankhamūn's tomb, in Central Asia, and in Chinese burials.[12] Through her knowledge of horses and a careful study of the iconographical evidence and archaeological finds, she had arrived at exactly the same conclusions about their function and purpose as Spruytte reached by practical experiments.

Another unusual item found in Tut'ankhamūn's tomb has helped to shed light on what appeared on some

[11] Littauer and Crouwel 1985: 88, 102, pls. LX, XLVIII.

[12] Littauer 1968.

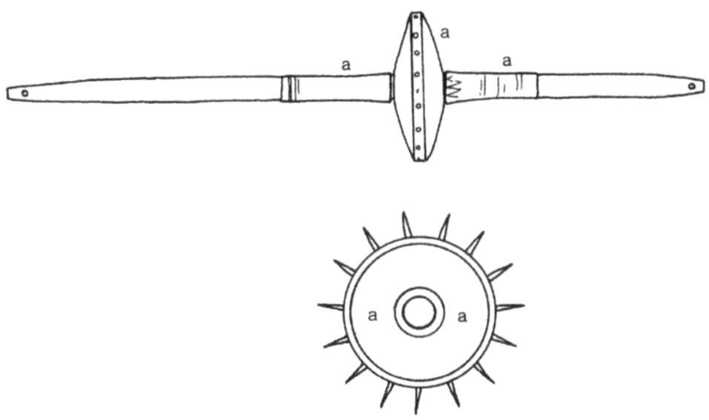

Figure 11. Check rowel from the tomb of Tutʿankhamūn, Obj. no. 620 (120). L. 27 cm., dia. of rowel (with spikes) 8 cm. After Littauer and Crouwel 1985, Fig. 18.

wall paintings and documents from the Amarna period (fourteenth century BCE) onwards to be an auxiliary rein with an attachment in its centre. This consists of a revolving disc with small bronze spikes mounted on a rigid element fitted between the yoke and the horse's bridle, which would prick the animal if it turned its head inwards towards its companion (Figures 4, 11). It has been proposed that it discouraged bickering between stallions[13] or helped them to move off in step.[14] By ensuring that both horses kept their necks bent slightly outwards, it may have helped to keep them close together under the yoke, and thus prevent the yoke saddle from slipping off sideways; nor could they swing their quarters out in the absence of traces. In initial experiments with inexperienced horses I found that if they moved apart, the yoke saddles could easily be pulled out of place. Tightening the *outside* rein has the same effect of turning the horses' heads out and making them bend their bodies inwards, bringing them closer to the pole, but this takes skill and experience and with two reins held in each hand would not have been easy to achieve. This may have been the purpose of the discs.

Greece

Yoke saddles were a standard element of chariot harness in Egypt, the Near East, and the eastern Mediterranean by the mid second millennium BCE (Figure 12). They were in use in the Aegean during the Bronze Age. In Knossos, the fourteenth century Linear-B ideograms picture them (Figure 13), while a mural clearly shows the upturned ends terminating in a knob,[15] fastened by a quick-release knot in a similar position to the holes for the breast strap and backing element at the lower end of the Egyptian saddle (Figure14). The formula *do-we-jo i-qo-e-qu* 'wooden horse-follower(s)' in the Linear–B tablets may refer to the yoke saddle.[16] The reins, however, run directly to the driver's hands instead of passing through loops or terrets near the horses' shoulders as they do further east (Figure 15).

Figure 12. Detail from an ivory gaming box from Enkomi, Cyprus, British tomb 58. British Museum, London, 97/4–1/996. After Crouwel 1981: Pl. 132, by permission of J. Crouwel.

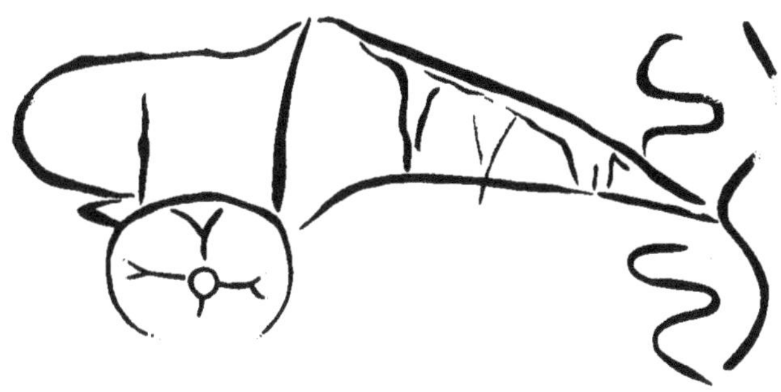

Figure 13. Ideogram *240, Linear B tablet Sc 238, from Knossos, Crete, 14th c. BCE. Ashmolean Museum, Oxford. After Crouwel 1981: Pl. 24, by permission of J. Crouwel.

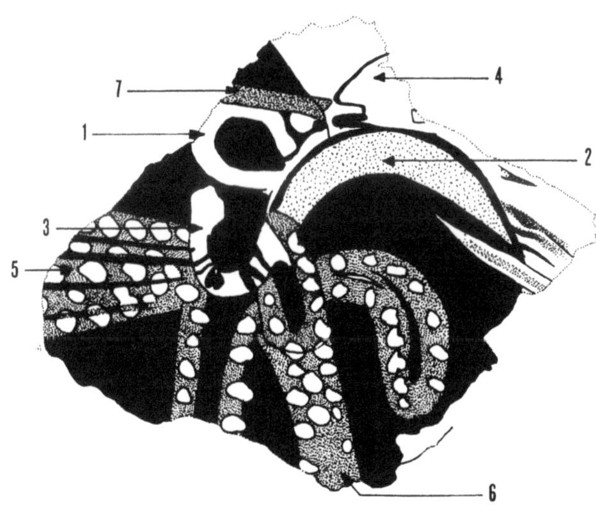

Figure 14. Wall painting from the palace at Knossos, Crete, 14th c. BCE. 1) Rein; 2) Yoke saddle; 3) Upturned terminal on leg of yoke saddle; 4) Finial of yoke saddle; 5) Neckstrap; 6) Backing element. Archaeological Museum, Heraklion, Crete. After Crouwel 1981: Pl. 108, by permission of J. Crouwel.

By the eighth century BCE, however, a different type of harness can be recognised in mainland Greece. The sharp upward curve of the pole visible in Geometric vase painting (Figure 16), combined with the central position of the axle, was associated with a dorsal yoke resting on the horses' backs instead of on their necks, and held in place by a surcingle.[17] Traction was by a breast band

[13] Littauer 1974: 523–24.
[14] Spruytte 1977: 47–52, 1983: 46–51.
[15] Crouwel 1981: 98 f.
[16] Crouwel 1981: 94 fn. 20.
[17] Crouwel 1981: 99; 1992: 39; Spruytte 1977: 53 ff.

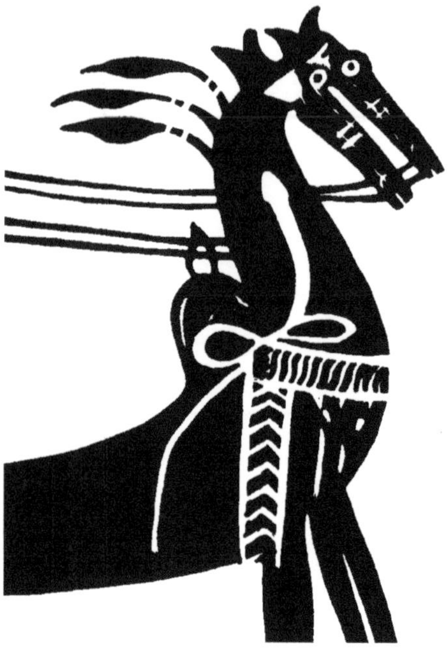

Figure 15. Detail of a vase painting (amphoroid krater), probably from Maroni, Cyprus. Metropolitan Museum of Art, New York, 74.51.964. After Crouwel 1981: Pl. 77 (detail), by permission of J. Crouwel.

Figure 16. Greek chariot with central axle and dorsal yoke. Detail of an Attic Geometric krater, 8th c BCE. Nicholson Museum, Sydney, Australia, 46.41. After Crouwel 1981: Pl. 144a, by permission of J. Crouwel.

lying across their shoulders and chests, lower than the throat strap envisaged by Lefebvre des Noëttes (Figure 17). From the seventh century BCE onwards, the light "high-front" chariot—usually shown with a central axle and chiefly used on the racetrack—became the standard type, drawn by two or four horses (Figure 18). In a team of four, so often represented in great detail on black- and red-figure vases, the outside horses had a long strap around the shoulders and breast with a single inside trace back to the vehicle. When depicting the war chariot of gods and heroes, the classical painter took for his model the racing *quadriga* of his own time.

Spruytte pointed out that the low breast band used with a dorsal yoke corresponds with the modern breast collar which is in common use today, while the neck yoke and

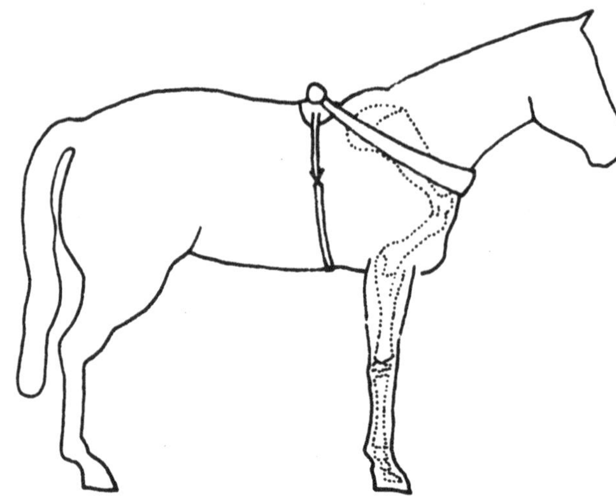

Figure 17. Dorsal yoke with shortened breast strap. After Spruytte 1983: p.14, by permission of A. Spruytte.

Figure 18. Detail of Attic black-figure neck amphora by the Piraeus Painter, from Piraeus, Greece, ca. 600–590 BCE. National Museum, Athens, 353. After Crouwel 1992: Pl. 9:1, by permission of J. Crouwel.

yoke saddles represent a form of shoulder traction, the forerunner of modern collar harness[18] (Figure 19).

I can find no evidence of a dorsal yoke anywhere other than mainland Greece. Images from the Eastern Greek settlements on the west coast of Asia Minor and the offshore islands show neck yokes like those that continued to be used throughout the Near East.[19]

Material from the rest of Europe does not allow conclusions to be drawn about details of harnessing other than the fact that the yoke lies on the animals' necks. From Central Asia to China, neck yokes with yoke saddles are evidenced by archaeological finds and by iconography until the Han period (second century BCE—second century CE), when the invention of shafts changed vehicle design, together with the introduction of a new harnessing system using a breastcollar.

[18] Spruytte 1983: 13, 124.
[19] Crouwel 1991: 71.

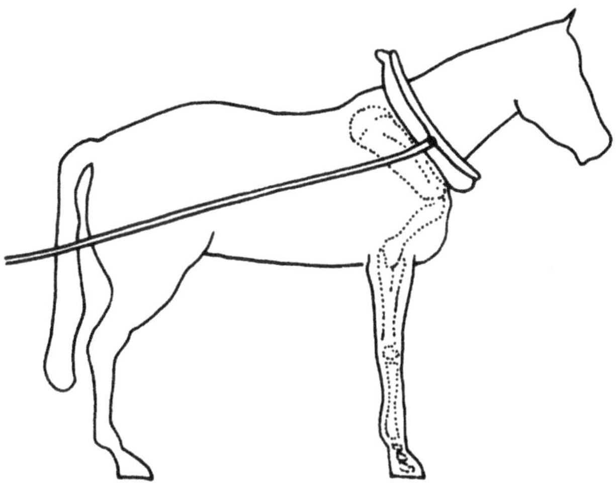

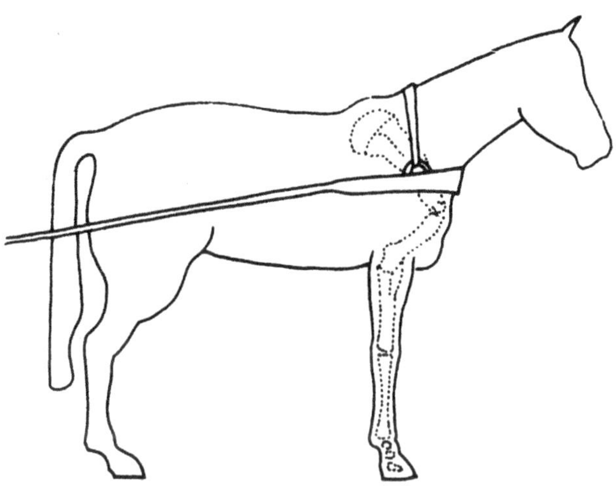

Figure 19. Modern harnessing: full collar (above) and breast collar (below). After Spruytte 1983: p. 16, by permission of A. Spruytte.

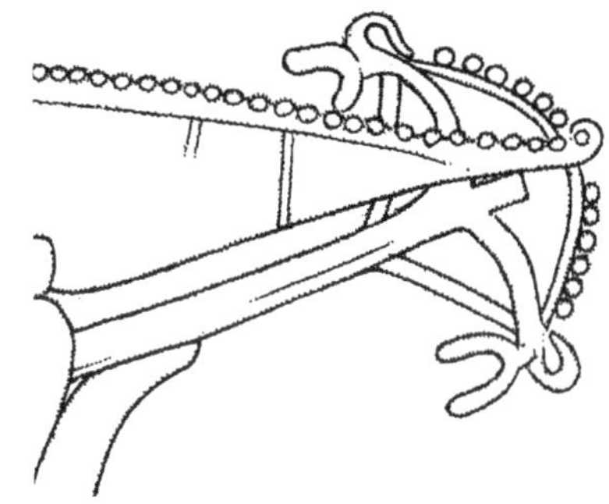

Figure 20. Yoke of an Assyrian chariot being transported on a boat. Detail of a relief of Ashurnasirpal II, N.W. palace, Nimrud, Iraq. British Museum, London, 118908. After Littauer and Crouwel 1979: Fig. 54 (detail), by permission of J. Crouwel.

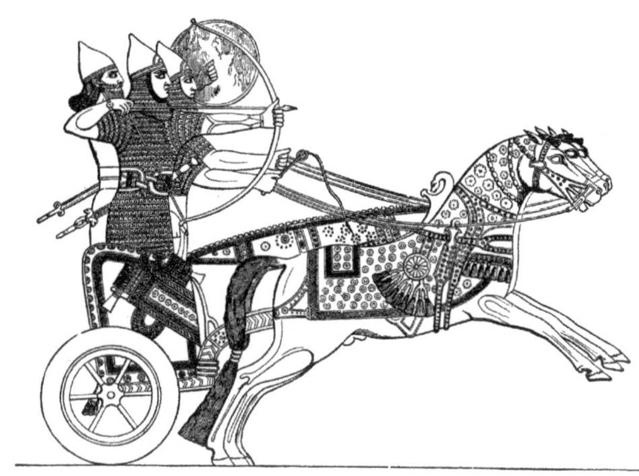

Figure 21. Assyrian warriors in battle. Detail of a relief of Ashurnasirpal II (885–859 BCE) from Nimrud, Iraq. After Layard 1849 vol. II: opp. p.269.

Assyria

Early Assyrian harness included yoke saddles[20] (Figure 20). Bas-reliefs from the royal palaces and the bronze reliefs on the great gates of Shalmaneser from Balawat, now in the British Museum, show that from the time of Ashurnasirpal II (883–859 BCE), a third horse—and sometimes a fourth on royal chariots—was added for extra pulling power, for display, or perhaps as a spare in case of injury (Figure 21). Since there is no evidence of a trace in the Greek manner, outriggers may have been attached to the yoke or to the adjacent horse.[21]

As in Egypt, chariots and harness were lavishly decorated. Wall paintings at Til Barsip had retained their bright colours when they were discovered in 1930, but, apart from fragments now in the Louvre and the Aleppo Museum, have now disintegrated on exposure to the air. Though the pigments have been lost from the stone reliefs that once decorated the palace walls, they too would once have been painted. Layard was greatly impressed by the sculptures of Ashurnasirpal II that he found during his excavations of 1845/7 at Nineveh:[22]

> "The harness and trappings of the horses were extremely rich and elegant. Plumes waved over their heads, or fanciful crests rose gracefully to an arch above the ears, and descended in front to the nostrils. To these ornaments were sometimes appended long ribands or streamers, which floated on the wind. Large tassels of wool or silk, dyed many colours, fell on the forehead, and were attached to many parts of the harness. The bridle generally consisted of a headstall, a strap divided into three parts connected with the bit, and straps over the forehead, under the cheeks, and behind the ears. All these details were elaborately ornamented. In the earlier sculptures we find the figures of winged bulls, and other symbolical devices, on parts of the head-furniture; in the latter, rosettes are more commonly introduced, frequently producing a very pleasing appearance.
>
> It is probable that the bits, as well as many ornaments of the bridle and trappings, were of gold and other precious materials.....

[20] Littauer 1968: 482.
[21] Littauer and Crouwel 1979: 116.
[22] Layard 1849, vol. 2: 272–74.

Around the necks of the horses were hung tassels, rosettes, and engraved beads. Three straps, richly embroidered, passing under the forepart of the belly, kept the harness and chariot-pole in their places. A breast band, adorned with tassels, was also supported by these straps. To the yoke was suspended a very elegant ornament, formed by the head of an animal, and a circle, in which was sometimes introduced a winged bull, a star, or some other sacred device. It fell on the shoulder of the animal, and to it were attached three clusters of tassels.

Embroidered clothes, or trappings, were frequently thrown over the backs of the chariot horses, and almost covered the body, from the ears to the tail. They were kept in their place by straps passing round the breast, the rump, and the belly.

The chariot horses of the later Assyrian period differed entirely in their trappings and harness from those of the earlier times. High plumes, generally three in number, and rising one above the other, waved over their heads. Frequently, an arched crest, and clusters of tassels, were placed between their ears. Similar tassels fell over their foreheads, and hung around their necks. The harness attached to the yoke, was more profusely ornamented with rosettes and fringes, than that of the earlier Assyrian chariots; but the ornaments showed less variety and taste.

The manes were either allowed to fall loosely on the neck, were plaited, or were cut short and stood erect. In the earlier sculptures, the tails of the horses were simply bound in the centre with ribands; in the later, the end was plaited, as is now the fashion in Persia and Turkey, and tied up in a bunch.

Each horse appears to have been guided by two reins, and the charioteer held three in each hand when driving three horses. He also carried a whip, which, like the Egyptians, consisted of a simple thong, attached to a loop at the end of a short handle. In the later Assyrian sculptures this thong was frequently divided into two or three lashes, the handle of the whip terminating in the head of a bull or lion."

Bas–reliefs of the ninth and early eighth century BCE show bits with rectangular cheekpieces, probably studded on the inside for directional control (Figure 21). During the next century, the influence of horse–riders from the north,[23] the increasing use of cavalry, and reduced need for precision steering for the less manoeuvrable four-horse chariots, brought changes in their design. Two short straps supported jointed bits with rod-shaped or curved cheekpieces of bronze, horn, or antler tines which passed through the ends of the mouthpiece (Figure 22). Rarely, under Sennacherib (704–681 BCE), decorative plaques in the form of a horse (Figure 23a) reflect the elaborate designs of earlier bits with long, straight mouthpieces from

Figure 22. Assyrian bridle with bit cheekpieces made from antler tines. Detail of a relief in the palace of Ashurbanipal (668–630 BCE), Nineveh, Iraq. British Museum, London 124875. After Littauer and Crouwel 1979: fig. 71, by permission of J. Crouwel.

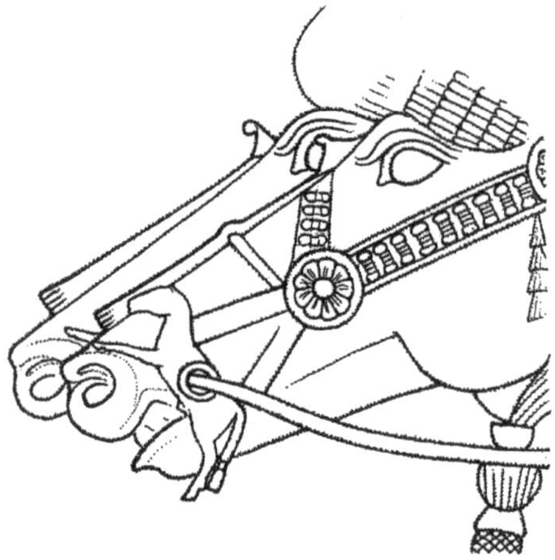

Figure 23a. Assyrian chariot horse wearing a bit with cheekpieces in the form of a horse. Detail of a relief in the palace of Sennacherib (704–681 BCE), Nineveh, Iraq. British Museum. London, 124783. After Littauer and Crouwel 1979: Fig. 72, by permission of J. Crouwel.

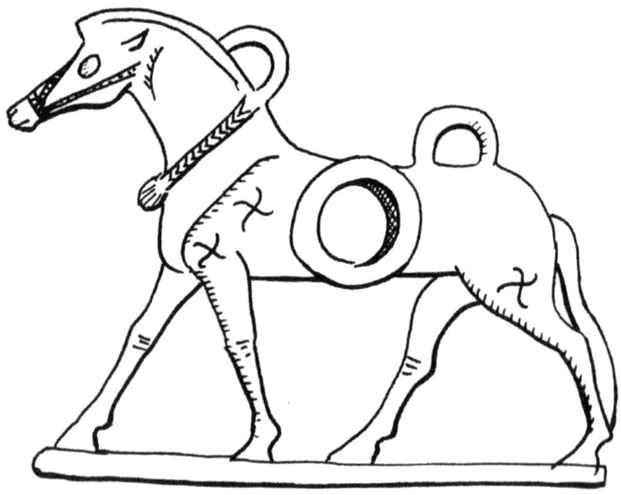

Figure 23b. Cheekpiece from Luristan (Iran) in the form of a horse. British Museum, London 1359–72. Author's drawing.

[23] Dietz 1998: 14 f., 182.

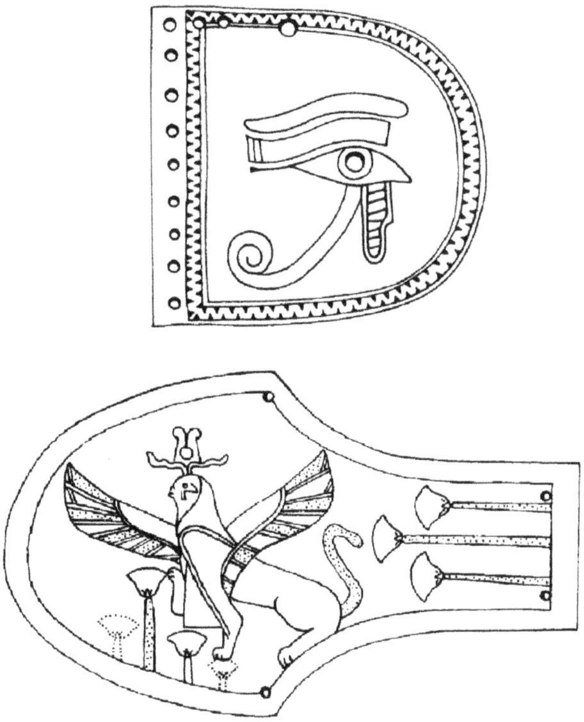

Figure 24. Top: Shield-shaped blinker from Room S.W. 37, Fort Shalmaneser, Nimrud, Iraq. Iraq Museum, Baghdad, 65201. L. 8 cm. (restored design). Adapted from Orchard 1967: Pl. I.1. Bottom: Spade–shaped blinker from Room S.W.37, Fort Shalmaneser, Nimrud, Iraq. L. 13.3.cm. (restored design). Iraq Museum, Baghdad. Adapted from Orchard 1967: Pl. XXIII:118.

Luristan in western Iran[24] (Figure 23b). Though actual examples of such cheekpieces have been found at Nimrud, Samos, and Rhodes, they are not typical: perhaps they had been taken as booty and formed part of the king's treasury, like the blinkers described by Homer when Menelaus is wounded and his blood flows

> "... as when some woman of Meonia or Caria
> stains white ivory with purple dye, making a
>   cheekpiece
> for a horse which, though many horsemen yearn to
>   bear it,
> lies in a storeroom as a king's treasure,
> alike an ornament for the team and its driver's glory."
> (Iliad IV, 140 ff.).[25]

A collection of blinkers, carved from ivory and gypsum in North Syrian and Phoenician style, which must have been acquired during campaigns in North Syria and the Levant, were excavated at Nimrud[26] (Figure 24). Shown very small on the reliefs where the browband joins the headpiece (perhaps the artist did not want to cover the horse's eye), blinkers are clearly depicted on a stone horse's head from Zinjirli in southeastern Turkey (Figure 25). By the seventh century BCE ornamental discs are worn at the browband

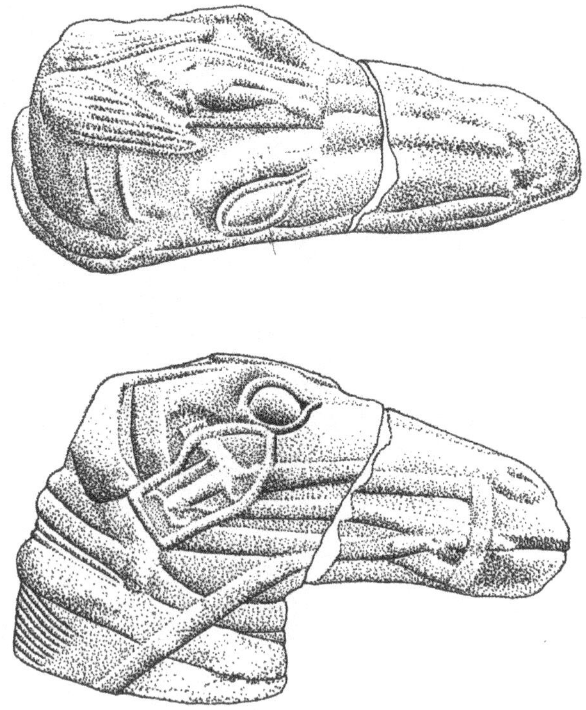

Figure 25. Stone horse head from Zinjirli, Turkey, Earlier first millennium BCE. Vorderasiatisches Museum, Berlin, 3004. After Littauer and Crouwel 1979: Fig. 64, by permission of J. Crouwel.

juction, as on driving bridles today, though sometimes there is an additional element which may be a vestigial blinker. Frontlets of metal or carved ivory, visible on the reliefs, have been found at Nimrud (Figure 26), in North Syria and Urartu, and also in Greece at Miletus and on the island of Samos.[27]

The Assyrian Four-Horse Yoke

During the eighth century BCE, mounted cavalry, first indicated in texts and on the reliefs of the previous century, seems to play an increasingly important part in the Assyrian army. Ridden horses can access places where chariots would have difficulty: marshy land, growing crops, woodland, or steep mountain slopes. Battles between opposing chariot forces must have taken place on appropriate terrain—open plains with room to manoeuvre—almost by prearrangement. The mobility and versatility of riders, whose role had originally been principally as messengers, would enable them to take over many of the functions of the earlier chariotry, including reconnaissance, flanking manoeuvres, harrying enemy troops, or pursuing fleeing soldiers. Larger horses, either imported or the result of selective breeding, are depicted on the Assyrian reliefs of the eighth and seventh centuries BCE. Suitable for riding, they were also able to draw bigger and heavier chariots, especially when harnessed in teams of four. Though less easy to manoeuvre at speed, these prestigious high-wheeled vehicles, now

24 Potratz 1966: 133–83.
25 Murray 1946: 163; quoted by Stephen Foltiny 1967: fn. 3.
26 Orchard 1967.

27 Barnett 1964; Kantor 1962.

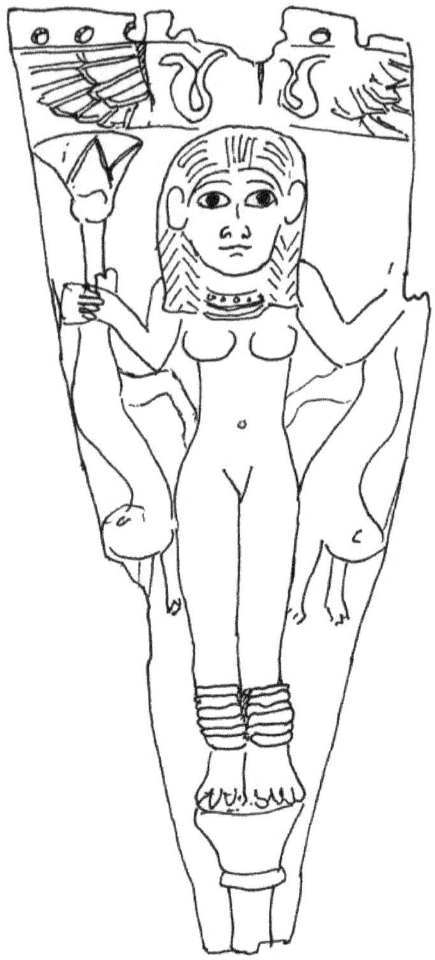

Figure 26. Ivory frontlet, 8th century BCE, from Room S.W. 37, Fort Shalmaneser, Nimrud, Iraq. H. 16 cm. Metropolitan Museum of Art, New York, 61.197.5. Author's drawing from Orchard 1967: Pl. XXX:137.

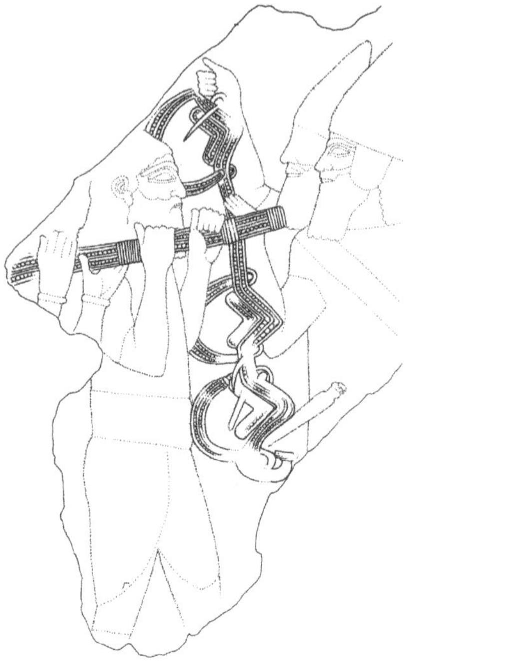

Figure 27. Four-horse fitted yoke on a relief of Sennacherib (704–681 BCE), Nineveh, Iraq. British Museum, London, 123339. After Littauer and Crouwel 1979: Fig. 61, by permission of J. Crouwel.

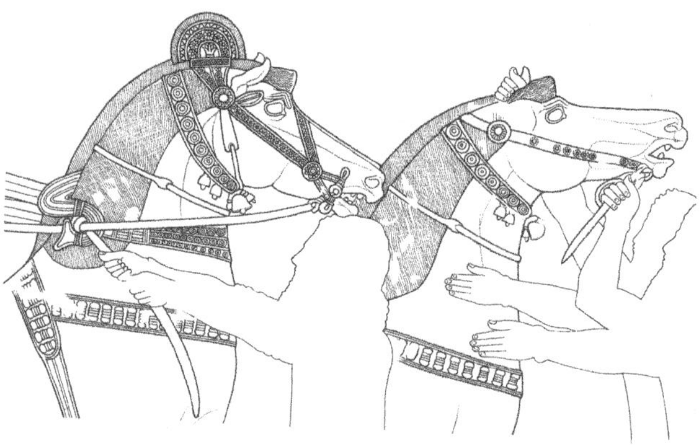

Figure 28. Horses being harnessed to a chariot. Detail from a relief of Ashurbanipal (668–630 BCE, Nineveh, Iraq. British Museum, London, 124859. After Littauer and Crouwel 1979: Fig. 62, by permission of J. Crouwel.

able to carry one or two additional passengers wielding shields or weapons, continued to be used as mobile firing platforms both in warfare and for hunting, as well as for ceremonial purposes. A new kind of yoke was developed to accommodate four horses. Instead of yoke saddles, it had bays shaped to fit over their necks in front of the withers (Figure 27). Broad crescentic neckstraps kept the yoke in place, tied with a quick-release knot, but the draught was still taken principally at the top of the shoulder (Figure 28). Two or four horses were harnessed under a fitted yoke in Assyria, while bas–reliefs celebrating campaigns by Ashurbanipal (668–630 BCE) show similar pair yokes in use with chariots and mule carts of the Elamites from western Iran (Figure 29).

Cyprus

In Cyprus, terracotta models dated to the 7th and 6th centuries BCE depict chariots drawn by two or four horses with fitted yokes (Figure 30). The quadrigas as well as the remains of a vehicle from Salamis[28] provide evidence for

Figure 29. Captured chariot from Elam (western Iran). Detail of a relief of Ashurbanipal (668-630 BCE), Nineveh, Iraq. British Museum, London. Author's drawing from Potratz 1966: Pl. XXIV:54.

[28] Crouwel 1987; Karageorghis 1969.

Figure 30. Terracotta model chariot from Ovgoros, Cyprus 7th c. BCE. Cyprus Museum, Nicosia. Author's drawing from Littauer and Crouwel 1977: Pl. 97c.

Figure 31. Royal Achaemenid chariot, on the Apadana stairway reliefs, Persepolis, Iran, 6th c. BCE. After Littauer and Crouwel 1979: Fig. 80, by permission of J. Crouwel.

two draught poles; the teams are sometimes harnessed under two pair yokes, sometimes under one long one.

These models, which have close Phoenician affinities, show that frontlets and blinkers continued to be used elsewhere after they seem to have gone out of fashion in Assyria. In the Salamis tombs, bronze blinkers and frontlets were discovered *in situ* on the heads of horses and donkeys that had been buried with the vehicles.[29]

Achaemenid Persia

By the sixth century BCE, the principal role of the chariot in the Near East remained that of fast personal transport, including hunting and ceremonial use. On the bas-reliefs at Persepolis, they are shown in procession, the disc which hung down over the horses' shoulders on the Assyrian reliefs replaced by a tassel, but no blinkers, facepiece, or protective blanket (Figure 31). Pairs of small horses are sometimes shown drawing the high-wheeled vehicle at gallop, though a more imposing team of four was generally considered appropriate for royalty.

Chariot Racing

As a prestige vehicle, the quadriga continued in use until the time of the Roman emperors, just as horse-drawn carriages form an important part of state processions even today, but the true descendant of the swift war machine was the racing chariot, which provided excitement in the circus until the end of the Byzantine Empire. The use of shafts and single harness had been developed in northern Europe around the second century CE for drawing utility vehicles carrying both goods and passengers, taking advantage of the strength and weight of the larger horse breeds to pull heavier loads on better roads. Transport in the arid Near East was dominated by the camel caravans to the extent that wheeled carriages virtually disappeared. Wherever there was racing, however, fast and agile horses were still yoked abreast, continuing a harnessing system that had lasted for more than two millennia (Figure 32).

Figure 32. Racing quadriga on a marble funerary stele from western Anatolia, Turkey. Roman period, ca. CE 200. Archaeology Museum. Istanbul. Author's drawing from www.photographersdirect.com/buyers/stockphoto.asp?imageid=1216761.

Acknowledgements

I have drawn so extensively from the seminal work by Mary Littauer and Joost Crouwel, *Wheeled Vehicles and Ridden Animals in the Ancient Near East* (1979), that I have refrained from inserting individual references to it. This book, together with discussions and correspondence with Mary herself, form the foundation for this study of chariots and related equestrian material in the Ancient Near East. I should like to thank Joost Crouwel for his encouragement and for permission to reproduce several illustrations from this and other publications.

Bibliography

Barnett, R. D. 1964. North Syrian and related harness decoration. In *Vorderasiatische Archäologie. Studien und Ansätze. Anton Moortgat zum fünfundsechzigsten Geburtstag gewidment von Kollegen, Freunden und Schülern*, edited by K. Bittel et al., 21–26. Gebr. Mann, Berlin.

[29] Crouwel 1987; Karageorghis 1969.

Bonomi, J. 1853. *Nineveh and its Palaces. The Discoveries of Botta and Layard, Applied to the Elucidation of Holy Writ.* 2nd edition Ingram, Cooke and Co., London.

Brownrigg, G. 2004. Schirrung und Zäumung des Streitwagenpferdes. Funktion und Rekonstruktion. In *Rad und Wagen. Der Ursprung einer Innovation. Wagen im Vorderen Orient und Europa*, edited by M. Fansa and S. Burmeister, 481–490. Beiheft der Archäologischen Mitteilungen aus Nordwestdeutschland 40. Von Zabern, Mainz.

Carter, H. 1972. *The Tomb of Tutankhamun* (abridged edition), Sphere Books, London.

Crouwel, J. H. 1981. *Chariots and other Means of Land Transport in Bronze Age Greece.* Allard Pierson Series 3. Allard Pierson Museum, Amsterdam.

Crouwel, J. H. 1987. Chariots in Iron Age Cyprus, *Report of the Department of Antiquities Cyprus,* 101–118. [Reprinted in Littauer and Crouwel 2002, 141–173].

Crouwel, J. H. 1992. *Chariots and other Wheeled Vehicles in Iron Age Greece.* Allard Pierson Series 9. Allard Pierson Museum, Amsterdam.

Dietz, U. L. 1992. Zur Frage vorbronzezeitlicher Trensenbelege im Europa. *Germania* 70, 17–36.

Dietz, U. L. 1998. *Spätbronze- und früheisenzeitliche Trensen im Nordschwarzmeergebiet und im Nordkaukasus.* Prähistorische Bronzefunde, XVI/5. Steiner, Stuttgart.

Kantor, H. J. 1962. A bronze plaque with relief decoration from Tell Tainat. *Journal of Near Eastern Studies* 21, 93–117.

Karageorghis, V. 1969. *Salamis in Cyprus. Homeric, Hellenistic and Roman. New Aspects of Antiquity.* Thames & Hudson, London.

Layard, A. H. 1849. *Nineveh and its Remains with an Account of a Visit to the Chaldean Christians of Kurdistan, and the Yezidis, or Devil-Worshippers. And an Enquiry into the Manners and Arts of the Ancient Assyrians,* 2 vols. Murray, London.

Lefebvre des Noëttes, R. 1924. *La Force motrice animale à travers les âges.* Berger–Levrault, Nancy, Paris and Strasbourg.

Lefebvre des Noëttes, R. 1931. *L'Attelage et le Cheval de Selle à travers les âges. Contribution à l'Histoire de l'Esclavage.* 2 vols. Picard, Paris.

Littauer, M. A. 1968. The function of the yoke saddle in ancient harnessing. *Antiquity* 42, 27–31. [Reprinted in Littauer and Crouwel 2002: 479–486].

Littauer, M. A. 1969. Bits and pieces. *Antiquity* 43: 289–300. [Reprinted in Littauer and Crouwel 2002: 487–504].

Littauer, M. A. 1974. An element of Egyptian horse harness, *Antiquity* 48:293–295. [Reprinted in Littauer and Crouwel 2002: 521–524].

Littauer, M. A. and J. H. Crouwel. 1977. Chariots with Y–poles in the ancient Near East. *Archäologischer Anzeiger* 1977, 1–8. [Reprinted in Littauer and Crouwel 2002: 238–245].

Littauer, M. A. and J. H. Crouwel. 1979. *Wheeled Vehicles and Ridden Animals in the Ancient Near East.* Handbuch der Orientalistik Abt. 7: Kunst und Archäologie, Bd. 1: Der alte Vordere Orient, Abschnitt 2: Die Denkmäler, B: Vorderasien, Lfg. 1. Brill, Leiden and Köln.

Littauer, M. A. and J. H. Crouwel. 1985. *Chariots and Related Equipment from the Tomb of Tut'ankhamūn,* Tut'ankhamūn's Tomb Series 7. Griffith Institute, Oxford.

Littauer, M. A. and J. H. Crouwel. 2001. The earliest evidence for metal bridle bits. *Oxford Journal of Archaeology* 20: 329–338.

Littauer, M. A. and J. H. Crouwel. 2002. *Selected Writings on Chariots and other Early Vehicles, Riding and Harness.* Culture and History of the Ancient Near East 6, edited by P. Raulwing. Brill, Leiden and Köln.

Murray, A. T. 1946. *The Iliad.* The Loeb Classical Library, Vol I. Cambridge/MA.

Orchard, J. J. 1967. *Equestrian Bridle-Harness Ornaments: Catalogue and Plates.* Ivories from Nimrud (1949–1963), Fascicule I, Part 2, edited by M. E. L. Mallowan and D. J. Wiseman. The British School of Archaeology in Iraq, London.

Potratz, J. A. H. 1966. *Die Pferdetrensen des Alten Orient.* Analecta Orientalia 41. Pontificium Institutum Biblicum, Roma.

Rommelaere, C. 1991. *Les chevaux du nouvel empire égyptien: Origines, races, harnachement.* Connaissance de l'Égyptien Ancienne 3. Connaissance de l'Égyptien Ancienne, Bruxelles.

Spruytte, J. 1977. Études expérimentales sur l'attelage: Contribution à l'histoire du cheval. Crépin-Leblond, Paris.

Spruytte, J. 1982. Démonstrations expérimentales de biges d'après quelques œuvres rupestres sahariennes. In *Les chars préhistoriques du Sahara: Archéologie et Techniques d'attelage.* Actes du colloque de Sénanque, 21–22 mars 1981, edited by G. Camps and M. Gast, 163–172. Laboratoire d'Anthropologie et de la Préhistoire des Pays de la Méditerranée Occidentale Université de Provence, Aix-en-Provence.

Spruytte, J. 1983. *Early Harness Systems. Experimental Studies. Contribution to the History of the Horse* [Translation of Spruytte 1977 by M. A. Littauer]. Allen, London.

7

That Strange Equid from Susa

Trudy S. Kawami

A small, unique shell image of a standing male equid was excavated at Susa in the Acropole/ Citadelle area in the late nineteenth or early twentieth century. Finding the place of its origin and explaining the location of its discovery has been difficult because the Susa equid belongs to a visually dynamic tradition of naturalistic, powerful, even numinous animal forms, rather than to the controlled, stratified world of the mid–third millennium of its find–spot where animals were not the central actors but were primarily modifiers of human activity. Seen in a new context, this essay reasons that the Susa equid is not an outlier of the mid–third millennium context, but part of the earlier Proto–Elamite animal world.

The Object

Between 1899 and 1902, a small (3 cm by 5.6 cm) shell inlay depicting a standing male equid (now Louvre Sb 5631) (Figure 1) was uncovered in the Acropole/Citadelle area (Figure 2) of Susa by G. Jéquier. The inlay shows the animal in profile to the right, head extended horizontally with mouth open as if neighing. It was described as ivory, and illustrated by a small line drawing, but nothing was published about its context.[1] It is possible that no additional information was ever recorded. At that time, de Morgan was excavating the Acropole with a spectacular cut of regular parallel trenches based on abstract geometry rather than the actual strata. The emphasis at the time was bulk clearing of earth to retrieve works of art and inscribed materials. Find spots, levels, or associated objects were not recorded, and intrusive burials were not recognized.[2] Even the beautifully painted archaic ceramics were given for analysis to a specialist in Greek ceramics who never visited Susa.[3] Two other small inlays are known from this early excavation, but neither have equid imagery.[4] The inlay is uniformly considered to date to the second half of the third millennium BCE.[5] No similar image has been discovered at Susa since de Morgan's excavations; the inlay is unique.

Figure 1. **Susa, Acropole area. Shell inlay, acc. no. Sb 5631. Photo by Hervé Levand courtesy of Réunion des Musée Nationaux/Art Resources, NY.**

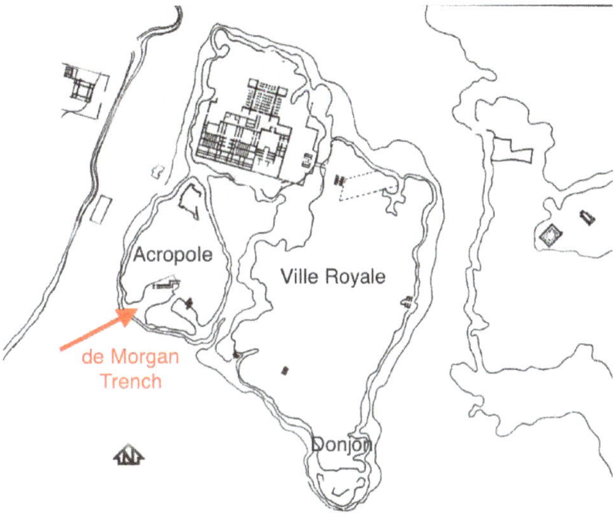

Figure 2. Plan of Susa with an indication of the location of the equid inlay.

A Wider Archaeological Context

A look at the material world of western Asia in the third quarter of the third millennium BCE provides a general context for the Susa find. The use of pale shell, stone, or bone inlays in composite panels (or mosaics) was a well–documented feature of tomb, temple, and palace furnishings

[1] Jéquier 1905: 26.

[2] Perrot and Ladiray 1989, Martinez–Seve 2004; Dahl 2005: 83–84; Pavel 2011.

[3] Le Breton 1957: 80 f.

[4] Dolce 1978, vol. 2, S1 and S3; p. 57 and pls. XVII and XLVIII.

[5] Amiet 1966: 194 f.; Littauer and Crouwel 1979: fig. 11; Dolce 1978, vol.2, 57 and 163, pl. XLVII, S2.

Figure 3. "Standard" of Ur, detail of equids, probably onager hybrids, on so–called war side. Photo courtesy of the Trustees of the British Museum.

in Mesopotamia and Syria during the Third Early Dynastic period (ED III, ca. 2450 BCE).[6] Excavations of the so–called Royal Tombs at Ur in southern Mesopotamia produced numerous examples, the best known of which is the "Standard of Ur" (Figure 3).[7] Other Mesopotamian examples come from Kish,[8] Khafaje,[9] Tell Chuereh,[10] Nippur,[11] and Tell al–Obaid,[12] while the Syrian sites of Mari[13] and Ebla[14] in modern Syria show the popularity of this luxurious décor extended far to the west. The Susa inlay is typical of Mesopotamian, rather than Iranian, artistic practices, but its presence at Susa can be explained by Mesopotamian influence. Susa was in the cultural, political, and perhaps military sphere of Mesopotamia in the mid–third millennium, though its population spoke Elamite rather than Sumerian or Akkadian.[15]

The Visual Tradition

Equid imagery is rare at Susa in the third millennium BCE. The only other depiction is a large painted jar from a funereal context, excavated between 1932 and 1939 from grave A 232 by R. de Mecquenem.[16] It shows two equids with short up–standing manes and tufted tails. They are not harnessed or led, and thus are not necessarily domesticates. The decorative geometric patterns on their bodies and the general abstracted decor of the pot stand in sharp contrast to the naturalism of the shell inlay, making direct comparison of both the style and the iconography difficult. In contrast, contemporaneous Mesopotamian and Syrian art has a rich inventory of equid images in both mosaics and relief sculpture. Equids pulling war wagons or straddle carts in combat scenes, or being led in parades of booty, are a standard component of elite art. The best–known examples occur on inlays like the Standard of Ur (Figure 3) and other small scale reliefs,[17] as well as works from Kish,[18] Khafaje[19] and Mari,[20] while larger, though now fragmentary, representations occur on monumental reliefs like the Vulture Stele from Tellō,[21] and in the round, like the cast copper model from Tell Agrab.[22]

Wagons and Carts

Susa has yielded no images of harnessed equids, but the remains of four–wheeled carts or wagons, and burial furnishings in at least three tombs, indicate that prestige wheeled vehicles formed part of the funeral panoply of the elite there as well. These vehicles were not necessarily equid–drawn, for a painted vessel shows a wagon pulled by bovids.[23] The Susa vehicles, like those in Mesopotamia and Syria, had pieced plank wheels with metal–reinforced rims. Two types of rims were documented at Susa. One type featured a rim edged with a narrow wooden tyre studded with copper nails or rivets. This not only reinforced the tread of the wheel against wear, but also formed an almost ornamental edge. Examples in two sizes were found in grave 280 in the Donjon area of Susa (Figure 4).[24] Numerous Mesopotamian representations of this wheel type are known,[25] as well as actual wheels.[26]

The second type of rim featured a segmented copper tyre with long tabs, or lapets, secured through the sides of the wheel with metal studs or pins. Two variants of this type appear at Susa. One type, a four–part tyre, was excavated from a late third millennium BCE tomb in the Ville Royale area by R. de Mecquenem (1943: pl. X, 1) (Figure 5). The other type, a six–part tyre (Figure 6), is known from a tomb in the Donjon area[27] and from Tomb K east of the

[6] Moorey 1994: 136–37; Dolce 1978.
[7] Aruz 2002, no. 52, pp. 97–100; Zettler and Horne 1998: 44, fig. 36a,b; Dolce 1978, pl. XXI.
[8] Aruz 2003, nos. 48–51, pp. 84–92; Dolce 1978: 20–34.
[9] Moorey 1994: 136; Dolce 1978: 57.
[10] Dolce 1978: 53–55.
[11] Dolce 1978: 37–45.
[12] Aruz 2003, no. 46, pp. 87 f.; Dolce 1978: 156–62.
[13] Aruz 2003: 96–100, 102–04, 156–62; Dolce 1978: 91–155.
[14] Aruz 2003, nos. 112, 114 f., pp. 173–177.
[15] Harper, Aruz and Tallon 1992: 81 f.
[16] Louvre Sb 5885; Amiet 1966:109, 150–51.
[17] Aruz 2003, no. 31, p. 72; Frankfort 1970: 74, fig. 76; Dolce 1978: pl. XXI, no. U172.
[18] Dolce 1978, pls. XV, K48 and K87.
[19] Frankfort 1970: 68, fig. 72.
[20] Dolce 1978: pl. XXXI.
[21] Aruz 2003: 190, fig. 52; Frankfort 1970.
[22] Frankfort 1970: 54, fig. 51.
[23] Amiet 1966: 46–147, no. 106; Le Breton 1957: 17, pl. XXVI.
[24] de Mecquenem 1943:123, fig. 89: 1–2 and pl. X:2; Amiet 1966: 143, no. 103; Littauer and Crouwel 1979, fig. 5.
[25] Littauer and Crouwel 1979: fig. 7; Dolce 1978: pl. XXXVI, M462–M463.
[26] Littauer & and Crouwel 1979: p. 19 and fig. 5.
[27] de Mecquenem 1943: 90, fig. 74:2 and 3.

Figure 4. Studded wooden wheel from Susa, Donjon area, Tomb 280. 1939 de Mecquenem photo courtesy of the Maison de l'Orient et de la Méditerranée, Université Lumière Lyon. Lyon.

Figure 5. Tomb 89, Donjon area, Susa. (Ville Royale area?); four–part copper tyre in situ and later reassembled. 1934 de Mecquenem photos courtesy of the Maison de l'Orient et de la Méditerranée, Université Lumière Lyon.

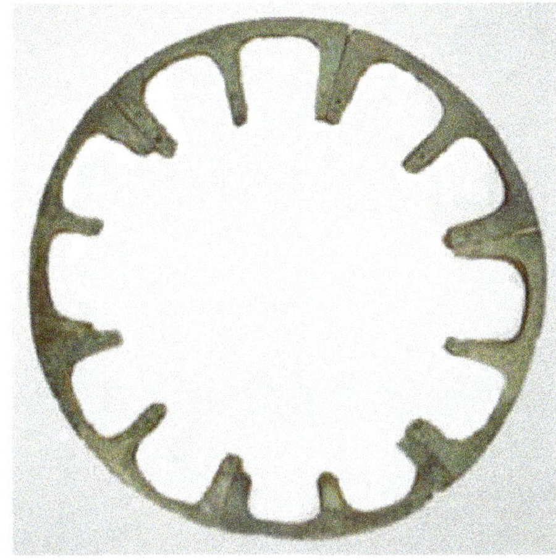

Figure 6. Tomb 89, Donjon area, Susa. Six–part copper tire (currently restored with spokes) Tehran, National Museum of Iran. Photo courtesy of Aaron Cripps (cyclinghistory. wordpress.com); visual editing by the author.

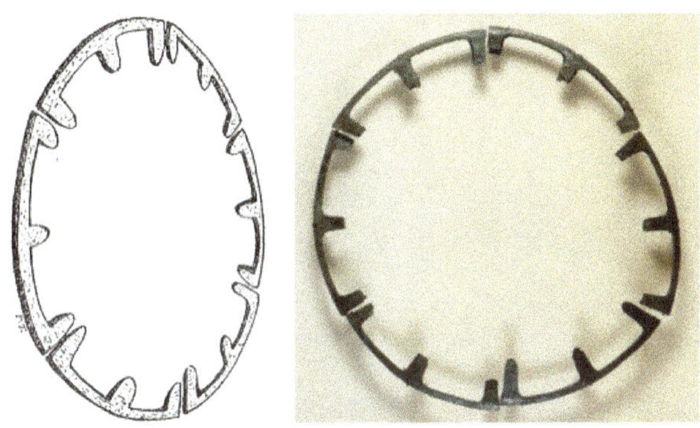

Figure 7. Tomb K, east of the Apadana, Susa. Six–part copper tyre; Louvre Sb 6829. Left: 1922 de Mecquenem photo courtesy of the Maison de l'Orient et de la Méditerranée, Université Lumière Lyon. Right: Photo by Franck Raux courtesy of Réunion des Musée Nationaux/Art Resources, NY.

Apadana (Louvre Sb 6829), where a single wheel and two cast copper linchpins were found placed above the remains of what was described as a sturdy "horse".[28] A second example in Tehran (Figure 7)[29] has been erroneously restored with spokes.

Wheels with segmented metal tyres have not been found, so far, in Mesopotamia or Syria, though one late third millennium BCE relief of unknown provenance may depict such a wheel.[30] Actual metal tyres are known in the Bactria–Margiana Archaeological Complex (BMAC) culture of the late third to early second millennium BCE in western Central Asia.[31] Tomb 3200 in the "Royal Necropolis" at Gonur–depe in the Merv oasis of modern Turkmenistan, contained among other furnishings, a four–wheeled wagon with a six–part metal tyre on each wheel (Figure 8). Tomb 3225 yielded another four–wheeled wagon with six–part tyres (Figure 9), although

[28] de Mecquenem 1922: 137 f. and fig. 14.
[29] Littauer and Crouwel 1979: fig. 19.
[30] Littauer and Crouwel 1979: 39, fig. 18:a, b.
[31] Kohl 2007: 196 f., 209.

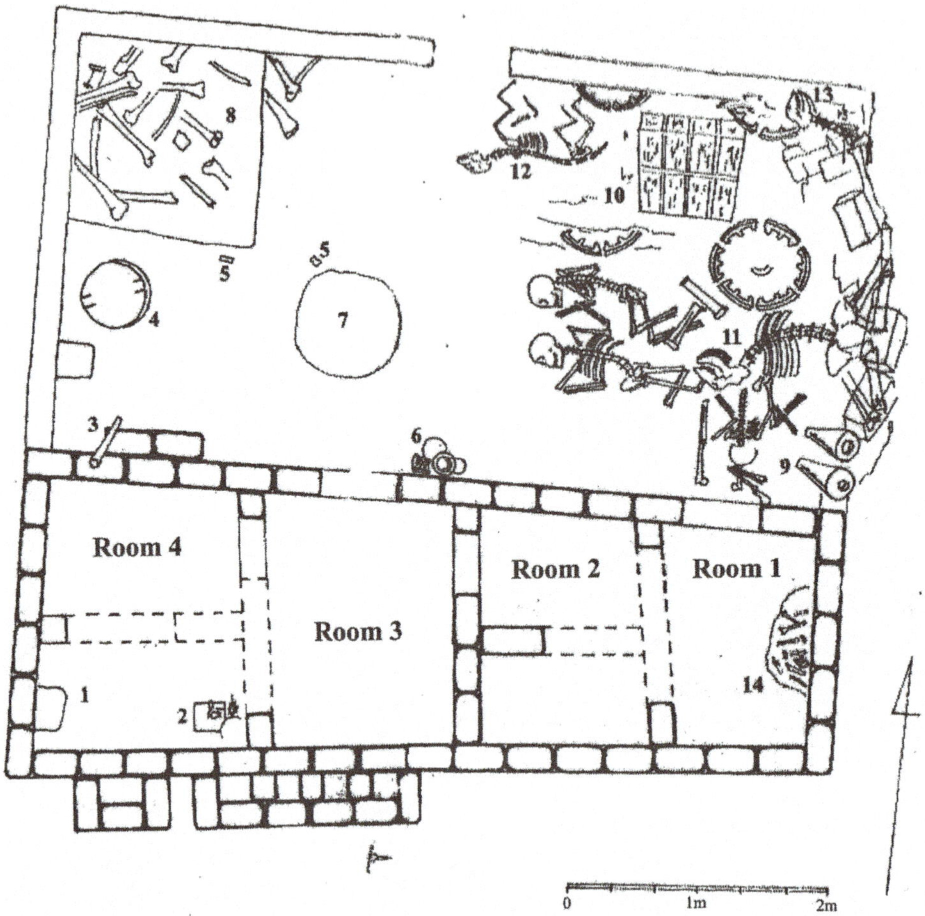

Figure 8. Gonur–depe, Tomb 3200; plan showing location of 6–part tires (After Kohl, fig. 5.11).

Figure 9. Gonur–depe, Tomb 3225; cart or wagon with 6–part tire (after Kohl 2007: fig. 5.10, p. 197).

Figure 10. Copper(?) horse–head terminal from Gonur–depe (after Sarianidi 2002, p. 237).

no two–wheeled "chariots," or straddle cars, like that of the Tell Agrab model have been discovered. In contrast to Mesopotamian practice, representations of wheeled carts or wagons are rare, as noted by, Kohl in a single example of a wagon pulled by bovids,[32] or in combat scenes or parades of booty. Excavated examples of equine imagery include a copper terminal from Gonur (Figure 10), and a pin from Khardchakhalif in Tajikistan,[33] as well as skeletal remains[34] and numerous unprovenanced items associated with the Bactria–Margiana Archaeological Complex (BMAC) culture.[35] These equids are shown with long, falling manes, small ears, and short heads with large eyes and small muzzles; they do not resemble the Susa equid in conformation.

[32] Kohl 2007: 213, fig. 5.18a.

[33] Kohl 2007: 209, fig. 5.15.

[34] Kohl 2007: 208; Sarianidi 2002: 238.

[35] Bothmer 1990: 43; Sarianidi 1986: 211.

The Mesopotamian and Central Asian finds establish a general context for the Susa inlay in an elite culture whose richly furnished tombs contained wheeled vehicles. Both regions had artistic traditions of equid imagery, but these traditions differed. The residences and tombs of Mesopotamia held images of equid–drawn four–wheeled vehicles in battle scenes, and some tombs contained actual carts or wagons.[36] In Central Asia, the carts were also part of the tomb furnishings, but no representations of the vehicles, or of war scenes for that matter, are known.

Identification of the Species

Przewalski Horse (*Equus przewalskii*)

The general archaeological context in both Mesopotamia and Central Asia is of little help in determining which species of equid the Susa inlay depicts. Bouman and Bouman (1994: 7) and Amiet (1966: 194 f., fig. 143) identify it as Przewalski horse (*Equus przewalskii*), (Figure 11) as did Ann Farkas, who considered it additional evidence of the connections between lowland Susa and the steppes of Central Asia,[37] connections that are now more clearly understood. The Przewalski horse has a stocky body, curving top line, short thick neck, an upstanding mane, large head, short ear, and long, though sparse, tufted tail.[38] In general, it resembles the Susa inlay. Przewalski horses have 66 chromosomes; domestic horses have 64. They are usually considered to be two different species, although their hybrid offspring with 65 chromosomes are viable and fertile.[39] The assumed range of the Przewalski horse was the Eurasian steppes, although the species is now found only in Mongolia.[40] Neither region is like the hot, humid lowlands of the Persian Gulf around Susa. In its behavior the Przewalski horse resembles the zebra, as in general it cannot easily be handled, haltered, nor mounted. When it needs to be transported or treated, it has to be tranquilized. In captivity, Przewalski horses are easily stressed, may react aggressively, and do not reproduce readily. Given the problems of handling and transporting the species in the twentieth century,[41] it seems unlikely that the Susa representation is a naturalistic depiction of a Przewalski horse driven or led far from its natural habitat.

Bouman and Bouman's (1994: 7) reference to an inlay with a neighing Przewalski horse at Tall-e Malyan on the Iranian plateau appears to be a confused reference to the Susa inlay. The Malyan excavation report documents only shell inlays in geometric patterns.[42]

Onager (*Equus hemionus *sp.)

Another West Asian equid species, the onager (*Equus hemionus* sp.) (Figure 12), still has a huge range of primarily dry open land from the Arabian Peninsula and Mesopotamia through Iran and across the southern sections of the Eurasian steppes to northwestern China. Onagers have a lighter, more gracile build than Przewalski's horses with a level topline and slender neck with dark upright mane. Their large heads have long ears; the tails are short and sparsely tufted. In temperament, the onager can be more biddable than the Przewalski's horse, and individually may be handled, trained, and used for traction. It can be crossbred with the domestic horse (*Equus caballus*) or the donkey (*Equus asinus*). The offspring of horse–onager crosses are not fertile, but those with donkeys are.[43]

Onagers are documented in the archaeological remains in Syria and Mesopotamia in the Early Dynastic period, although difficulties in distinguishing onagers from asses remain substantial.[44] A center for crossbreeding onagers and donkeys has been identified at Abu Salabikh.[45] The equids pulling the Mesopotamian and Syrian wagons in the mosaics appear onager–like, with large heads, long

Figure 11. Przewalski's horse at the Smithsonian's facility, The Wilds, Ohio. Photo by Jeff Kubina.

Figure 12. Persian onagers. Photo by Hamed Moshiri courtesy of Plan4land.org.

[36] Littauer and Crouwel 1979: 18–19; Zettler and Horne 1998: 32, fig. 29; Woolley 1934: pls. 33a, b.
[37] Ann Farkas, personal communication, 1974.
[38] Groves 1974.
[39] Boyd and Houpt 1994.
[40] Bouman and Bouman 1994; Groves 1994.
[41] Bouman and Bouman 1994; Levine 2003; Boyd and Houpt 1994: 226.
[42] Sumner 2003: 69–71, 83, 160–63, and fig. 39.
[43] Clutton–Brock 1992: 88.
[44] Vilà 2006.
[45] Postgate 1992.

Figure 13. Chased décor on a silver vessel from Maikop. Hermitage, St Petersburg. After M. I. Rostovetzeff, *Iranians and Greeks in South Russia*. The Clarendon Press, Oxford. 1922., p. 23, fig. 2.

ears, upright manes, tufted tails, and a generally gracile build. It is unlikely that these are true onagers, which are not reliably biddable; they are most probably hybrids.[46] Given the close connections between Susiana and Mesopotamia and the broad natural range of the onager, there is no problem hypothesizing it or hybrids based on it at Susa. However, the conformation of the onager and onager–hybrids[47] does not resemble the Susa inlay.

*Tarpan (*Equus ferus ferus *or* Equus caballus ferus*)*

The closest parallel to the Susa equid appears on a silver vessel (Figure 13) from Maikop to the northwest of the Caucasus. Now dated securely to the mid–late 4th millennium BCE—antedating the Susa, Mesopotamian, Syrian, and Central Asian equid images by centuries—the vase was one of many precious items excavated in 1897 from a kurgan, or burial mound, in the town of Maikop.[48] The Maikop kurgan finds should not be confused with the "Maikop Treasure" of later date.[49] The Maikop vase shows eight wild animals in two bands of four animals each. The widest band around the sides of the vessel depicts a confronted pair of horned bovids, followed by a male equid and a large felid with a bird perched on its back. The Maikop equid has a stocky body, a thick neck with what appears to be an upright mane, a large head with a small muzzle and ears, and a tufted tail very much like the Susa inlay. This equid has been identified as a tarpan (*Equus ferus ferus* or *Equus caballus ferus*) (Figures 14a and 14b), the now–extinct hypothesized ancestor of the domestic horse.[50]

The exact appearance of the tarpan is unclear. An early 19th century drawing, (Figure 14a) supposedly from nature, still bears the stylistic stamp of its time, showing the young animal with a thick curving neck, concave topline, and substantial hindquarters, yet delicate elongated legs with tiny hoofs and a small muzzle. The vague landscape—an

Figure 14a. 1845 drawing of 5 month–old Tarpan colt (foal?) by Borisov. After Clutton-Brock, p.30, fig. 2.2.

Figure 14b. Photograph of "The Tarpan of Kherson" captured in 1866 and kept at the Moscow Zoo. Date of photo unclear. Photo via Wikimedia Commons.

English park or the Crimean steppes—suggests that the image is at some remove from a naturalistic depiction. An earlier eighteenth century drawing[51] is similarly stylized. The detailed descriptions provided by Heptner et al. (pp. 1037–1049) are based on earlier written sources and not on the authors' direct observation. The only known photograph of a tarpan (Figure 14b) shows an equid that

[46] Clutton–Brock 1992: 43 f., 87–90.
[47] Clutton–Brock 1992, p. 42, fig. 3.1, right.
[48] Kohl 2007: 72–85; Aruz 2003: 290–96.
[49] Leskov 2008.
[50] For a discussion of the problems in nomenclature see The Extinction Website http://www.petermaas.nl/extinct/speciesinfo/tarpan.htm.

[51] Clutton–Brock 1992: 30, fig. 2.1.

could easily pass for a delicate, well–bred small horse, not the stocky animal shown on the Maikop vase. Thus some have questioned whether the so–called Kherson tarpan is indeed a purebred tarpan and not a cross–bred feral animal (The Extinction Website). The range of the tarpan included the steppes of Eastern Europe and western Central Asia as well as the steppe–forest zone of these regions,[52] an area that encompassed the western Caucasus, where the Maikop vase was found. In behavior, the tarpan seems to have been more biddable than the Przewalski's horse and less flighty than the onager. When crossbred with domestic horses (*E. caballus*), its offspring were fertile. The range and temperament of the tarpan, if that is the species shown on the Maikop vase, present it as the most likely candidate species for the Susa inlay. But the lack of absolute identification for the now–extinct tarpan precludes any certainty.

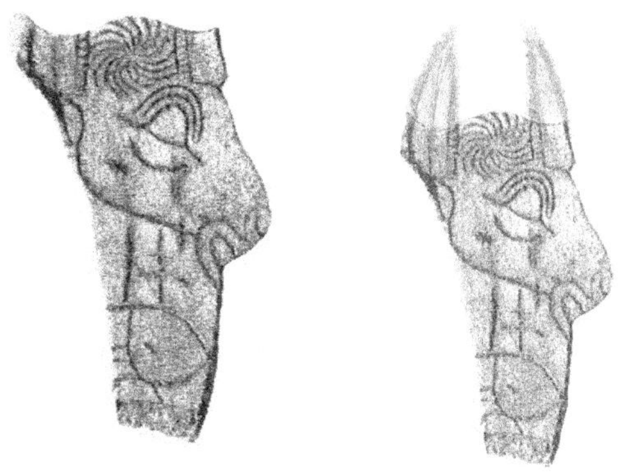

Figure 15. Left: Susa, shell inlay, no. 4899. Musee du Louvre, Paris, acc. no. Sb 5630. Right: Partial reconstruction of Sb 5630 by the author. Photo by Raphaël Chipault © Musee du Louvre, Dist. RMN-Grand Palais/ Art Resource NY.

Domestic Horse (*Equus caballus*)

The Susa equid has also been called a true domestic horse (*Equus caballus*) by Littauer and Crouwel (1979: 26), Zarins (1976: 280 ff.; 2014: 131 f., 252), and Moorey (1967: 107), though the sparse tufted tail and general conformation argue against this. The Susa equid is not shown as domesticated, but as a wild animal without harness, halter, or evidence of handler. In the absence of an immediate context for the inlay, it is difficult to pursue this identification.

Some Observations and Speculations

Having surveyed the material culture of the period to which the shell inlay has been attributed and looked at the possible species that it may depict, we can make several observations. The first is that while the inlay is similar to objects from Mesopotamia and Syria, it is stylistically distinct. The Mesopotamian and Syrian equids have angular forms with straight lines, a lack of interior modeling, and almond eyes with central dots. Neither does the inlay depict any animal known in works from those regions. Thus the Susa inlay was not imported from Mesopotamia or Syria. Nor does it resemble anything made at or near Susa, i.e. Old Elamite Susiana. Its very distinctiveness is a salient characteristic. The second observation is that the species shown is visually far closer to a Przewalski's horse or a tarpan, if that is indeed the equid shown on the Maikop silver vase. The Susa inlay is clearly not an onager or onager–hybrid. The third observation is that the equid shown appears to be wild, not a domesticate. If the piece is accepted as an import from a region other than Mesopotamia or Syria, the identification of a Przewalski's horse or a tarpan is less problematic.

Consideration of another inlay fragment apparently found, or at least published with, the equid inlay may help further refine the possibilities. This second shell fragment, now Louvre Sb 5639 (Figure 15), shows an herbivore, head in profile to the right with a frontal whorl of hair at the top of the head where one might expect horns or ears to be. (The identification of the animal as a bovid by Dolce (1978: 57), will be considered below.) The elegance of the engraving, the curving line around the muzzle and the conventions of the eye: an arching upper lid, flat lower lid and separate curved line beneath the lower lid, indicate the two fragments, the equid and the herbivore head, came from the same workshop, if not the same hand. The style of these two "orphan" inlays is quite unlike the Mesopotamian parallels. They have more modeled, naturalistic forms, and, in the case of the equid, realistic positions of the surviving legs and head. The pronounced difference in style and subject matter with the third inlay known from Susa, an awkwardly schematic human figure in the Mesopotamian style,[53] suggest that the equid and the herbivore were part of the same assemblage.

The herbivore head has been called bovine, but close observation of the stubby "horns" on either side of the hair whorl shows that the smoothly recurving edge that gives the impression of stubby horns is in fact merely the worked edge of the shell. The engraved vertical lines that rise on either side of the whorl do not resemble the natural forms of bovine horns, and have finely hatched or fringed edges suggestive of hairy or furry ears. The proper left "horn" also has a single straight line running vertically between the two "hairy" edges, another visual element not found on cattle horns. If these lines are extended along their apparent orientation, (Figure 15, right) one sees instead a bovine head, the head of a donkey *(Equus asinus)*. The drooping lower lip, highly unusual in a bovid, then becomes the pendulous lower lip of a donkey at rest, perhaps explaining the presence of the horizontal lower leg and hoof below the jaw. The whorl of hair between

[52] Groves 1974: 39–47; Boyd and Houpt 1994.

[53] Dolce 1978: 57, S1, and pl. XVII.

the ears would then be the short stiff forelock typical of the donkey especially in winter coat. Donkeys and donkey images are well attested in early Iran, and there is no difficulty with its presence at Susa.[54]

If the identification of the head as that of a donkey is correct, the object that once bore the two Susa inlays showed a scene of wild equids, or at least equids not under immediate human control, rendered in a style not known at Susa, nor in Mesopotamia or Syria in the latter half of the third millennium BCE. Given the lack of close stylistic and iconographic parallels plus the chaotic character of the original Susa excavations, there is no reason to assume a mid–third millennium BCE date for the Susa inlay. Indeed the excavations that produced the two animal inlays also produced a large number of Proto–Elamite tablets (ca. 3200–2800 BCE), suggesting that levels far earlier than the mid to late third millennium BCE had been reached by the early trenches.[55] The similarity of the Susa "horse" to the equid on the well–dated Maikop vase further supports a late 4th millennium date. One contemporaneous Proto–Elamite tablet, Louvre Sb 6320, bears pictographs of several types of equids. Some have large heads, thin necks and sparse upstanding manes, while others have small heads and thicker, heavier necks.[56]

It is also in this period, the latter portion of the fourth millennium, that some of the most astonishingly naturalistic animal images in Near Eastern art were produced. Most relevant to our discussion is the silver gazelle, an archaeological orphan now in the Metropolitan Museum of Art, NY.[57] Its technical sophistication and realistic style stymied attempts to date it for decades and provoked chronologies as late as the Achaemenid period until its artistic commonality with Iranian works like the Guennol lioness, the Metropolitan Museum's silver bull, and other works dating to the latter part of the fourth millennium was noted.[58] The Susa equid, (Figure 1) and its probable companion piece (Figure 15), belong to this visually dynamic tradition of naturalistic, powerful, even numinous animal forms, rather than to the controlled, stratified world of the mid–third millennium where animals are not the central actors but mere modifiers of human activity. Seen in this new context, the Susa equid is no longer an outlier of the mid–third millennium, but part of the earlier Proto–Elamite animal world.

Bibliography

Abdi, K. 2003. From *écriture* to civilization: Changing paradigms of proto–elamite archaeology. In *Yeki bud, yeki nabud: Essays on the Archaeology of Iran in honor of William M. Sumner,* edited by N. F. Miller and K. Abdi, 140–151. Cotsen Institute of Archaeology, University of California, published in association with the American Institute of Iranian Studies and the University of Pennsylvania Museum of Archaeology and Anthropology, Los Angeles.

Amiet, P. 1966. *Elam*. Archée, Auvers–sur–Oise.

Aruz, J. 2002. Power and protection: A little proto–elamite silver bull pendant. In *Leaving no Stones Unturned. Essays on the Ancient Near East and Egypt in Honor of Donald P. Hansen,* edited by E. Ehrenberg, 1–14, Eisenbrauns, Winona Lake/IN.

Aruz, J. ed. with R. Wallenfels. 2003. *Art of the First Cities: The Third Millennium B.C. from the Mediterranean to the Indus*. Metropolitan Museum of Art, New York and Yale University Press, New Haven.

von Bothmer, D. 1990. *Glories of the Past. Ancient Art from the Shelby White and Leon Levy Collection.* Metropolitan Museum of Art, New York. [Distributed by H. N. Abrams].

Bouman, I. and J. Bouman. 1994. The history of Przewalski's horse. In Boyd and Houpt 1994, 5–38.

Boyd, L. and K.A. Houpt eds. 1994. *Przewalski's Horse. The History and Biology of an Endangered Species.* SUNY Series in Endangered Species. Albany State University of New York Press, Albany/NY.

Le Breton, L. 1957. The early period at Susa, Mesopotamian relations. *Iraq* 19:79–124.

Clutton–Brock, J. 1992. *Horse Power. A History of the Horse and the Donkey in Human Societies*. Natural History Museum Publications, London, and Harvard University Press, Cambridge/MA.

Dahl, J. 2005. Animal husbandry in Susa during the Proto–Elamite Period. *Studi Micenei ed Egeo Anatolici* 47, 81–137.

Dolce, R. 1978. *Gli intarsi mesopotamici dell'epoca protodinastica.* Istituto di studi del Vicino Oriente, Università, Rome.

Frankfort, H. 1970. *The Art and Architecture of the Ancient Orient*. 4th edition, Pelican History of Art. Yale University Press, New Haven and London.

Groves, C. P. 1974. *Horses, Asses, and Zebras in the Wild.* David and Charles Publications, Newton Abbot and London.

Groves, C. P. 1994. Morphology, Habitat, and Taxonomy. In Boyd and Houpt 1994, 39–60.

Harper, P. O., J. Aruz, and F. Tallon eds. 1992. *The Royal City of Susa. Ancient Near Eastern Treasures in the Louvre*. The Metropolitan Museum of Art, New York. [Distributed by H. N. Abrams].

Heptner, V. G., A. A. Nasimovich and A. G. Bannikov. 1988. Tarpan: *Equus (Equus) przewalskii. Mammals*

[54] Potts 2011.
[55] Dahl 2005: 84. For an overview of the term Proto–Elamite see Abdi 2003.
[56] Dahl 2005: 97, fig. 11 and p. 125; Amiet 1966: 97, no. 50B and Le Breton 1957: 104, fig. 19.
[57] Accession Number 47.100.89.
[58] Aruz 2002; Harper, Aruz and Tallon 1992: 68–70.

of the Soviet Union. Vol. 1. ATIODCYLA and PERISSODACTYLA. Smithsonian Institution Libraries and The National Science Foundation. Washington, D.C., 1037–1058.

Jéquier, G. 1905. Fouilles de Suse. 1899–1900, 1900–1901, 1901–1902. Description des monuments. *Mission Archéologique de Perse. Memoires de la Delegation Archéologique en Perse.* Vol. 7, Leroux, Paris, 2–42.

Kohl, P. L. 2007 *The Making of Bronze Age Eurasia.* Cambridge World Archaeology. Cambridge University Press, Cambridge and New York.

Leskov, A. M. 2008. *The Maikop Treasure.* University of Pennsylvania Museum of Archaeology and Anthropology, Philadelphia.

Levine, M., C. Renfrew and K. Boyle eds. 2003. *Prehistoric Steppe Adaptation and the Horse.* McDonald Institute Monographs. McDonald Institute for Archaeological Research, Cambridge.

Littauer, M. A. and J. H. Crouwel. 1979. *Wheeled Vehicles and Ridden Animals in the Ancient Near East.* Handbuch der Orientalistik Siebente Abteilung. Kunst und Archäologie. Erster Band. *Der Alte Vordere Orient.* Zweiter Abschnitt. *Die Denkmäler.* Brill, Leiden and Köln.

Martinez–Seve, L. 2004. Mecquenem, Roland de (1877–1957). In *Encyclopaedia Iranica.* http://www.iranicaonline.org/articles/mecquenem-roland-de. (Accessed 15 January 2017).

de Mecquenem, R. 1922. Fouilles de Suse. Campagnes des années 1914–1921–1922. *Revue d'assyriologie et d'archéologie orientale* 19, 109–140.

de Mecquenem, R. 1943. Fouilles de Suse, 1933–1939. *Mémoires de la Mission archéologique de Perse* 29, 3–161.

Moorey, P. R. S. 1967. Some aspects of incised drawing and mosaic in the Early Dynastic Period. *Iraq* 29, 97–116.

Moorey, P. R. S. 1994. *Ancient Mesopotamian Materials and Industries. The Archaeological Evidence.* Clarendon Press, Oxford.

Pavel, C. 2011. *Describing and Interpreting the Past: European and American Approaches to the Written Record of the Excavation.* Editura Universităţii din Bucureşti, Bucureşti (Romania).

Perrot, J. and D. Ladiray 1989. La Haute–Terrasse et la nécropole du IVe millénaire. *Dossiers Histoire et Archéologie.* No. 138, 38–41.

Postgate, J. N. 1992. *Early Mesopotamia. Society and Economy at the Dawn of History.* Routlegde, London [etc.].

Potts, D. T. 2011. *Equus asinus* in Highland Iran: Evidence old and new. In *Between Sand and Sea. The Archaeology and Human Ecology of Southwestern Asia. Festschrift in Honor of Hans–Peter Uerpmann,* edited by N. J. Conrad, P. Drechsler and A. Morales, 167–176. Kerns, Tübingen.

Sarianidi, V. 1986. *Die Kunst des alten Afghanistan.* Seemann, Leipzig.

Sarianidi, V. 2002. *Marguş. Murgap derýasynyň köne hanasynyň aýagyndaky gadymy gündogar şalygy / Drevnevostčnoe carstvo v staroj del'te reki Murgab / Ancient Oriental kingdom in the Old Delta of the Murgab River.* [Turkmen, Russian and English]. Türkmendöwlethabarlary, Aşgabat, Turkmenistan.

Sumner, W. M. 2003. *Early Urban Life in the Highlands of Anshan: Excavations at Tal-e Malyan in the Highlands of Iran.* University Museum Monograph 117 / Malyan Excavation Reports 3. University of Pennsylvania Museum of Archaeology and Anthropology, Philadelphia.

Vilà, E. 2006. Data on equids from late fourth and third millennium sites in northern Syria. In *Equids in Time and Space. Papers in honour of Véra Eisenmann,* edited by M. Mashkour, 101–123.Proceedings of the 9th Conference of the International Council of Archaeozoology, Durham, August 2002. Oxbow, Oxford, 2006.

Woolley, C. L. 1934. *The Royal Cemetery. Ur Excavations* Vol. 2. *Plates.* Publications of the Joint Expedition of the British Museum and of the University Museum, University of Pennsylvania, Philadelphia to Mesopotamia. British Museum Press, Oxford.

Zarins, J. 1976. *The Domestication of Equidae in Third Millennium B.C. Mesopotamia.* Vols. I–II. PhD Dissertation The University of Chicago.

Zarins, J. 2014. [with the assistance of R. Hauser]. *The Domestication of Equidae in Third Millennium B.C. Mesopotamia.* Cornell University Studies in Assyriology and Sumerology 24. CDL Press, Bethesda, Maryland [update of *The Domestication of Equidae in Third Millennium B.C. Mesopotamia,* PhD Thesis University of Chicago 1976].

Zettler, R. and L. Horne eds. 1998. *Treasures from the Royal Tombs of Ur.* University of Pennsylvania, Museum of Archaeology and Anthropology, Philadelphia.

8

The Use of Equines in Archaeological Research: A New Methodology Combining Ancient DNA and Trace Element Analysis Tested on Israeli Equines from ca. 900–1200 BCE

Gila Kahila Bar–Gal, Lia Hadas*, Hadas Motro, Amit Reem, Reuven Amitai, Ronnie Ellenblum, Michael Haahr Friis, Kurt Villads Jensen and Kaare Lund Rasmussen*

**These authors contributed equally to this work.*

In medieval times, equines were of great importance in transportation, agriculture, and war. In the archaeological assemblages, they are rare due to the way in which their corpses were normally disposed. Therefore, traditionally, equines have played only a very minor role in archaeological reconstructions of medieval events. We present a new methodology in the study of equine remains that encompasses the combined use of ancient DNA and trace element analysis. The genetic data yield information pertinent to the origin of equines. The trace element composition gives, analogous to human bones, dietary information, which in the case of equines is related to the geographical/geological locations of pastures. This methodology has been applied to a set of equine remains from Israel, amongst them two unique equine assemblages Vadum Iacob (October 1178–August 1179 BCE) and the Armenian monastery in Jaffa (Muslim period; ninth–tenth century BCE). We believe that the equines, horses and mules, originated from three different sources.

Introduction

Historical sources testify to the importance of equines, i.e. horses (*Equus caballus*), donkeys, (*Equus asinus*) and their hybrid, the mule, in medieval Europe, the Middle East, and Asia for transportation, warfare, agriculture, economy, and status symbols (Bennett 2001; Clutton–Brock 1992; Hyland 1999; Kelekna 2009). Improvements in horse equipment enabled the utilization of horsepower more effectively and raised the importance of equines in the European economy. Furthermore, the establishment of empires and other political developments of the Middle Ages reinforced the horse's status, as most of these empires were heavily dependent on the beast. The expansion of the Islamic state also played a role in this process, and Muslim armies always incorporated a large cavalry component. Later in the crusaders' period, horses from Europe were transported to The Holy Land via land and sea. Historical texts testify to the hardships of the journey and adjustments to the conditions in the new land, which eventually led to the death of many horses (Clutton–Brock 1992; Hyland 1999; Kelekna 2009). Therefore, the Southern Levant during the medieval period served as a major junction of interaction among Eastern and Western civilizations in which the horse and its relatives played a major role.

Morphological studies of equine skeletal remains can determine the species (horse, donkey), or hybrids based on several indicative bones (Eisenmann 1980; 1981; 1986). Bone morphometric also contribute to the general knowledge of the individual animal (age, sex, body size, weight, health, and utilization), enabling a better understanding of the roles of equines in human society. However, at present, the understanding of the characteristics of horses in the Southern Levant and the places of their origin are based solely on descriptions and interpretations of historical and artistic sources without much supporting scientific evidence. In archaeological contexts, equine remains are rare due to sanitary disposal, as big animals were usually buried or dumped away from settlements in order to prevent unpleasant odors and the spread of diseases (Motro 2004). The unprotected carcasses were often eaten and dispersed by carnivores, resulting in only isolated and incomplete equine specimens left to be unearthed by archaeologists.

With the absence of morphologically distinctive bones, most of the equine remains cannot be properly identified, and the possibilities for in–depth studies are therefore often very constricted. A molecular genetics approach using ancient DNA (aDNA) analysis based on the recovery of DNA from archaeological remains may enable identification of species (Kahila Bar–Gal et al. 2003; Leonard et al. 2007; Pääbo et al. 2004). The quality and quantity of the aDNA varies due to degradations processes (Herrmann and Hummel 1994), which have been shown to be dependent on the state of preservation of the sample and to a lesser degree on the age of the sample (Höss and Pääbo 1993; Poinar and Stankiewicz 1999). Despite these drawbacks, once aDNA can be recovered it provides a window into the past, presenting unique, quantitative data about the genetic links between extant and extinct species (Yang 1997).

Trace element concentrations in human bones have been interpreted as reflecting the diet (e.g. Schutkowski et al. 1999, Mauerer et al. 2011; and Cucina et al. 2011). It is likely that this could also be the case for equines, although little work has been done on the subject. In the present study, three trace–elements, lead, nickel, and chromium, have been measured in the bones of ancient equines in order to attempt to group them according to differences in trace element concentrations, which hypothetically are supposed to vary in water, plants, and/or soil of the areas on which the animals have grazed for most of their lives. Lead (Pb), nickel (Ni), and chromium (Cr) were selected because they form divalent cat ions, which substitute willingly for Ca^{2+} in the bone apatite and are not normally present in large amounts in the surrounding soil, and therefore not likely to have participated in large–scale diagenetic processes. Lead, nickel, and chromium are also known not to be metabolically–controlled, like zinc (Zn), for example. The choice of metabolically–controlled elements would have prevented the trace element distribution in equine bones to reflect the distribution in the crop, soil, or water. The large size of the equine bones limits the probability of diagenetic processes. Such processes are also less likely to have occurred in the relatively arid climate in Israel compared to many other archaeological sites.

Archaeologically, the present study was made possible by the discovery of two unique and large equine assemblages in Israel. First, over the course of a salvage excavation led by the Israeli Antiquities Authority in the Armenian monastery in Jaffa (AMJ), a set of equine skeletons dated to the early Muslim period was found under the floor of the seventeenth century monastery (Reem Unpublished). Secondly, an excavation at the Crusader fortress at *Vadum Iacob* in northern Israel revealed a unique equine assemblage from a siege that took place in August 1179 BCE (Ellenblum 2007). This assemblage represents the first finding of complete medieval equine skeletons in a battlefield from the eastern Mediterranean. Both assemblages consist of relatively complete articulated skeletons of dozens of equines.

The goals of this study were to reconstruct the roles of equines in medieval historical events through the determination of their species, differences in their genetic origin, and differences in their dominant grazing areas. Each technique in itself is not sufficient, but with a combination of knowledge about genetic profile and differences in grazing areas, it is indeed possible to obtain new archaeological and historical insight.

Table 1. Information summary of contemporary specimens included in the study

Lab code	Species		Breed
226	*E. caballus*	Horse	Arabian
228	*E. caballus*	Horse	Arabian
233	*E. caballus*	Horse	Arabian
241	*E. caballus*	Horse	Quarter
330	*E. caballus*	Horse	Warmblood
354	*E. caballus*	Horse	Thoroughbred
417	*E. caballus*	Horse	Pony
456	*E. caballus*	Horse	Thoroughbred
519	*E. caballus*	Horse	Andalusian
596	*E. caballus*	Horse	Pony
D52	*E. asinus*	Donkey	–
D53	*E. asinus*	Donkey	–
Mu4	*E. caballus x E. asinus*	Mule	–

Material and Methods

Contemporary Equine Samples

Blood samples of 13 modern equines, including ten domestic horses, two donkeys and one mule, from Israel were used as references in the genetic analysis. The domestic horses in the sample were of different breeds including Arabian, Quarter Horse, Warm–blood, Thoroughbred, Pony, and Andalusian (Table 1).

Equine Remains from Archaeological Sites

Equine remains from four different archaeological sites excavated in Israel were studied. Three sites were dated to the Muslim period (eighths–tenth century): the Armenian Monastery in Jaffa, the Olē Ziyyon Street in Jaffa, and Ramla (Table 2, Figure 1). One sample from Ramla was dated to the thirteenth to fourteenth century (Table 2). The fourth site was *Vadum Iacob* dated to the Crusade period (1179 AD) (Table 2, Figure 1).

The Frankish castle of *Vadum Iacob* ("Jacob's Ford," or according to its modern Hebrew name, Metzad Ateret), is located in Northern Israel in a strategic place, guarding the only crossing point over the Jordan River between its sources and the sea of Galilee, on the road between Acre and Damascus (Figure 1). The castle was constructed by Crusaders and guarded by Templar knights, and was populated for only 11 months (October 1178 to August 1179 AC). The castle was conquered and destroyed by Saladin and his soldiers and was never rebuilt. Based on historical sources, Saladin's army slaughtered some 800 Crusaders, dumping bodies of humans and horses alike into a deep cistern (Mitchell, et al. 2006). In excavations led by the Hebrew University of Jerusalem, 10 articulated equines were uncovered with distinct arrowheads between their skeletons, which indicate their cause of death during

Table 2. Information summary of archaeological equid remains included in the study

Excavation	Date	Individual / Lab code	Skeletal element
Armenian Monastery	9th–10th Century	Eq.1	Humerus, Skull
Armenian Monastery	9th–10th Century	Eq.10	Tooth, Mandible
Armenian Monastery	9th–10th Century	Eq.11	Mandible
Armenian Monastery	9th–10th Century	Eq.12	Tooth (LM & P4)
Armenian Monastery	9th–10th Century	Eq.13	Tooth (LM), Skull, Mandible
Armenian Monastery	9th–10th Century	Eq.17	Mandible, Tooth (LM)
Armenian Monastery	9th–10th Century	Eq.18	Tooth (L-P4)
Armenian Monastery	9th–10th Century	Eq.4	Tooth (M2 & M3)
Armenian Monastery	9th–10th Century	Eq.7	Skull
Armenian Monastery	9th–10th Century	Eq.8	Mandible
Armenian Monastery	9th–10th Century	Eq.9	Mandible
Armenian Monastery	9th–10th Century	Eq.21	Mandible
Armenian Monastery	9th–10th Century	Amj9	1st phalanx
Armenian Monastery	9th–10th Century	Amj10	2nd phalanx
Armenian Monastery	9th–10th Century	Amj16	Humerus
Armenian Monastery	9th–10th Century	Amj17	Humerus
Armenian Monastery	9th–10th Century	Amj19	Radius/Ulna
Armenian Monastery	9th–10th Century	Amj20	Humerus
Armenian Monastery	9th–10th Century	Amj21	Humerus
Armenian Monastery	9th–10th Century	Amj22	Tibia
Armenian Monastery	9th–10th Century	Amj23	Tibia
Armenian Monastery	9th–10th Century	Amj24	Femur
Armenian Monastery	9th–10th Century	Amj25	Tibia
Armenian Monastery	9th–10th Century	Amj26	Humerus
Armenian Monastery	9th–10th Century	Amj28	Radius
Armenian Monastery	9th–10th Century	Amj29	Scapula
Armenian Monastery	9th–10th Century	Amj30	Scapula
Armenian Monastery	9th–10th Century	Amj31	Vertebra
Armenian Monastery	9th–10th Century	Amj32	Pelvis
Armenian Monastery	9th–10th Century	Amj33	Scapula
**Armenian Monastery	9th–10th Century	AmjT1	Mandible
**Armenian Monastery	9th–10th Century	AmjT2	Skull
**Armenian Monastery	9th–10th Century	AmjT3	Skull
**Armenian Monastery	9th–10th Century	AmjT4	Bone fragment
**Armenian Monastery	9th–10th Century	AmjT5	Bone fragment
**Jaffa, Olē Ziyyon St	8th–10th Century	MF14	Scapula
**Jaffa, Olē Ziyyon St	8th–10th Century	MF15	Femur
**Jaffa, Olē Ziyyon St	8th–10th Century	MF16	Bone fragment
**Jaffa, Olē Ziyyon St	8th–10th Century	MF17	Mandible
**Jaffa, Olē Ziyyon St	8th–10th Century	MF18	Bone fragment
**Ramla	8th–10th Century	MF19	Femur
**Ramla	8th–10th Century	MF20	Femur
**Ramla	13th–14th Century	MF21	Skull
Vadum Iacob	1179 CE	AT1	Molar
Vadum Iacob	1179 CE	AT2	Molar, patella
Vadum Iacob	1179 CE	AT3	Metatarsal
Vadum Iacob	1179 CE	AT4	Astragalus calcaneum
Vadum Iacob	1179 CE	AT5	Ulna

Table 2 *continued*

Excavation	Date	Individual / Lab code	Skeletal element
Vadum Iacob	1179 CE	AT 6	Molar, radius-ulna, metatarsal
Vadum Iacob	1179 CE	AT7	Molar, tibia
Vadum Iacob	1179 CE	AT8	Bone fragment
Vadum Iacob	1179 CE	AT9	Molar, carpal tarsal
Vadum Iacob	1179 CE	AT10	Molar
Vadum Iacob	1179 CE	AT126	Metacarpal
Vadum Iacob	1179 CE	AT113	Metacarpal
Vadum Iacob	1179 CE	At2019	Metatarsal
Vadum Iacob	1179 CE	AT333	Molar

** For these samples only trace element analysis was carried out. No DNA was extracted.

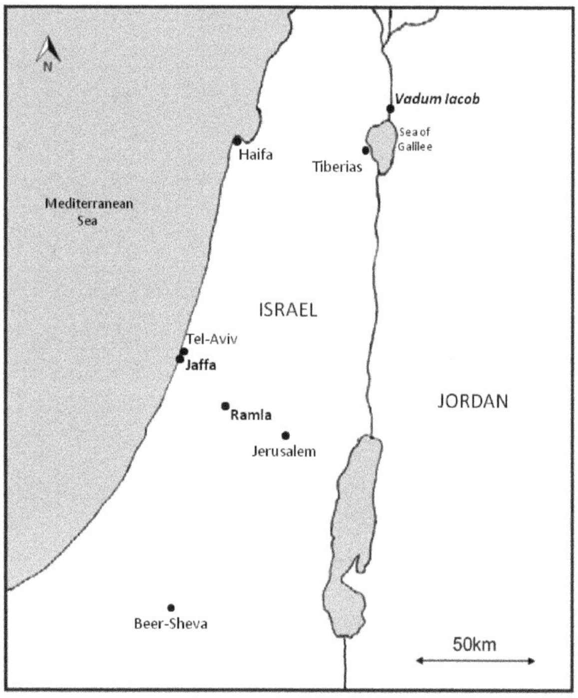

Figure 1. Map of geographical location of archaeological sites. 1: Vadum Iacob; 2: Jaffa (Armenian Monastery, Olē Ziyyon St); 3: Ramla.

the aforementioned battle. In addition, equine bone remains that could not be attributed to specific individuals were excavated, among them, four that were sampled for the study.

The Armenian Monastery in Jaffa was unearthed during a salvage excavation conducted by the Israel Antiquities Authority (November 2005–January 2006) initiated by the renovation of a building (Figure 1). The equine remains were found in a wide pit, sealed by a floor of a seventeenth–eighteenth century Ottoman building that re–used the Armenian Monastery floor. The pit contained dozens of articulated individual equines, some of which still had their horseshoes in place. On top of and in between the equine skeletons were layers of lime. A few bones showed signs of burning, and charcoal was found between some of the bones. The equine remains were radiocarbon dated to the early Muslim period (nineth–tenth century BCE) (Dr. Elisabetta Boaretto, Weismann Institute, unpublished information).

Forty–four specimens were sampled and studied for aDNA from the two major sites *Vadum Iacob* (10 identified individuals, labelled AT1 through AT10, plus four miscellaneous remains) and the Armenian Monastery in Jaffa (12 identified individuals, Eq., plus 18 miscellaneous remains, Amj). Some individuals were sampled more than once from different and/or the same bone. Based on availability, different skeletal elements were sampled for DNA analyses including teeth, skulls, femurs, humeri, phalanges, and vertebrae (Table 2). Teeth were preferred over bones, because the enamel protects the DNA from degradation and external damage (MacHugh et al., 2000). Twenty–nine of these samples were also analyzed for trace elements. In addition, three samples from Ramla, five samples from another site in Jaffa (Olē Ziyyon Street) and five remains (AmjT1–AmjT5) from the Armenian Monastery in Jaffa were included and studied only for trace elements. One sample from Ramla, was dated after sampling to the thirteenth to fourteenth century, the Ottoman period. Samples for the trace element analyses were taken only from the skeletal parts.

Genetic Analysis

DNA Extraction

Contemporary samples of blood from the modern equines were extracted using the phenol–chloroform method (Mathew, 1984) in a laboratory designated for contemporary DNA research.

DNA from ancient samples was extracted from the equine remains in a dedicated aDNA laboratory following strict guidelines for aDNA sample preparation procedures (Cooper and Poinar 2000; Hermon et al. 2011). The DNA was extracted using the Guanidine Thiocynate solution (GuSCN) (Boom, et al. 1990) followed by the silica beads method (Höss and Pääbo 1993). This method has been shown to be the most useful method for aDNA extraction

(Rohland and Hofreiter 2007). Each extraction included a negative control sample for ruling out contamination during the extraction.

DNA Amplification

A 105bp long nuclear Short Interspersed Nucleotide Element (SINE), named Equine Repetitive Element–1 (ERE–1), was amplified to determine the quality of the DNA extracted from the ancient samples (Walker et al. 2004).

Three primer sets were designed using Primer3 software (MIT) to identify the equine species based on a single nucleotide polymorphism (SNP) that vary among equine species. The screening for target sequences/amplicons focused on class 1 SNP's, since they more clearly present differences in their melting curves (Liew et al. 2004). Two primer sets targeting single point substitution along the mitochondrial genes, cytochrome B (CytB) and cytochrome oxidase I (COI), were designed to distinguish between horses and donkeys. The primer sets were designed based on a database that included horse and donkey sequences obtained from GenBank (ID numbers: EU433685, EU433686, EU4433661, X97337, EU257216, EU257215, EU834862, X97337.1, NC_001788, EF165072.1, EF165077.1, EF165078.1, EU887260.2, EF127815.1) and from sequences of local Israeli horses obtained by Dr. Glantz (unpublished data). The size of the amplicons was optimized for aDNA work, not exceeding 150bp (Table 3). The amplicon of the CytB included one point mutation 99C>T (EU433686), and the COI amplicon included three point mutations: 342G>A, 345A>G and 369G>A (EU834862). In addition, one nuclear gene, Prion Protein (PRNP), was also chosen in order to enable identification of horses/donkeys/mules and/or hinnies. The primer set was designed based on sequences obtained from GenBank (EF165072.1, EF165077.1, EF165078.1, EF127815.1, EU887260.2) (Table 3). The amplicon includes one point mutation 506G>A (EU887260.2). All primer sets were designed for maximum specificity to equines.

The extracted DNA was amplified using Polymerase Chain Reaction (PCR) assay and Real Time PCR followed by High Resolution Melting (HRM) curve analysis.

PCR Assay

Each reaction had a final volume of 25μL, containing 7μL of the sample DNA, 10X buffer, 0.3mM each dNTP, 3.5mM MgCl, 0.4μM each primer set, and 0.5–1U Taq polymerase [Amplitaq Gold or Amplitaq Gold 360 (Applied Biosystems)]. DNA was amplified using Biometra thermocycler with a touchdown cycle program: 10 min 95°C for initial denaturation, 18 cycles of 15 s at 95°C for denaturation, 45s at temperatures varying from 60°C to 50°C for primer annealing (3 cycles for each temperature) and 30s at 72°C for extension, 21–26 cycles of 15s at 95°C for denaturation, 45s at 48°C for primer annealing and 30s at 72°C for extension, 10 min at 72 °C for final extension. The amplified fragments were separated on a 1.5% agarose gel to determine successful amplification. In case inhibition was suspected, the extracted DNA was filtered using Microcon® YM–50 (Millipore™) following the manufacturer's protocol and PCR assay was performed.

Real–Time PCR and HRM Analysis

To determine the equine species three separate Real Time PCR targeting CytB, COI, and PRNP were conducted. Real Time PCR was performed on Corbett's Rotor–Gene 6000 Thermocycler. Amplification mixture was either purchased (Themo–Start PCR master mix, ABgene) or prepared separately for every reaction. The prepared mixture included 5–10ul of DNA sample, 10X PCR buffer, 35mM MgCl, 2.5 mM dNTPs, 10μM of each primer, 0.5 U Amplitaq Gold, 1.5 mM SYTO9 (Invitrogen). DNA amplification cycles resembled those described for the standard PCR assay above. After amplification, the samples were heated from 50° to 99°C with a rise in 1°C per second. Following this melt was a hybridization step in which the samples were cooled back to 50°C with a decrease of 1°C per second. The HRM analysis was conducted according to the melting step with a gradual rise in temperature of 0.1–0.4°C per second (Wittwer et al. 2003). Fluorescence acquisition was measured at the end of each extension step in the amplification cycles and during the HRM step with automatic optimization. HRM and melting curves were analyzed using Rotor–Gene 6000 software (Corbett).

Table 3. Primer sets used to determine equine species

Gene	Primer name	Primer sequence	Reference (Accession no.)	Horse *	Donkey **	Fragment size
Cytochrome B	Eq CytoB-F	5`-CCTCAAACATTTCATCAT-`3	EU433686	99C	T	60
	Eq CytoB-R	5`-GGAGGATTAGGCAGATTC-`3				
Cytochrome Oxidase I	Eq COI-F	5`-GGAGGATATACGGTTCAG-`3	EU834862	342G 345A 369G	A G A	109
	Eq COI-R	5`-CCCGAATAAACAACATAA-`3				
Prion Protein	Eq PRNP-F	5`-CTATGAGGACCGTTACTATC-`3	EU887260.2	506G	A	92
	Eq PRNP-R	5`-AGTTGTTCTGGTTGCTGTA-`3				

Single base substitutions were found between E. asinus and E. caballus and the primers flanking each substitution. *Position of SNP and horse genotype. **Donkey genotype.

Each Real Time PCR analysis included positive controls of DNA samples belonging to modern horses, donkeys, and mules. The melting curves of these samples were used as references for genotyping the ancient samples. It is important to emphasize that the modern samples were never introduced to the ancient laboratory, and were extracted and handled only in a separate modern laboratory to prevent carry–over or cross–contamination.

Sequencing

PCR products showing successful amplification on the agarose gel, or at the Real Time PCR analysis, were direct sequenced for both strands. Prior to sequencing, PCR products were purified using Bioneer's DNA purification kit or Exonuclease Shrimp Alkaline Phosphatase (HDV Pharmacia). Sequencing was carried out on an ABI 3700 (Applied Biosystems) by the Centre for Genomic Technologies at Hebrew University.

Genetic Data Analysis

The sequences obtained were initially verified as equine genes (ERE–1, COI, and PRNP) using the CBI BLAST algorithm (http://www.ncbi.nlm.nih.gov/blast/Blast.cgi). Chromatograms were individually examined in order to confirm the quality of sequences and visually screened for errors, using Sequencher 4.9 (Gene Codes Corp.). Both strands (sense and antisense) of each successful PCR product were direct sequenced and assembled into a contig in Sequencher. For each individual contig, a final consensus sequence was generated for further analyses. Sequences with poor quality chromatograms were excluded from the study. Sequences were compared to local modern equine sequences included in this study.

Trace Element Analysis

Analyses for Pb, Ni, and Cr were done by graphite furnace atomic absorption spectroscopy (GF–AAS). In order to remove any contamination deposited on the bone surface in the soil, during excavation or during curatorial handling, a ca. 0.5 cm^2 surface area of the bone was removed by an electric drill operated at low velocity. Following the decontamination, a sample of ca. 50 mg was procured with the drill. Approximately 15 mg of sample was placed in a flask and, 1 mL analytical grade conc. HNO_3 and 1 mL conc. analytical grade H_2O_2 were added. The flask was sealed and placed on a shaking table for 24 hours. 4 mL of 0.2 wt.% ammonium–di–hydrogen phosphate was added to the solution as a matrix modifier and the flask filled to 100 mL with 1% analytical grade HNO_3. Pb, Ni, and Cr were measured by the standard addition technique on a 2380 graphite furnace atomic absorption spectrometer (GF–AAS) from Perkin Elmer equipped with a HGA–400 graphite furnace and using hollow cathode lamps as light sources. The detection limits of the measurements were typically 0.8 µg Pb per gram of bone, 2 µg g^{-1} for Ni and 0.9 µg g^{-1} for Cr. Standard deviations were calculated based on three independently repeated measurements and also included the uncertainty originating from the calibration curve.

Results

DNA Preservation

Overall 44 specimens were sampled for DNA analysis (Table 2). DNA was extracted from each specimen, at least once independently. For each extraction, several PCR attempts ($2<N<6$) were conducted with all primer sets to increase the probability of successful amplification. To verify the identification of the sample as an equine, and to determine its DNA preservation, all samples, modern and ancient, were first amplified using the ERE–1 primer set. Sequences of contemporary Israeli horses (including one donkey, D52) were identical to each other, but differed by two point mutations: 94G>A (D26756.1) and 100A>G (D26756.1) along the ERE–1 sequence from the reference sequence of a domestic Thoroughbred horse (accession number D26567.1) (Table 4a). These findings indicated that the primer sets identify equines, but do not distinguish between breeds and species. The differences between the Israeli Thoroughbred horse samples and the reference sequence suggest that the variation is probably random.

Eleven successful ancient sample amplifications were achieved, although several amplification attempts were conducted for each extraction. The successful amplifications represent nine samples from AMJ, and two from *Vadum Iacob*. All positive amplifications were successfully sequenced for both strands (sense and antisense) using direct sequencing. The sequences representing the same specimen, different extractions and different amplifications were aligned together in one contig, and a consensus sequence was determined for each specimen. Alignment of the consensus ancient sequences with a sequence of a contemporary local horse (an Arabian horse, sample 226) revealed high variability: 23 positions along the ERE–1 sequence (Table 4b). The nine AMJ sequences differed from each other. Overall, a total of 20 substitutions were found among the AMJ sequences. Some

Table 4a. Summary of the base substitutions found in the ERE-1 region among the studied contemporary equids from Israel

Variable nucleotide position			
Sample	**94**	**100**	**Species**
D26567.1	A	G	*E. caballus*
226	G	A	*E. caballus*
228	G	A	*E. caballus*
241	G	A	*E. caballus*
330	G	A	*E. caballus*
354	G	A	*E. caballus*
496	G	A	*E. caballus*
419	G	A	*E. caballus*
D52	G	A	*E. asinus asinus*

Table 4b. Summary of the base substitutions found in the ERE-1 region among archaeological equid remains: Armenian monastery (AMJ), Vadum Jacob (AT), W=A or T; Y=T or C; M=A or C; R=A or G; K=T or G; N=Any; X=Position with no coverage, :=Gap.

Variable nucleotide position																																
Sample	**1**	**3**	**4**	**6**	**11**	**13**	**17**	**18**	**19**	**21**	**22**	**25**	**26**	**27**	**29**	**30**	**31**	**32**	**33**	**34**	**35**	**36**	**37**	**38**	**39**	**40**	**41**	**42**	**43**	**45**	**52**	**58**
226	G	C	G	G	T	G	T	G	C	C	A	A	A	A	C	A	C	G	C	T	G	A	G	G	C	A	G	C	G	C	G	A
Amj28																																
Amj17	T						C	A	A				G	G	T			A	T						T	G					A	
Eq.8	X	X	X	X	X	X	X												T												C	
Eq.17						A					G	G	G	G								G			T	G						
Amj10		W			Y				M				G	G			T												A			
Amj20			A																T												C	
Amj29												G		G																		
Amj32			A		C		N			A	G		:	G	K	N	N	N	N	N	N	N	N	N	T		N					M
Amj33													G																			
AT9	X	X	X	X	X	X	X	X	X	X	X	X	X	X	X	X									T	G		T		T		C
AT10	X	X	X	X	X	X																		A								

of them were shared between two to five individuals, such as substitutions at position 26 and 27 (Table 4b). The sequences of the two samples from *Vadum Iacob* differed from the contemporary horse reference sequence and also between them (Table 4b). Moreover, they differed from the ancient samples of AMJ. The quantitative differences among the consensus ancient sequences may indicate the low preservation of DNA in the samples, and can be interpreted as nucleotide misincorporations (C <> T and G <> A), which are often seen in ancient DNA (Hofreiter et al. 2001; Gilbert et al. 2007). In our study, the consensus sequence of a specimen represents sequences obtained from different extractions and amplifications supporting its authenticity and suggesting the possible genetic diversity among the individuals. Although variations were found among all ancient sequences, they were all found to be equines (57–93% homology), thus supporting the morphometric identification (Motro 2004, Motro et al. 2014).

Species Identification

Species identification of the contemporary equines was determined using Real Time PCR assay with the HRM method supported with direct sequence of the amplicon (Figure 2). The genotype of the horse was clearly distinguished from that of the donkey with all primer sets (CytB, COI, and PRNP). The PRNP primer set presents discernible curves for the three equines: horse, donkey, and mule (Figure 2c). Combination of the data available from all the primer sets together enabled the identification of the mule samples.

Relying on the success of the DNA amplification with the ERE–1 primer set as an indication of the DNA preservation level, we amplified the DNA of various ancient equine

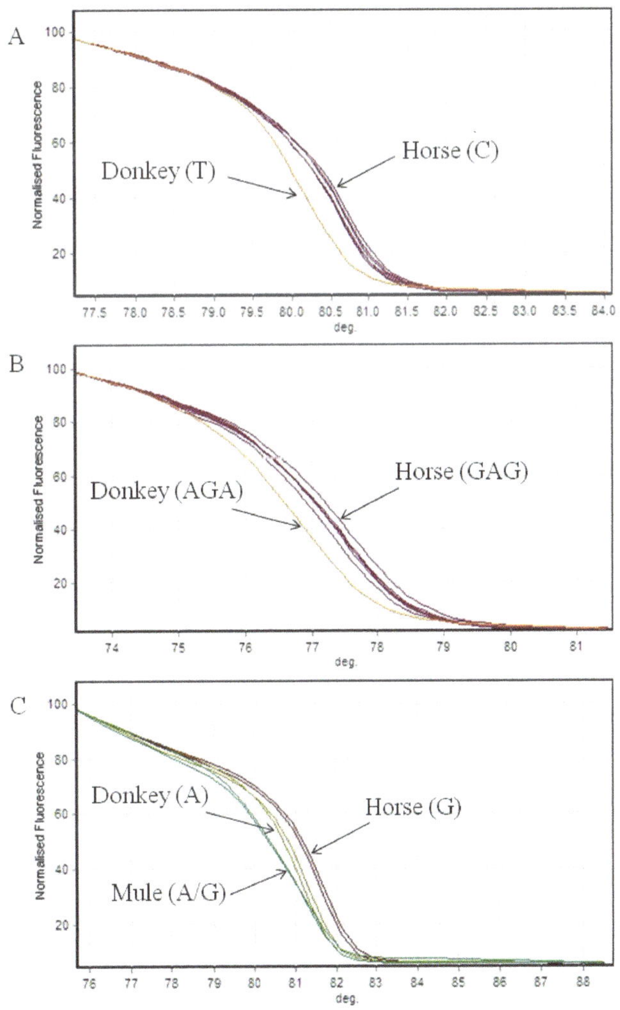

Figure 2. Real Time PCR with HRM analysis of contemporary equid samples. HRM is shown in the normalization function. (A) CytB primer set. (B) COI primer set. (C) PRNP primer set.

Table 5. Identification of the equid species among the ancient studied specimens

Sample	Site	Gene	aDNA Result	Morphological Result
AT113b	Vadum Iacob	COI	Horse/Mule	
AT126	Vadum Iacob	COI	Horse/Mule	
AT2019	Vadum Iacob	COI	Horse/Mule	
AT6	Vadum Iacob	PRNP	Horse	Horse/Mule
Amj29	Armenian Monastery	CytoB	Donkey	
Eq.10	Armenian Monastery	PRNP	Mule	Mule
Eq.8	Armenian Monastery	COI	Horse/Mule	Mule
Eq.11	Armenian Monastery	COI	Horse/Mule	Horse/Mule

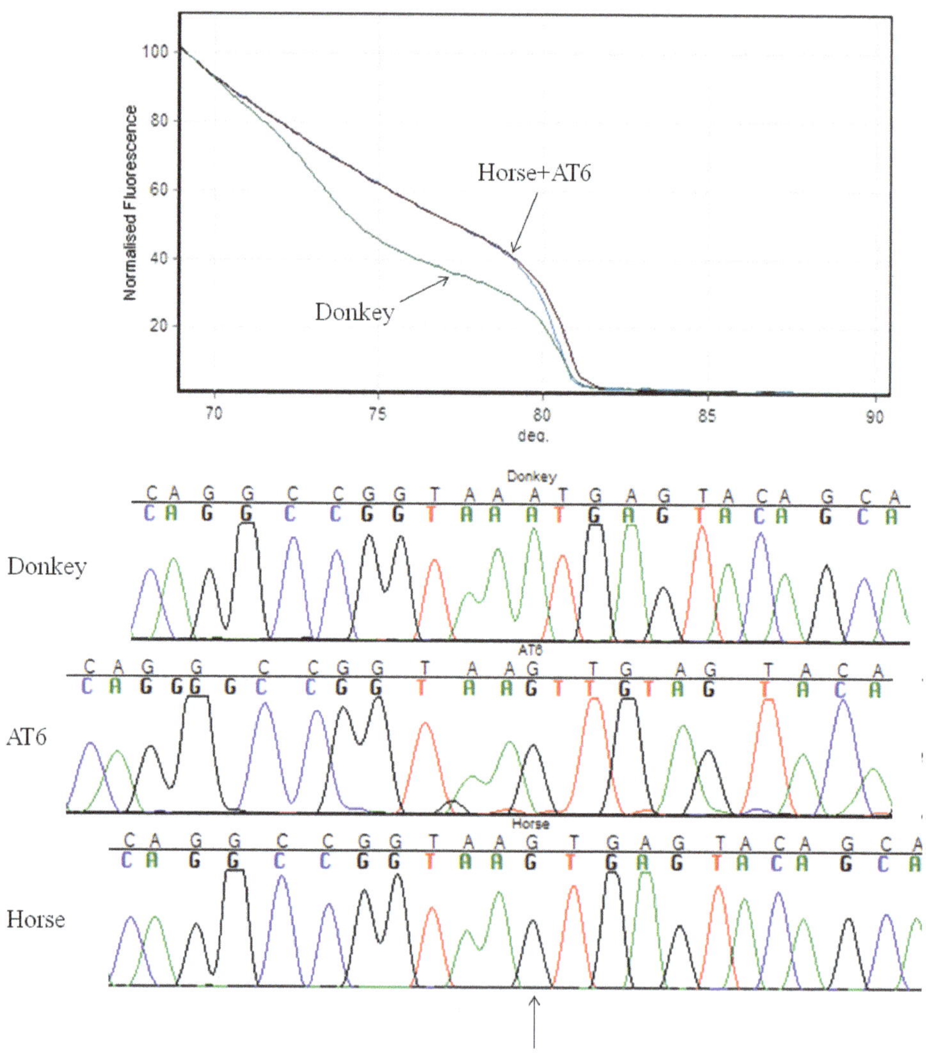

Figure 3. Identification of species among ancient equid remains using the Real Time PCR with HRM analysis and direct sequencing. Genotyping of AT6 from Vadum Iacob using PRNP primer set with HRM analysis and direct sequencing.

samples with the three primer sets using similar Real Time PCR conditions for the contemporary samples. Overall, eight samples were successfully amplified by one primer set (Table 5). The nature of the aDNA, low quantity and quality (degraded and frequently severed), explains the inconsistency of amplification success among all primer sets. Moreover, it indicates the inconsistency to amplify different genes due to random degradation. The successful amplifications indicated the presence of horses, donkeys, and mules among the ancient archaeological remains (Table 5, Figure 3). All Real Time PCR assays showed no amplification of the negative controls samples (both negative extraction and negative amplification), and, therefore, contaminations can be ruled out. It is important to stress that the Real Time PCR assay was designed to differentiate between species, using one SNP along the short amplified region of the degraded DNA. Therefore, the ancient positive amplifications that were sequenced were found to be identical to the contemporary horse sequence, except for one sample (Amj29) that was identical to the contemporary donkey sequence.

Among the equine ancient samples for which the species was unknown, we were able to determine the exact species of three specimens, and, for five specimens, we could rule out that they were donkeys, meaning they were domestic horse or mule (Table 5). In one specimen (AMJ Eq.1), both morphometric and genetic results identified the second phalanx as mule. The species of two specimens that could not be determined by morphological criteria were identified by the genetic analysis, indicating that one individual (AT6) was a horse, and the other (Amj29) was a donkey. For the other specimens (such as AT113b, AT126, AT2019), the genetic analysis minimized the options and identified them as possible horse or mule. The genetic identification of sample AMJ Eq.8 as horse/mule was supported by the morphological identification of a mule; therefore, we suggest that sample AMJ Eq.8 is indeed a mule.

Trace Element Analysis

The results of the trace element determination are listed in Table 6. Interpreting the trace element data was done by principal component analysis (PCA) using the Unscrambler© software. The data points from six samples were categorized as outliers. It was seen that principal component number 1 (PC–1) is almost solely dependent on Pb, while most of the variation in PC–2 is shared between Ni and Cr (Figure 4a). Seven samples had such different and unique trace element concentrations that they were singled out as outliers in the PCA analysis (AMJ–Eq8, AMJ–28, AMJ–T5, AT–3, AT–4, AT–5, and AT–2019). The trace element compositions were found to be effective in dividing the investigated equines into three distinct groups A, B, and C (Figure 4). The group in the lower left quadrant of Figure 4 consists of five of the equines from *Vadum Iacob* (AT2, AT6, AT7, AT7a, AT7b), one from Ramla (R–MF20), one from Olē Ziyyon Street in Jaffa (J–MF18), and two from the Armenian Monastery in Jaffa (AMJ–Eq.12 and AMJ–T4). This distinct group is tentatively interpreted as Crusader horses due to the fact that the majority of the samples are from *Vadum Iacob.* The group of red points in the lower right consists of data from the Armenian Monastery in Jaffa (AMJ–Eq4, AMJ–Eq9, AMJ–Eq10, AMJ–Eq11, AMJ–Eq17, AMJ–Eq18) and one individual from Ramla (R–MF21). The R–MF21 specimen is dated to a later period, the thirteenth to fourteenth century—the Muslim period. This group is tentatively interpreted as of specific Muslim provenance. It is interesting to note that no samples from *Vadum Iacob* were found in this group. Finally, the large and most diverse group in the upper part of the diagram consists of equines from a variety of archaeological sites, and due to the diverse nature of this collection, it is interpreted as another separate group of equines of Muslim provenance originating from different geographical area. From the mid–seventh century until the coming of the Crusaders by the end of the eleventh century, and then again from the later thirteenth century, the Muslims were the hegemonic group in the country. Horses would have been raised and used mainly by the military and political elite, and thus we can assume that these horses were indeed of "Muslim provenance" (Amitai 2010).

Discussion

The study of the ERE–1 SINE element proved effective as a marker to estimate the preservation state of the DNA in the samples, and also to support the published morphological identification of the samples as equines (Motro 2004, Motro et al. 2014). The low success rate (only 25%) of amplifying nuclear DNA gene fragment (ERE–1 SINE) indicated low preservation and low quality of DNA extracted from the archaeological specimens. The significant difference in the success rate between the samples (36% of the AMJ and 14% from *Vadum Iacob)* probably implies the differences in the general state of preservation of the DNA, although the samples are dated to approximately the same period. It is possible that some of the differences are caused by type 2 transition cytosine deamination, which has been shown to be a result of DNA damage in archaeological samples (Gilbert et al. 2007). But, as some of the mutations were found from different extraction and amplification of the same specimens, we believe that they are authentic, representing the differences between samples. The differences could be due to the different equine species or breed, geographical locations, climate, and soils (Pääbo et al. 2004, Poinar and Stankiewicz 1999). The climate in Israel is subtropical. The northern and coastal regions have a Mediterranean climate, characterized by hot dry summers and cool rainy winters, conditions which are not optimal for DNA preservation (Herrmann and Hummel 1994; Hofreiter et al. 2001).

The finding that several ancient samples represent the same substitution, as expected, further supports that the results are authentic probably representing differences in origin. For example: two samples, AMJ–Eq.10 and AMJ–Eq.8, both excavated at the Armenian Monastery in Jaffa and both identified as mules, differed from each other in the ERE–1 sequence. These two samples were found to differ also in the trace element content: AMJ–Eq.10 belongs to group C (Muslim equines mainly found at the Armenian Monastery), while AMJ–Eq.8 has a very different trace element concentration pattern and belongs to group B, probably of Muslim origin. (Table 6, Figure 4).

The results of the Real Time PCR, followed by HRM analysis, proved advantageous in amplifying short fragments (50–110bp), which are suitable for degraded ancient DNA, and also for determining the equine species. As expected, the higher success rate of positive amplifications was among the mitochondrial genes due to the number ratio of nuclear and mitochondria DNA in

Table 6. Results of the Atomic Absorption Spectroscopy (AAS) analyses

Site	Sample	Pb	1σ	Ni	1σ	Cr	1σ
		μg g.1	μg g-1	μg g.1	μg g-1	μg g.1	μg g-1
Armenian Monastery	AMJ-Eq7	15.5	0.56	23.72	0.91	1.18	0.27
	AMJ-Eq 8	82.75	1.38	18.07	1.26	16.33	0.55
	AMJ-Eq9	42.21	0.44	15.34	0.42	20.71	0.8
	AMJ-Eq10	21.6	0.34	15.87	0.63	15.82	0.52
	AMJ-Eq11	20.17	0.4	18.78	0.53	23.73	0.78
	AMJ-Eq4	24.61	0.36	13.35	0.97	20.92	0.7
	AMJ-Eq17	36.16	0.43	10.88	0.54	21.28	0.7
	AMJ-Eq18	19.56	0.49	14.76	1.09	22.05	0.76
	AMJ-Eq21	12.53	0.84	20.9	0.6	1.24	0.26
	AMJ-Eq1	8.31	0.31	13.68	0.47	13.74	0.48
	AMJ-28	126.31	2.26	15.19	0.83	46.45	2.4
	AMJ-T1	8.03	0.33	32.59	0.99	4.48	0.12
	AMJ-Eq12	4.59	0.29	4.14	0.12	<0.42	0.14
	AMJ-T2	25.04	0.49	29.03	0.41	6.73	0.31
	AMJ-T3	14.89	0.39	17.66	0.57	6.37	0.14
	AMJ-T4	14.18	0.55	1.91	0.51	5.37	0.29
	AMJ-Eq4	19.21	1.07				
	AMJ-17	18.82	0.72	27.01	1.01	8.22	0.16
	AMJ-T5	81.65	3.08	33.43	0.37	6.75	0.15
Jaffa - Olē Ziyyon St	J-MF14	10.65	0.48	12.66	0.57	<1.44	0.48
	J-MF15	10.34	0.44	22.46	0.98	<1.32	0.44
	J-MF16	7.83	0.32	19.28	0.79	<0.96	0.32
	J-MF17	9.96	0.32	19.42	0.62	6.19	0.58
	J-MF 18	1.2	0.09	2.5	0.41	12.17	0.32
Vadum Iacob	AT-2	1.41	0.12	7.8	0.28	<1.14	0.38
	AT-3	<0.30	0.1	2.69	0.44	<1.17	0.39
	AT-4	<0.30	0.1	<1.35	0.45	<1.53	0.51
	AT-5	<0.33	0.11	4.67	0.41	<1.23	0.41
	AT-6	0.88	0.14	3.95	0.82	6.73	0.59
	AT-7	1.11	0.1	3.15	0.51	<1.68	0.56
	AT-7a	2.33	0.11	7.91	0.68	<1.92	0.64
	AT-7b	2.78	0.09	6.92	0.36	<2.16	0.72
	AT-8	13.32	0.64	20.83	1.02	14.71	0.35
	AT-9	14.1	0.48	27.39	0.97	3.34	0.33
	AT-9a	7.95	0.66	16.22	1.36	22.27	0.48
	AT-10	12.11	0.8	22.63	1.51	<1.17	0.39
	AT-113	15.5	0.56	23.86	0.92	<0.87	0.29
	AT-126	12.5	0.25	35.27	1.66	4.9	0.46
	AT-2019	144.43	9.89	29.72	0.88	7.38	0.24
Ramla	R-MF19	8.78	0.32	18.85	0.7	<1.44	0.48
	R-MF20	7.71	0.19	2.66	0.47	4.26	0.39
	R-MF21	27.7	0.33	7.02	0.52	29.98	0.72

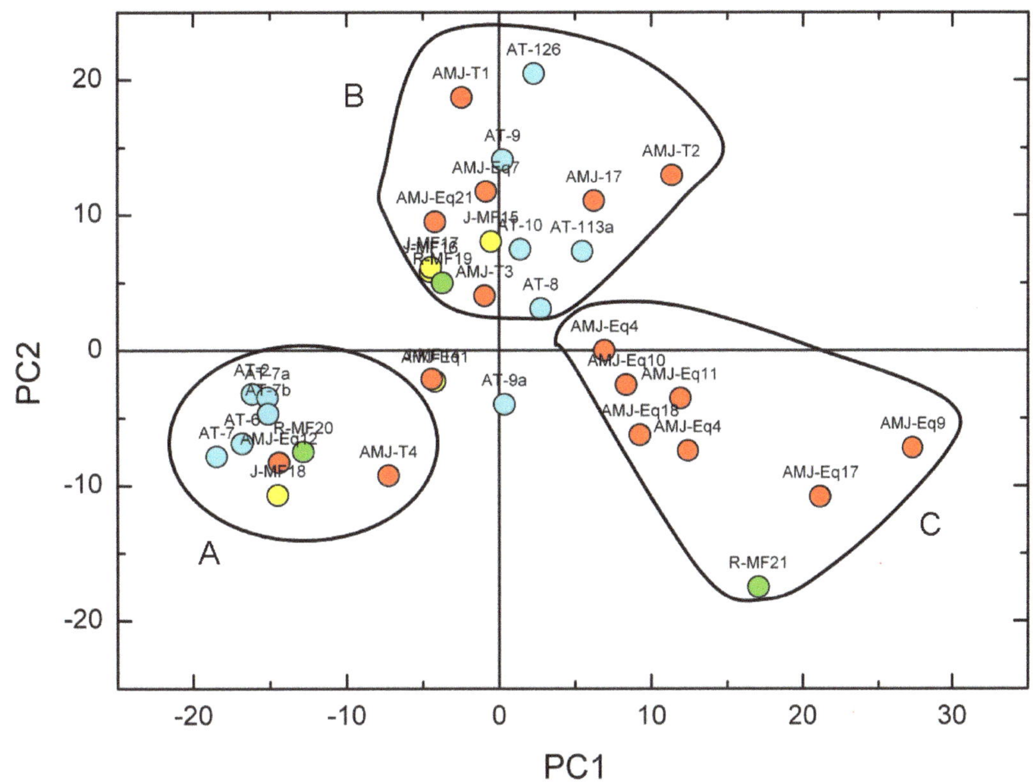

Figure 4. Principal Component Analysis results. The factor loadings for the Principal Component Analysis: Factor number 1 (PC-1) is almost solely dependent on Pb, while the variation in PC-2 is shared between Ni and Cr. The color codes represent four sites: cyan: Vadum Iacob, red: Jaffa—Armenian Monastery, yellow: Jaffa—Olē Ziyyon St, and green: Ramla.

every cell. This phenomenon is the general reason most ancient DNA studies are based on mitochondrial genes (Hofreiter et al. 2001). Unfortunately, we cannot base the species identification on the mitochondrial markers alone, because mules are hybrids between a female horse and a male donkey, and the mitochondria are maternally inherited. The low success rate in amplification of the Prion Protein (PRNP) gene, a nuclear gene, prevented the determination of the exact equine species, but nonetheless we were able to rule out the possibility of a donkey. Although, in this study, the success rates determining the equine species were low; the established method was found to be efficient and can be used on other equine archaeological assemblages.

The interpretation that we impose on the trace element data is that they reflect the trace element chemistry of the pasture area. Whether this is the trace element chemistry of the soil, the water, or the crop on which the equines feed is not of essence. The knowledge of the equine species, its place of origin, and its prevalent pasture site shed light on questions pertinent to human activities such as long–range transportation of the horses. Did the animals travel with the conquering army from the homeland in Europe, or were they local horses looted, bought, or raised locally by the Crusaders in the Holy Land and the surrounding regions? Our results seem to sustain both possibilities. The tentative interpretation of the trace element concentration data is that three distinct groupings can be recognized: A: a group of Crusader's horses (lower left quadrant); C: a rather well defined group of Muslim equines mainly found at the Armenian Monastery in Jaffa (lower right quadrant) which have probably been grazing at a rather well defined location characterized by the highest Pb load of them all; and finally, B: a group probably of Muslim origin (upper half in Figure 4). The latter group is found at all the sites investigated. Since these equines are found in so many different places, it is likely that this origin area B is the most prolific in raising, and in particular, in exporting equines in the Near East, the major junction between Eastern and Western horse breeds, as is indicated in the historical records of the Medieval Levant (Pryor 1982). Although we are not able to pinpoint a specific geographical position to group B, it is our hope that either research in historical sources, or more measurements of both genetics and trace elements in equines from the Middle East, eventually will reveal the identity of group B.

At *Vadum Iacob* the finding of both horses and mules in the same context fits well with the hypothesis that the castle was still under construction when it was destroyed on 30 August 1179 BCE (Ellenblum 2007). The mules could have been used to carry stones from the quarry during the building of the castle and the Templar Knights could have used the horses, both for regular transportation and for fighting the Muslims (Barber 1998; Ellenblum 2003). Concerning the main areas of pasture for the equines from *Vadum Iacob,* trace element analyses indicates there are several European horses (AT2, AT6, AT7, AT7a, AT7b), whereas five seem to be local—perhaps looted or purchased from the Muslims at source B (AT8, AT9,

AT10, A–113a, and AT–126), or raised by the Crusaders from such stock.

There are no direct historical sources describing the purpose of the Armenian Monastery in Jaffa in medieval times, but the city is known to have served as a major port for the Holy Land since Biblical times (Galili et al. 2000). Among the equine remains, we identified horses, donkeys, and mules. In a busy port, the donkeys and mules could have been an integral part of everyday life, carrying travelers and merchandise from place to place. The horses could have been imported on boats arriving in the port or bred locally. In any event, the horses were likely to have been used for human transportation inland. If any horses were transported across the sea, it is most likely they were imported into Jaffa from overseas, rather than locally bred horses being exported from Jaffa port. The individual equines from the Armenian Monastery in Jaffa can be divided into three groups based on the trace element concentrations: two equines of supposedly European origin (AMJ–Eq12 and AMJ–T4) as they cluster with the Crusader horses, seven from the distinct Muslim site C (AMJ–Eq4 (two samples), AMJ–Eq9, AMJ–Eq10, AMJ–Eq11, AMJ–Eq17, AMJ–Eq18), and six from the dispersed Muslim source area B (AMJ–Eq7, AMJ–Eq21, AMJ–17, AMJ–T1, AMJ–T2, and AMJ–T3). The reason we designate B and C as Muslim in origin is because of the archaeological date. One sample from Olē Ziyyon Street is of European origin (J–MF18), while three are from the prolific Muslim source B (J–MF15, J–MF16, J–MF17). Thus, all lines of evidence, the morphology, places of origin as reflected by the genetics and variation in trace element concentrations reflecting the grazing areas, point to Jaffa as a busy melting pot–type harbor city with equines coming from all corners of the medieval world.

The present study shows that a combined investigation using morphology, aDNA, and trace element analysis on equine remains gives a coherent picture that can enhance our understanding of archaeological contexts containing equine remains. The aDNA analysis was found successful in identifying equine species, even when the classical morphological methods fail to determine the species. It was shown that trace element analysis has the potential to distinguish different areas of pasture for the equines independent of their genetic origin.

Acknowledgements

The study in Israel was partially funded by Israel Science Foundation, Bikura, Grant no. 1431/07: "Medieval Human History of the Levant Written on Horseback".

Bibliography

al–Athir, Ibn. 1872. Extraits de la chronique intitulée Kamel–Altevarykh. In *Recueil des historiens des croisades. Historiens orientaux*, edited by C. Barbier de Meynard [l'Académie royale des inscriptions et belles–lettres], vol. 4, 636. Imprimerie royale, Paris.

Amitai, R. 2010. Armies and their economic basis in Iran and the surrounding lands, ca. 1000–1500 C.E. In *The New Cambridge History of Islam*, Vol. 3: *The Eastern Islamic World Eleventh to Eighteenth Centuries*, edited by D.O. Morgan and A. Reid: 539–560. Cambridge University Press, Cambridge.

Barber, M. 1998. Frontier warfare in the Latin Kingdom of Jerusalem: The campaign of Jacob's Ford, 1178–79. In *The Crusades and Their Sources Essays Presented to Bernard Hamilton Ashgate*, edited by J. France and W.G. Zajac, 9–22. Ashgate, Aldershot (Hampshire) and Brookfield/VT.

Bennett, D.G. 2001. Bits and bitting: form and function. *Proceedings of the American Association of Equine Practitioners* 47: 130–137.

Boom, R., C. J. Sol, M. M. Salimans, C. L. Jansen, P. M. Wertheim–van Dillen, and J. van der Noordaa. 1990. Rapid and simple method for purification of nucleic acids. *Journal of Clinical Microbiology* 28(3): 495–503.

Clutton–Brock, J. 1992. *Horse Power. A History of the Horse and the Donkey in Human Societies*. Natural History Museum Publications, London and Harvard University Press, Cambridge.

Cooper, A. and H. N. Poinar. 2000. Ancient DNA: do it right or not at all. *Science* 289(5482): 1139.

Cucina, A., V. Tiesler, T. S. Sosa, and H. Neff. 2011. Trace–element evidence for foreigners at a Maya Port in Northern Yucatan. *Journal of Archaeological Science* 38(8): 1878–1885.

Eisenmann, V. 1980. *Les chevaux (Equus sensu lato) fossiles et actuels: crânes et dents jugales supérieures*. Cahiers de paléontologie. Éditions du Centre national de la recherche scientifique, Paris.

Eisenmann, V. 1981. Étude des dents jugales inférieures des Equus (Mammalia, Perissodactyla) actuels et fossiles. *Palaeovertebrata* 10(3–4): 127–226.

Eisenmann, V. 1986. Comparative osteology of modern and fossil horses, half–asses, and asses. In *Equids from the Ancient World*, vol. 1, edited by R. H. Meadow and H.–P. Uerpmann, 67–116. Tübinger Atlas des Vorderen Orients, 19/1. Reichert, Wiesbaden.

Ellenblum, R. 2003. *Frankish Rural Settlement in the Latin Kingdom of Jerusalem*. Cambridge University Press, Cambridge.

Ellenblum, R. 2007. *Crusader Castles and Modern Histories*. Cambridge University Press, Cambridge.

Galili, E., J. Sharvit, and B. Rosen. 2000. Symbolic engravings on Byzantine sounding leads from the Carmel coast of Israel. *International Journal of Nautical Archaeology* 29(1): 143–150.

Gilbert, M. T., L. P. Tomsho, S. Rendulic, M. Packard, D. I. Drautz, A. Sher, A. Tikhonov, L. Dalén, T. Kuznetsova,

P. Kosintsev, P. F. Campos, T. Higham, M.J. Collins, A.S. Wilson, F. Shidlovskiy, B. Buigues, P. G. Ericson, M. Germonpré, A. Götherström, P. Iacumin, V. Nikolaev, M. Nowak–Kemp, E. Willerslev, J. R. Knight, G.P. Irzyk, C. S. Perbost, K. M. Fredrikson, T. T. Harkins, S. Sheridan, W. Miller, and S. C. Schuster. 2007. Whole–genome shotgun sequencing of mitochondria from ancient Hair shafts. *Science* 317(5846): 1927–1930.

Hermon, D., R. Gafny, A. Zamir, L. Hadas, M. Faerman, and G. Bar–Gal. 2011. The genetic signature of a shrunken head. *Archaeological and Anthropological Sciences* 3/2: 223–228.

Herrmann, B., and S. Hummel. 1994. *Ancient DNA: Recovery and Analysis of Genetic Material from Paleontological, Archaeological, Museum, Medical, and Forensic Specimens*. Springer, Heidelberg.

Hofreiter, M., V. Jaenicke, D. Serre, A. von Haeseler, and S. Pääbo. 2001. DNA sequences from multiple amplifications reveal artifacts induced by cytosine deamination in ancient DNA. *Nucleic Acids Research* 29(23): 4793–4799.

Höss, M., and S. Pääbo. 1993. DNA extraction from Pleistocene bones by a silica–based purification method. *Nucleic Acids Research* 21/(1), 3913–3914.

Hyland, A. 1999. *The Horse in the Middle Ages*. Sutton, Stroud/UK.

Kahila Bar–Gal, G., P. Ducos, and L. Kolska Horwitz. 2003. The application of ancient DNA analysis to identify neolithic caprinae: a case study from the site of Hatoula, Israel. *International Journal of Osteoarchaeology* 13(3): 120–131.

Kelekna, P. 2009. *The Horse in Human History*. Cambridge University Press, Cambridge.

Leonard, J. A., O. Shanks, M. Hofreiter, E. Kreuz, L. Hodges, W. Ream, R. K. Wayne, and R. C. Fleischer. 2007. Animal DNA in PCR reagents plagues ancient DNA research. *Journal of Archaeological Science* 34/9: 1361–1366.

Liew, M., R. Pryor, R. Palais, C. Meadows, M. Erali, E. Lyon, and C. Wittwer. 2004. Genotyping of single–nucleotide polymorphisms by high–resolution melting of small amplicons. *Clinical Chemistry* 507: 1156–1164.

MacHugh, D. E., C. J. Edwards, J. F. Bailey, D. R. Bancroft, and D. G. Bradley. 2000. The extraction and analysis of ancient DNA from bone and teeth: A survey of current methodologies. *Ancient Biomolecules* 3(2): 81–102.

Mathew, C. 1984. The isolation of high molecular weight eukaryotic DNA. *Methods in Molecular Biology* 2: 31–34.

Maurer, A., M. Gérard, A. Person, I. Barrientos, P. Del Carmen Ruiz, V. Darras, C. Durlet, V. Zeitoun, M. Renard, and B. Faugère. 2011. Intra–skeletal variability in trace elemental content of Precolumbian Chupicuaro human bones: The record of post–mortem alteration and a tool for palaeodietary reconstruction. *Journal of Archaeological Science* 38(8): 1784–1797.

Mitchell, P., Y. Nagar, and R. Ellenblum. 2006. Weapon injuries in the 12th century Crusader garrison of Vadum Iacob Castle, Galilee. *International Journal of Osteoarchaeology* 16(2): 145–155.

Motro, H. 2004. *Equid Skeletons from the Frankish Castle of Vadum Iacob: Archaeozoological Evidence of a Massacre*. M.A. Thesis. Hebrew University of Jerusalem, Jerusalem.

Motro, H., 2014. *Medieval Equids in the Southern Levant: Methods for Characterizing Species and Breeds*. Ph.D thesis, The Hebrew University of Jerusalem, Israel.

Pääbo, S., H. Poinar, D. Serre, V. Jaenicke–Després, J. Hebler, N. Rohland, M. Kuch, J. Krause, L. Vigilant, and M. Hofreiter. 2004. Genetic analyses from ancient DNA. *Annual Review of Genetics* 38(1): 645–679.

Poinar, H. N., and B.A. Stankiewicz. 1999. Protein preservation and DNA retrieval from ancient tissues. *Proceedings of the National Academy of Sciences* 96(15): 8426–8431.

Pryor, J. H. 1982. Transportation of horses by sea during the era of the Crusades. Part II: 1228–1285: With particular reference to Louis IX and Charles of Anjou. *The Mariner's Mirror: The Journal of the Society for Nautical Research* 68(2): 103–125.

Rohland, N., and M. Hofreiter. 2007. Ancient DNA extraction from bones and teeth. *Nature Protocols* 2(7): 1756–1762.

Schutkowski, H., B. Herrmann, F. Wiedemann, H. Bocherens, and G. Grupe. 1999. Diet, status and decomposition at Weingarten: trace element and isotope analyses on early mediaeval skeletal material. *Journal of Archaeological Science* 26(6): 675–685.

Walker, J. A., D. A. Hughes, D. J. Hedges, B. A. Anders, M. E. Laborde, J. Shewale, S. K. Sinha, and M. A. Batzer. 2004. Quantitative PCR for DNA identification based on genome–specific interspersed repetitive elements. *Genomics* 83(3): 518–527.

Wittwer, C. T., G. H. Reed, C. N. Gundry, J. G. Vandersteen, and R. J. Pryor. 2003. High–resolution genotyping by amplicon melting analysis using LCGreen. *Clinical chemistry* 49(6): 853–860.

Yang, H. 1997. Ancient DNA from Pleistocene fossils: Preservation, recovery, and utility of ancient genetic information for quaternary research. *Quaternary Science Reviews* 16(10): 1145–1161.

Part III

The Horse's Arrival in North Africa and Arabia

9

Northern Africa: Equestrian Penetration of the Sahara and the Sahel and its Impact on Adjacent Regions

Pita Kelekna

The diffusion of the domesticated horse worldwide has wrought many changes on all six continents. This article examines the impact of the horse across the centuries on the arid and semi–arid sections of Northern Africa. Whereas certain innovations accompanying the introduction of the horse to Africa were similar to occurrences elsewhere in the world (Kelekna 2009), others were specialized responses to the environmental challenges of the African deserts and savannas. Clearly, the Arabian horse was integral to the rapid Islamic expansion east and west out of Arabia. But other types of horses were additionally operative in the evolutionary development of technology, economy, trade, metallurgy, social organization, warfare, and religion among different ethnic groups across Northern Africa. These advances, described in this paper, span a period of 4,000 years (Up to the point of, but not including, modern European maritime expansion) and involved horses both imported to and domestically reared in the Sahara and Sahel (a semi-arid zone of grass and thorny scrubland, transitioning between the sands of the Sahara to the north and the moister savanna to the south).

Initial Arrival of the Domesticated Horse in the Ancient Near East, Egypt, and Arabia

While wild asses and zebras are indigenous to the African continent and primary domestication of the donkey took place there, the horse arrived from elsewhere. From domestic horse bones reported in Khabur, northern Syria, and the first appearance of a word for "horse" in cuneiform texts of the Ur III period, it is known that by the end of the third millennium BCE horse culture had penetrated the Near East from the Eurasian steppes (Oates 2003: 117). The true horse, "ass of foreign mountain countries," quickly replaced its donkey and onager predecessors and their hybrids throughout the region in battle (Shaughnessy 1988: 211). Actual conquest with horse and chariot was initially evidenced among the Hittites, Hyksos, Kassites, and the Mitanni–ruled Hurrians (Gurney 1990: 86). By the beginning of seventeenth century BCE, the Hyksos, probably a western Semitic people, had introduced the horse–drawn chariot along with the war goddess Astarte to the land of Pharaohs, to dominate for over a century the eastern delta and Middle Egypt (Cotterell 2004: 96). As the two–horse chariot was deployed across the Levant, Mesopotamia, and up the Nile valley, chariot speed in combat created a revolution in warfare. With the newly acquired horse in Eighteenth–dynasty Egypt, the pharaohs Thutmose III and Amenhotep II captured large numbers of slaves on expeditions to Nubia and the Sudan (Beachey 1976: 2). In military campaigns, horse–riding was initially undertaken by scouts and messengers, albeit at first assuming the rear "donkey seat" on the rump (Raulwing and Clutton–Brock 2009: 9 f.). Fully competent mass cavalry, however, would not be achieved until the first millennium BCE, with the saddle, probably first invented on the Eurasian steppes, making its appearance in the Near East toward the end of the era.

In the funerary depictions of pharaonic tombs, Egyptian horses appear bedecked with ostrich plumes, their slender bodies and small heads showing a marked resemblance to the modern Arabian. Similarly, remains of horses from seventeenth–century–BCE Buhen (Upper Nubia) and fifteenth–century–BCE Thebes also exhibit the same fine limbs characteristic of horses in Egypt and Arabia today. And both skeletons share a feature in common with the modern Arabian horse, namely five lumbar vertebrae in contrast to the customary six or seven of other domestic horse breeds (Clutton–Brock 1992: 81–83). By the close of the second millennium BCE, possibly even earlier, a petroglyph of a two–horse chariot indicates horse presence at Jubbah in northern Saudi Arabia (Olsen 2010: 54). Actual burials of distinctive Arabian–type horses date to 2,000 years ago at Mileiha (UAE) where they were elaborately interred with bejewelled trappings (Bin–Sultan al–Nahyan 1998: 7). On the Arabian Peninsula, scrupulous inbreeding of mounts produced an animal of unrivalled stamina. The slighter build of these southern horses, which contrasted to the stockier physique of the earlier northern steppe domesticates, likely was an adaptation to dry, hot desert conditions. Such differences in physical type reflect variation in morphology, evidenced in many mammalian species of wide distribution, whereby in warmer climes there is a tendency to develop a gracile body with fine limbs (Bergmann's rule) and longer extremities (Allen's

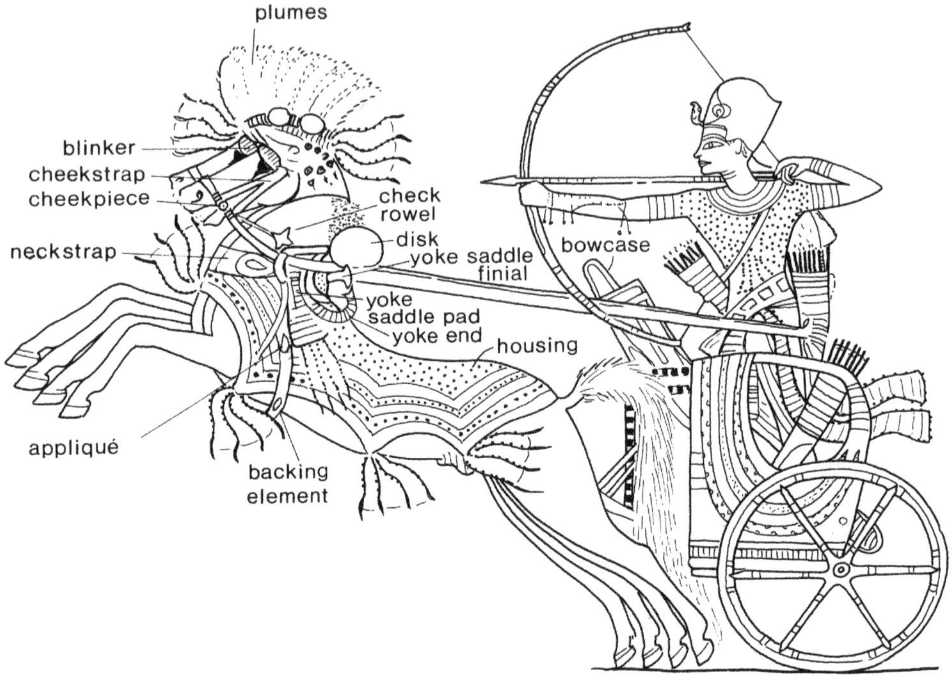

Figure 1. Sketch of the Egyptian chariot of pharaoh Tut'ankamum (Littauer and Crouwel 1985: Fig. 2). Courtesy of Joost Crouwel.

rule) (Clutton–Brock 1992: 61). Vertebral reduction too may have been an adaptive response by the evolving proto–Arabian horse to the stresses of southern latitudes.

As illustrated in Figure 1, by the reign of Tutʿankhamūn in the fourteenth–century BCE, the Egyptian chariot was a formidable war machine. Its six–spoked wheels featured an integral nave with the axle positioned to the rear edge of the rider's box, to endow the vehicle with great fore–aft stability. Also, lashed to the yoke, was an important chevron–shaped wooden device, designed to adapt the yoke to the narrow equid neck.[1] This "yoke saddle" basically prevented the yoke from slipping down behind the withers, thereby enabling the shoulders to exert a strong forward pull (Littauer 2002: 479–84; Spruytte 1983: 30 f.). The driver, clad in scale armour for protection, with reins tied around his waist, wielded a bow and arrow. Affixed to the side of the chariot was a javelin case.

In keeping with the technological complexity of the light, spoked–wheeled war chariot, from Kikkuli's Manual for the training of chariot horses, recorded on clay tablets written expressly for the Hittite kings of Anatolia, we are aware of the extraordinary attention accorded the training of chariot horses in the middle of the second millennium BCE. The most famous of these texts is named after Kikkuli, who introduces himself in his manual as *aššuššanni*, the "horse trainer" from the land of Mittani,[2] a Hurrian–speaking state in northern Syria and southeast Anatolia 1500–1300 BCE. Kikkuli's training instructions were found in the capital of the Hittite kings in Hattušša. Three training manuals survived. The first one (CTH 284) named after Kikkuli Kikkuli's manual was composed in fifteenth century BCE; however, the original clay tablet was not found in the Hittite archive, and only a copy from the thirteenth century BCE survived. The second one (CTH 285), a training instruction with a ritual introduction dated in the fourteenth century BCE, and the third one (CTH 286) written in Hittite from thirteenth century BCE.[3] Several kings of Mittani also had Indo–Aryan names in addition to their Hurrian throne names (Mayrhofer 1996; 1974; Raulwing 2006). Some of those Indo–Aryan names, together with specific *termini technici* of a hippological sphere, have been interpreted as proof of a great esteem of horses and chariots in Mittani in the second half of second millennium BCE.[4] The Kikkuli Text employed "interval training" techniques over a time span of at least seven months comparable to modern equine sport principle of progression, peak loading systems, electrolyte replacement, fartlek conditioning, and repetitions (Horn 1995, tables; Raulwing 2006; Nyland 2009: 38). The first few days of Kikkuli's course comprised a series of trials, designed from the outset to eliminate horses unfit for strenuous training. The Indo–Aryan compounds for "rounds" build with *–wartanna* (Vedic *vartaní* "round")— *aika–* (Vedic éka– "one"), *tera–* (Vedic *trī* "three"), *panza* (Vedic *páñca–* "five"), *satta* (Vedic *saptá–* "seven"), *na*

[1] The yoke, originally invented for ox–draft, was ill suited to horse anatomy and therefore required this additional device for more efficient traction.

[2] Raulwing and Schmitt 1998: 698 reconstruct: Old Indo–Aryan *ásva–* "horse" + Old Indo–Aryan root *sami* "to come to rest, to become exhausted" = "the one who drives the horse (during training or race) to exhort the last reserves, to become exhausted".

[3] For an overview on the historical context, languages used in the training manuals and etymology of Kikkuli's profession title see Raulwing and Schmitt 1998; Raulwing 2006 on hippological interpretations of the Kikkuli Text.

[4] EWAia, s.v; Raulwing 2004: 519 ff. and 518, Table 1 with terms which have been accepted by most scholars as Indo–Aryan; Mallory 1996: 37 f.

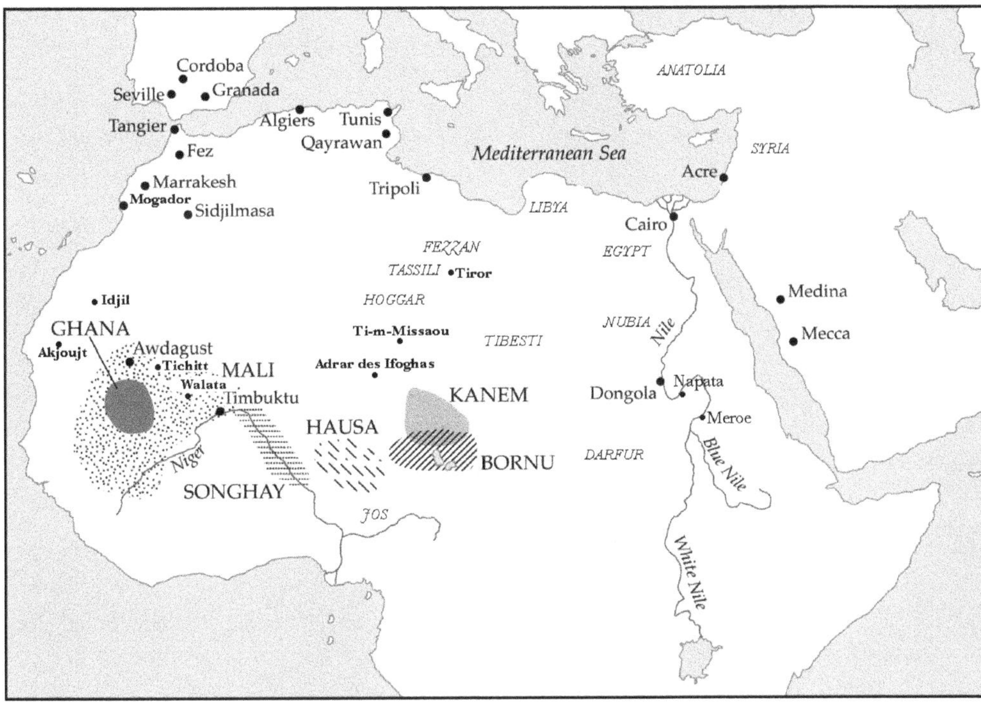

Figure 2. Map of North Africa, showing locations of early chariot and equestrian penetration, later Arab and Almoravid expansions, and Sahel cavalry kingdoms of the first- and second-millennia CE.

(haplolocally shortened by omitting the second syllable; Vedic *nava* "nine") and *wašanna* (Gen. Sg.) "track, 'stadion'"—represent an older dialect than Vedic as the diphthong in *aika* (Vedic éka) indicates.[5] Other cuneiform archives in the ancient Near East have yielded further Indo–Aryan linguistic traces such *termini technici* for horse colors in Nuzi (Kirkuk)[6] and a collar decoration for horses in the Amarna Tablets.[7] Such was the Mitanni reputation for fine horsemanship that Egypt maintained diplomatic correspondence with that dynasty, requesting both horses and chariots. The marshes of Lower Egypt were not great horse–breeding country, as a consequence fine horses had to be continually imported from Asia (Epstein 1971: 428) (Figure 2).

Diffusion of the Domesticated Horse to the Sahara and the Sahel

Thirteenth–century–BCE Egyptian records further indicate horse presence elsewhere in North Africa, as when the pharaoh Merneptah was attacked from the west at the Battle of Perire by a coalition of Libyan, Achaean, and Cretan charioteers. These latter invading "Sea Peoples," well experienced in distant wars, possibly were assimilated

Figure 3. Saharan rock drawing of a two-horse chariot, the horses depicted in the famous "flying gallop." Cave of Tamadjert, Tassili-n-Ajjer (Spruytte 1996: Fig. 8). Courtesy of Alain Spruytte and Franck David.

into North African populations. Across Saharan oases, over 200 rock drawings of chariots, dating from mid–second millennium BCE onward have been discovered, some representations showing certain affinities with the Mycenaean art style, the horses in the distinctive "flying gallop" (Hyland 2003: 31; Lhote 1953: 1141–43, 1158) (Figure 3). Thus, it would appear there were multiple early entrances of the warhorse into North Africa. These chariots of the Sahara were very light, seemingly without wooden or wicker sides. French scholar, Jean Spruytte, from his decades–long research in the Sahara, constructed first models then full–scale experimental replicas of the rupestrian chariot–drawings he had photographed. A working replica of the light Saharan chariot is displayed in Figure 4, its horses breaking into a gallop. Note, attached to the very end of the draught pole, the traction bar that

[5] In a treaty between the Hittites and the Mitanni, the Mitanni king first swore an oath by Hurrian gods then by the Indo–Aryan deities in CTH 51, ^{d}*In–da–ra*, ^{dMEŠ}*Mi–it–ra–aš–ši–il*, ^{dMEŠ}*Na–aš–at–ti–ya–an–na*, and *Váruṇa*, corresponding to *Mitrá–*, *Váruṇa–*, *Índra–* and the Nāśatyā in Rigveda 10.125.1 (Mallory and Mair 2000: 257; Raulwing 2013: 254 f.). See Thieme 1960 for an interpretation of the religious background specific to the Indo–Aryan development of these gods.

[6] Indo–Aryan *p/babru–nnu* (Vedic *babhrú–* reddish brown, brownbraun"), *p/barita–nnu* (Vedic *palitá–* "grey", *p/binkara–nnu* (Vedic *pingalá–* "reddish brown, reddish yellow, greenish yellow).

[7] *mani–nnu* (Vedic *maní–*).

Figure 4. Working replica of Saharan chariot, approximating the "flying gallop." Note the frontal traction bar-pole harness (Spruytte 1996: Fig. 82). Courtesy of Alain Spruytte and Franck David.

Figure 5. Close-up, showing details of working replica of Saharan two-horse chariot (Spruytte 1996: Fig. 13). Courtesy of Alain Spruytte and Franck David.

passes in front of the horses' breasts. A close–up of chariot construction is shown in Figure 5. The sideless platform consisted of tightly woven leather strips (Spruytte 1996: 20). The wheel, intricately modelled on an ancient Egyptian wheel (Brooklyn Museum; Littauer and Crouwel 1979), comprises 20 separate pieces.

In former times, much of the Sahara was still open parkland, home to plentiful wild game, such as antelope, ass, elephant, giraffe, mouflon, rhinoceros, hippopotamus, lion, and leopard. Hunting was widely practiced, as was fishing near lakes and perennial streams. Rock art of the region shows hunters of both Afro–Mediterranean and Negroid types,[8] suggesting a mingling of races in these southerly latitudes. Also depicted were long–and short–horned cattle, sheep, and goats. But due to encroaching desertification, stockbreeding was transhumant, complemented by the gathering of wild cereals as herders moved between seasonal pastures. Cultivation of wheat and barley, staples of the Mediterranean littoral, was not possible in the Sudanic belt, with its climate of uniform hours of daylight and pattern of summer rainfall. Continuing climatic deterioration, however, led to experimentation with wild grasses and, by the first millennium BCE, the eventual domestication of millets and sorghums (Oliver and Fagan 1975: 12–14, 48).

In 450 BCE, Herodotus reported (2007: 4.183) that the Berber Garamantes of the Fezzan, driving four–horsed chariots, used to raid for slaves "Ethiopian troglodytes" in the Hoggar and Tibesti massifs of the central Sahara. These may have been the Teda, who still today use caves as refuges. From rock drawings of Saharan four–horse chariots, Spruytte constructed a working model (Figures 6 and 7). The harnessing encountered in the Sahara differed significantly from the Eurasian technique. In the latter

Figure 6. Rock drawing of a Saharan chariot with multiple draught poles joined by frontal traction bar (Spruytte 1996: Fig. 20). Courtesy of Alain Spruytte and Franck David.

Figure 7. Constructed replica of a Saharan chariot with multiple draught poles joined by frontal traction bar (Spruytte 1996: Fig. 21) Courtesy of Alain Spruytte and Franck David.

[8] These Afro–Mediterraneans spoke languages of the Afro–Asiatic family—Coptic, Berber, Chadic, and Cushitic; they early utilized both copper and bronze. The Negroes of the south spoke languages of the Niger–Congo, Saharan, Central Sudanic, and Nilotic families (Oliver and Fagan 1975: 3).

case, the ancient quadriga utilized a central draught pole and yoke, to which the two interior horses were harnessed by yoke saddles, the two outspanners fastened merely by traces. Contrastingly, the North African quadriga featured multiple draught poles connected at the tip by a traction bar to be located in front of the horses' breasts (frontal traction bar also shown in Figure 4). Spruytte attributes this distinctive harnessing method to Phoenician influence in North Africa, since similar materials are attested in Cyprus during the first millennium BCE; nothing comparable is known from Egypt (Spruytte 1996: 112 f.).

Herodotus also commented upon chariot travel undertaken deep into the interior by Libyans, who reached a great river infested by crocodiles—probably the Niger. In fact, across the Sahara, numerous rock drawings of horse–drawn chariots have been discovered that cluster along two major routes connecting perennial waterholes while avoiding the worst areas of shifting sand: one passing over the Hoggar and along the ridge of the Adrar des Ifoghas toward the gold deposits of the Niger bend; a second running far west to the copper mines of Akjoujt in southwestern Mauritania (see Figure 2) (Oliver and Fagan 1975: 10). Ox–drawn wagons and carts are also featured in these early engravings of the central and western Sahara (Lhote 1953: 1162). In the Tichitt valley, a natural line of communication between the salt mines of Idjil, the Akjoujt copper zone of Mauritania, and the Niger bend, excavations by American archaeologist P. J. Munson at Dhar Tichitt–Walata have revealed a thousand–year–long occupation by hunting and farming cultures dating from 1500 BCE. Exploitation of aquatic resources and wild cereals marked the earliest sites. By 1200 BCE, herding of cattle and goats, pottery, and incipient cultivation of bulrush millet were evident, with lakeshore villages similar in type to those of the Mande savanna peoples to the south. But nomads, herding animals on the fringes of the desert, posed a threat to these settled farmers, in times of drought raiding their southerly villages in search of water and pastures. Around 1000 BCE, there were both developed agriculture and significant population increase; sedentary settlements, however, were now located on high escarpments encircled by defensive walls. By mid–first millennium BCE, rock engravings of horse–drawn vehicles along the Tichitt valley suggest Berber penetration from the north, probably for Mauritanian copper and for gold from the western Sahel. Copper arrows, axes, and spearheads lying in the desert resemble forms from the Maghreb during this epoch. But between 600 and 300 BCE, heavier fortifications of the oasis settlements signalled serious conflict. And by the latter part of the first millennium BCE, all traces of indigenous farmers had disappeared from Tichitt, to be replaced by *tifinar* inscriptions, Libyo–Berber tombs, and paintings of mounted warriors, indicating Berber exploitation of Mauritanian copper mining and metallurgy from approximately 500 BCE culminating in final equestrian occupation. The retreat toward more humid lands by indigenous sedentary agriculturalists, due to gradual desiccation of the Sahara, was thus accelerated by the Berber advance south (Garlake 1990: 119; Levtzion 1980: 5 f., 12 f.) (Figure 8).

Figure 8. Saharan rock drawings of horsemen from Ti-m-Missaou, guiding the horse with a withe around the neck and taps of a rod. (Giraffe heads are more ancient). (Lhote 1953: Fig. 20). (Courtesy of IFAN, Dakar).

Conflict had occurred elsewhere, far to the east in Nubia, where c. 727 BCE the kings of Kush had conquered and ruled Egypt as the XXV dynasty. Defeated by the Assyrian Ashurbanipal in 664 BCE, the dynasty retreated south first to Napata on the Fourth Cataract, then to Meroe above the Fifth. Like Egyptian nobles, the Kushites raised fine horses for chariotry and riding and in fact had long exported them North, to Lower Egypt. In funerary pomp, their steeds, measuring 15 hands at the withers, were buried with elaborate trappings of bronze, silver, shell, and faience; an ostrich plume in a gold holder was worn on the head (Olsen 2010: 48; Raulwing and Clutton–Brock 2009: 54–56, and Doxey in this volume). Unlike the Egyptians, Kushite pastures, though, were not limited to the narrow Nile floodplain, but watered by annual rains extended across wooded savannas with plentiful grass, where horses and other stock thrived. In addition to an abundant supply of wood, Meroe possessed much iron. Rich in these two commodities, iron ore and the fuel to smelt it, by 300 BCE Meroe had acquired great urban wealth.

Figure 9. From the region south of Lake Chad, a Musgu rider of the mid-nineteenth century, armed with a large spear and two iron throwing axes. The pony appears to have no saddle and there is only a single bridle without a bit (Barth 1857).

Also, active in hunting wild elephants from horseback, Meroe achieved international prominence in the luxury trade of ivory, slaves, ostrich feathers, rare skins, ebony, and gold from inner Africa to the Mediterranean and east to Arabia, Persia, and India (Oliver and Fagan 1975: 5 f.). From rock art in the Sahel of the Chad area, it is also evident that from Meroe by 300 BCE, if not earlier, iron had diffused westward, manufactured as large–bladed spears and wielded as lances by horse riders. Traces of ancient ironworking in sites long abandoned in the desert are very plentiful, suggesting that "smiths accompanied the horsemen from the time of their first appearance in these regions." Thus, across the Sudanic bush country, 10° north of the equator, iron was regularly used at this time. However, these horses of the savanna were quite different from the proto–Arabian horses of the north; they were very small, measuring only 10 hands. Furthermore, these ponies were always ridden, never driven (Oliver and Fagan 1975: 62). Blench (1993: 88 f.) has remarked that adverse environmental conditions commonly produce domestic animals of diminutive size and believes these small ponies were derived from larger northern horses, possibly of different breeds. In Africa's extremely hot climate, various other livestock species, cows, pigs, sheep, and goats are all known to have undergone dwarfing; in fact, there appears to be a general correlation between degrees of latitude south and animal height. Also, in antiquity Nubian donkeys, settled on the island of Socotra, 300 km off the Horn of Africa, in feral state were subsequently reduced to a stature of 10 hands (Clutton–Brock 1992: 40). In the case of these Sahel ponies, therefore, in addition to displacement southward, feralization might have been one further contributing factor to height reduction.

In Kushite expansion out of the Nile valley, Meroitic hunting of wild elephants, slave–capturing, and escort of rich caravans through alien territory were quasi–military operations, in which armed horsemen ranged much further afield than transhumant pastoralists. During these expeditions, it is entirely possible that domestic horses may have escaped across the scrub savanna to survive and proliferate in the wild. At a much later date, Arab geographer Leo Africanus in fact confirmed the presence of feral horses in the Sahara (Blench 1993: 91). Identical developments, of course, occurred on other continents, where escaped Spanish horses formed the nuclei of the great feral mustang/mesteño herds of North and South America. In Australia, the brumbies of the outback were a comparable phenomenon. In every case the feral horse was smaller than its domestic predecessor (Hyland 2003: 30). As is well known, on the prairies and the pampas, re–domestication of feral horses with minimal apparatus revolutionized the hunting economies and military potential of the indigenous inhabitants. So too, in the Sahel, it appears that in the latter part of the first millennium BCE small feral horses, re–domesticated by transhumant herders of different ethnic groups, developed the capacity to survive on a broadly–based diet. Deployed in a variety of local usages, they were largely responsible for the diffusion of iron technology across a broad belt of the African savanna, later even extending to the edge of tropical forests, where these equines acquired partial resistance to trypanosomiasis[9] (Figure 9).

Geographer and trader accounts over the centuries attest to the continuing presence of ponies across the Sahara and Sahel. They were used in trade and the collective spear hunting of savanna game, but also militarily, both

[9] There are various forms of trypanosomiasis transmitted to humans and animals by different species of tsetse fly. For horses, the most dangerous form is *Trypanosoma brucei*, which causes fatality within three months; less severe rates of mortality resulting from other infections range from six months to two years (Law 1980: 77–79). Pony resistance was highly geographically specific. Although immune in their home range, local stock transported elsewhere would die rapidly from other fly–borne diseases (Blench 1993: 91).

in mounted raids to enslave neighbours and as defence against enemy slavers (Blench 1993: 94–96). Dated from the first millennium BCE, a brass figurine shows a warrior mounted without saddle or stirrups on a very small blunt–nosed horse, controlled by a single–reined, bitless bridle. Unfortunately, not a great deal is known about these ancient riders. Today though, we may perhaps glean a hint of their technology from remote non–Muslim regions of Northern Africa, particularly the Jos plateau. There, during the twentieth century, Ron tribesmen still rode bareback with a simple rope knotted around the horse's nose. Alternatively used was a noseband, comprising two curved iron bars joined on one side by an iron ring and by a linking chain on the other. The bar placed under the horse's chin was commonly spiked to exert a severe curb pressure whenever the rein was pulled. In lieu of saddlery, another Jos custom was to cut the centre of the horse's back, so that the raw and bleeding surface, like cement, would afford the rider a grip which otherwise would be lacking. This incision ultimately formed a callous pad that protected the horse's spine. (Law 1980: 93–95, 112, 127). These ponies were widely valued for their agility in rocky environments. They served as bridewealth payments and were paraded at the investiture of a chief. Horse sacrifice was also an important component of funeral obsequies. Upon a chief's death, ponies were clubbed to death, the deceased buried wearing the skin of his favourite mount as a shroud (Blench 1993: 94, 97).

With the expansion of horsepower, there arose in the Sahel the important kingdom of Wagadu, its capital Kumbi Saleh strategically located along the north–south trade route, midway between the desert sources of salt and the Bambuk gold fields of the Senegal and Falem drainage. Control over the sources of gold was essential to Wagadu's economic prosperity and political control. Fabulously wealthy, the Mande ruler was known as *Kaya maghan* "king of gold," or *Ghana* "war chief" (Garlake 1990: 118 f.). The king deployed cavalry as a striking force in his armies, military superiority asserted through the use of iron. With horses, iron swords and spears, the Soninke elite extended Wagadu territories, raiding and enslaving their southern neighbours, who possessed no iron and fought only with ebony staves (Levtzion 1980: 13 f.). In the trans–Saharan trade, kola nuts, hides, ostrich feathers, ivory, gold, and slaves travelled north. Slaves were assigned to heavy or dangerous work in portage, mining, cutting salt, and agriculture. While gold was exported, Saharan salt and Akjoujt copper were imported south, the latter not for weapons or tools, but to make elaborate regalia for kings and the wealthy, although also as rods to be used later as currency (Garlake 1990: 119 f.). During the first millennium BCE, Phoenician mariners had settled the southern Mediterranean coast. As their wealth increased with copper, iron, and silver shipments from Spain, their colony Carthage developed commerce with the Garamantes of the Fezzan, who controlled the trade from the Saharan interior. But as early as 490 BCE, the Carthaginian Hanno had attempted to circumvent Berber monopoly of this lucrative trade and to gain direct access to the valuable gold of the southern savannas by establishing trading posts along the Atlantic coast of northwest Africa. The most southerly of these outposts was the island of Kerne, identified with Mogador or Herne. This maritime venture however was opposed by Pharusian nomads, who rode across the desert with water skins strapped under the bellies of their horses. Berber middlemen in the gold trade, the Pharusians relentlessly attacked these competitor Punic colonies along the Atlantic coast (Levtzion 1980: 125 f.; Strabo 1996: 17.3.7). According to Strabo (1996: 17.3.19), around this time, the still verdant Mediterranean littoral of North Africa was home to vast herds of horses that yielded 100,000 foals a year (Hyland 1990: 177). Herodotus (2007: 7.86) recounts how these Barbary horses—known to the Greeks as Libyans and later to the Arabs as Barbs—had famously been conscripted by Xerxes as chariot horses for the 480 BCE Achaemenid invasion of Greece (Hyland 1990: 24). The Barbary horse, its convex profile widely depicted in rock paintings and engravings (Lhote 1953: 1204), also provided military escort to caravans carrying iron tools from Carthage southward. During the Punic invasion of Italy, Hannibal's armies exceeded by far in ratio of horse to infantry anything the Romans had ever fielded in battle, enlisting the formidable Garamantes cavalry of the Fezzan. But the Numidians were the most feared of Hannibal's cavalries, inflicting dire defeats on Rome at Trebia, Trasimene, and Cannae (218–216 BCE). These North Africans from the Aures mountains rode their swift, nimble horses bareback without a bridle, guiding their horses with body movements, a withe around the neck, and taps of a rod (similar to rock depiction in Figure 8). They were merciless in skirmishing manoeuvres, harrying, and tearing the fleeing Romans from their horses' backs (Hyland 1990: 173–175; Lhote 1953: 1156; Polybius 2011a: 3.115–117). Survivor of Cannae, Scipio Africanus finally countered Punic might by emulating Carthaginian tactics with Iberian horse to defeat Hasdrubal at Ilipa, Spain in 206 BCE, then to forge an alliance with Masinissa, the famed Numidian general. Aided by 6,000 Numidian cavalry, Scipio in North Africa finally annihilated Hannibal's army at Zama in 202 BCE (Polybius 2011b: 15.12–14), a victory which catapulted Rome from a regional to an international power.

Early in the first millennium CE, yet another important development had taken place. The one–humped camel or dromedary, originally domesticated in Arabia during the second millennium BCE, had been introduced into the Sahara via the Sinai. Capable of carrying burdens of up to 200 kg at a pace of 2–3 km per hour, its bulk–transport capability, primarily as a pack animal[10] soon displaced wheeled vehicles in Northern Africa (Köhler–Rollefson 1996: 282–286, 291). The camel's extraordinary efficiency in the desert environment would facilitate a highly specialized form of mobile pastoralism across the central and western Sahara, but also inhibit transmission of the wheel southward. With the advent of the dromedary,

[10] The camel was also widely used in plowing and lifting water.

the volume of trade across the Sahara rose significantly. Imperial Rome, in turn, would be beneficiary of this valuable commerce, obtaining from the African interior wild animals for its famed circuses—and the slaves and gold brought from the Sahel by Berber caravan traders (Oliver and Fagan 1975: 53).

Equestrian Islamization of the Sahara and the Sahel

In the seventh century CE, however, Roman dominion in North Africa was challenged by Bedouin warriors from the Arabian Peninsula. First reaching south into Nubia, a treaty was concluded at Dongola in 642 that laid the foundation for the Muslim slave trade in Africa. The treaty stipulated that a mosque be maintained in the center of town and required each year the delivery of 360 male and female adult slaves (Heers 2003: 27). In the Islamic western expansion (643–647) out of al Fustat (Arabic "tent" or Cairo of the future), both the camel and Arabian horse were to play important roles. In raiding, the dromedary could be ridden over distances of 65–80 km per day. Yet in combat, even at a gallop, the camel could never deliver the momentum and impact of the cavalry charge. While fighting with lance from camelback was practiced, it was not the easiest mode of combat, since the camel was notoriously unresponsive to rider command (Bulliet 1990: 99). In armed hostilities, everywhere the warhorse was the preferred steed. Thus, in the waterless desert, the dromedary came to fulfil the important function of supporting horses, which to conserve energy during travel were tied to the cinch of the camel's saddle and mounted only at time of attack (Hill 1975: 34, 36). The Arab's camel could be bled to provide 5 L of blood per month. It transported containers of water for both human and horse consumption. It also provided as much as 10 kg of milk per day for horse nourishment *en route* (Köhler–Rollefson 1996: 286). Additionally, in time of dire emergency, its stomach could be slit to access its reserve of water. This was accomplished by initially withholding liquid, then forcing the thirsty camels to drink copious amounts of water. The camels' mouths were tied to prevent their chewing cud and contaminating water in the stomach. On campaign, each day a specific number of camels would be slaughtered, their stomachs slit to provide water to men and warhorses (Donner 1981: 121 f.).

Acquiring the stirrup from the Sasanians during the conquest of Persia, the Arabs introduced this important device to the Sahara (Lhote 1953: 1200–02). In those early wars, the Arabs quickly gained advantage over their enemies through the superior mobility afforded by dromedary support of the horse. By using the desert for passage, raiding base, and refuge, they were able to cover long distances rapidly and to concentrate their forces at the point of greatest danger. Their central stratagem was to mount surprise attacks across the desert borders. In situations of adversity they could retreat back into the desert, without fear of Byzantine pursuit, to await a more favourable opportunity to rally. With these tactics, their rear and lines of communication were safe from enemy interference and their dispatch of reinforcements was free of hazard (Hill 1975: 41 f.).

Well accustomed to the precariousness of the desert, the Arab's horse showed a hardiness and grace under the most trying conditions. With a long–arched neck, concave forehead, straight nasal profile, and large somewhat protruding eyes, the Arabian horse had a deep body culminating in powerful quarters; the hocks well set down gave the horsepower and ability to manoeuvre. Renowned for its intelligence, the Arabian was the swiftest and toughest of horses. It remained healthy on little fodder, achieved the fastest speeds, and displayed extraordinary stamina (Bin–Sultan al–Nahyan 1998: 8 f. Hyland 1990: 24–27, 209). Its distinctiveness was intensified and maintained by long inbreeding in an environment demanding toughness and alertness. The desert tribes' respect for endogamy was mirrored in their close supervision of horse–breeding practices. An *asil* (purebred) mare was only ever covered by an asil stallion. When raiding into enemy territory, the Arab sewed up the mare's vagina to prevent her from mating with an inferior stallion. The Bedouin believed in telegony. Should the mare be covered by a non–asil horse, it was feared that all future foals would be contaminated by impure blood. In birthing, the cherished asil foal was not allowed to fall to the ground but caught and caressed as an infant. Mare and foal shared the men's tent, the mare's neck often a pillow for its master. After a month, the foal was weaned then fed on camel milk and later on barley and dates. (Amirsadeghi 1998: 18 f.).

Backed by the water–carrying, bulk–transporting dromedary, the cavalry forces of Islam spread rapidly across North Africa reaching the Atlantic in 680. Arab victories could not be attributed to their masses, for they were vastly outnumbered by their enemies. Arab strengths were their religious fervour and their astonishing military mobility, afforded by the Arabian warhorse. Both men and horses were accustomed to living off the meagre diet of Bedouin existence. Hardy and resourceful, without supply train, they traversed barren and inhospitable lands, traveling at night with the bright light of the desert stars—even fighting at night (Kennedy 2007: n. 371). Poised between the Atlantic and the Mediterranean, the tide of Islamic conquest turned north. In the east, Byzantium would continue to defy Muslim might for many centuries to come, but in the west Spain furnished an alternate route to assault Christian Europe. In 711 at Gibraltar, a Muslim army commenced the invasion of Visigothic Spain, a land internally divided. By 720 all provinces of the peninsula had been subjugated; albeit in Asturias, the Christian revolt of 722 would repulse the Arabs at Covadonga to form the embryonic core of the Spanish *Reconquista.*[11] The tide of Islamic equestrian conquest continued its advance north, finally to be repulsed on the plains of Poitiers in 733 by a Frankish army headed by Charles Martel. And so began the great era of Arabic learning, which would extend

[11] Christian reconquest of the Iberian Peninsula from Islam.

across three continents and achieve prodigious heights in art, architecture, literature, science, mathematics, and medicine.

At the beginning of the eighth century in Africa, the Arabian horse also played an important role in the conquest of southern Morocco. Lured by the fame of the Sudanese gold, the Umayyad governor in 734 ordered an expedition to *Bilad al–Sudan* (Lands of the Black People), where large quantities of gold and slaves were seized. Realizing, however, that such raids would not secure a steady supply of the precious metal, the Arabs encouraged Sanhadja nomad trade across the desert, by digging wells along the trails between oases. By the second half of the century, independent Kharijite principalities ruled over the northern gateways to the southern goldfields, promoting the development of private enterprise by Muslim merchants (Levtzion 1980: 126). Al Bakri of Cordoba later would report how the king of Wagadu, now referred to as Ghana by the Arabs, appeared at audiences in the capital bedecked in a golden headdress and surrounded by ten horses, small in stature, but caparisoned in gold. Behind stood ten slaves bearing swords and shields mounted with gold (Trimingham 1962: 53 f.). According to Soninke tradition, 22 monarchs, who exercised both secular and religious authority, had ruled before Muslim contact. The king's residence was in a palisaded grove; nearby, a pagan priesthood guarded the sacred groves surrounding the tombs of past kings. Arab Muslims lived in a separate town, six miles away, as did the Zanata and Sanhadja nomads engaged in the trans–Saharan trade (Oliver and Fagan 1975: 163 f.).

Gold, together with silver, constituted the basis of the Muslim monetary system. From the mid–eighth century onward, the flow of gold from the Sahel to the Maghreb increased dramatically. Rival caliphates competed for control of this commerce. By 909 the Fatimids succeeded in dominating Ifriqiya, where Sudanese gold financed their economic and political expansion and subsequent conquest of Egypt. Upon the caliph's entry into Cairo, he was presented with "500 horses with saddles and bridles encrusted with gold, amber, and precious stones" (Culbertson 2010: 79). The early eleventh century saw a major migration of Arabs with their flocks out of Upper Egypt across the Maghreb. Displacing Berbers and spreading their distinct Arabic dialect as far west as the Atlantic, these nomads, the Banu Hilal and Banu Sulaim, destroyed the irrigation works and ruined the agricultural base of the North African littoral (Trimingham 1962: 18 f.). In response to Shiite Fatimid rule of former Sunni populations, there also occurred a zealous revival of orthodox Sunni Islam. Because in the westernmost corner of Islam there existed considerable religious heterodoxy, the missionary Abdallah ibn Yasin was dispatched to indoctrinate the equestrian Sanhadja of the western Sahara. *Ribat* (fortified religious centres) had been established along the Islamic frontiers in Africa, where devout Muslims could honour Allah and extend the kingdom of Islam by fighting infidels. Under Ibn Yasin's stern tutelage, the nomads developed into a tough disciplined fighting force. Known as *al–murabitun* (ribat fighters of jihad), later as Almoravids, they rallied many desert tribes to their cause. As a Sunni reform movement, the Sanhadja between 1054 and 1059 succeeded in vanquishing their Zanata Berber political rivals, capturing Sidjilmasa in the north and Awdagust in the south. By strengthening the staging posts along the connecting routes between Morocco and Senegal, Almoravid equestrian militarism succeeded in integrating the different economies of the region into the political unity of the Maghreb; many elements of Arab riding gear also diffused south to the Sahel at this time. With the two termini of the trans–Saharan gold route under their authority, the Almoravids controlled all the trade of northwest Africa. Ibn Yasin was succeeded by Abu Bakr who led his tribal cavalries and cameleers south (Hrbek and Devisse 1988: 347 f., 361). What followed was essentially a racial war between Berbers and Negroes, in which farming communities were destroyed and marginal agricultural land permanently ruined by pastoralist over–grazing. After final defeat in 1076, Ghana's formal acceptance of Islam and payment of tribute were exacted by the Almoravids (Oliver and Fagan 1975: 166).

While Muslim power was being consolidated in the south, Sanhadja cavalry aggressively attacked in the north, subduing semi–Islamized Berber groups and establishing a military camp at Marrakesh, which grew rapidly to become the Almoravid capital. By 1083, all of Morocco and western Algeria were conquered and the Gibraltar strait secured. In al–Andalus, the Hispano–Umayyad caliphate had collapsed in civil war with rebel Berbers and been replaced by various Taifa kingdoms, which due to incessant internecine feuding failed to gain significant political power. In response to Taifa pleas for aid in repelling the northern Christian armies, the Almoravid Yusuf ibn Tashfin, with 15,000 frontier fighters, invaded the Iberian Peninsula just as Toledo was being recaptured in 1085 by Alfonso VI of Castile and Leon. In 1086 the Christian forces were spectacularly defeated at the battle of az–Zallaqa (Sagrajas) near Badajoz. Continuing Almoravid victories on the battlefield kept the Christians at bay in the north but failed to regain Toledo. Earlier, Arab conquest of Morocco had given the trans–Saharan commerce a fresh impetus. Now, as the Almoravids wielded extraordinary power, their territories extending from the Senegal river in the south to the Ebro river in northern Spain, the gold trade rose to a new peak. Under Almoravid rule, the economy boomed; increased supplies of copper, iron, and silver were exported from Morocco and vast quantities of African gold reached al–Andalus across the straits (Hrbek and Devisse 1988: 349–51, 363). Almoravid dinars were of pure gold and therefore in greatest demand throughout Europe. The Almohads, another revivalist group who challenged the Almoravids in North Africa and invaded al–Andalus in 1147, continued to maintain the same high standard of gold coins (Levtzion 1980: 129). Ghana, long accustomed to its fabulous wealth in the gold trade, continued to flourish as an independent Muslim kingdom throughout the twelfth century, imposing tolls, custom duties, market dues and fees for adjudication of commercial disputes. By mid–thirteenth century, however, Ghana was in decline. As gold exploitation shifted to Bure on the headwaters of the Niger and further south to Akan on the fringes of the rainforest, Mali emerged a dominant power (Levtzion 1980: 155 f.). Muslim merchants continued to establish trading centres throughout the region, so that conversion to Islam became essential for peoples of the savanna wishing to participate in this prosperous commercial network. Muslim religious leaders regularly consulted with the different rulers, who as early recipients of Islamic teachings greatly expedited the process of islamization (Levtzion 1999: 476–81).

During the Fatimid and Ayyubid caliphates, nomadic Arabs had always constituted a turbulent element in Upper Egypt. The Egyptian Mamluks, however, who had a passionate love for purebred Arabian horses, greatly expanded the earlier *furusiyya* treatises on horsemanship, chivalry, and veterinary medicine and built imposing hippodromes for military exercises that often included the game of polo—imported from Persia (Culbertson 2010: 78). Under their sultan Baybars, Arab uprisings were systematically put down by the better–trained and equipped Mamluk cavalries. As a consequence, many nomadic tribes migrated southward, creating an explosive situation in the Nubian borderlands. In response to Nubian aggression, Baybars in 1276 launched a punitive expedition against Dongola to defeat the Nubian army, this event contributing directly to the decline of Nubian Christianity and deterioration of Coptic Christian status in Egypt. Subsequent disintegration of Nubia facilitated the great Arab breakthrough to pasturelands between the Nile valley and Lake Chad, resulting in the islamization of the Nilo–Chadic Sahel. In 1324 on his legendary *hajj* pilgrimage from the Niger to Mecca, Mansa Musa, king of Mali, visited Cairo displaying his wealth and generosity by distributing as gifts enormous amounts of gold, with the result the price of the metal significantly depreciated. In response, the sultan presented him Arabian horses equipped with Egyptian saddles, bridles, and stirrups. Musa also purchased 30 Mamluks as his bodyguard to return with him to Mali. Every year, as great caravans travelled between Egypt and Mali, peoples across the Sudan increasingly looked toward Cairo as the commercial centre of the Muslim world. Reporting on the Mali court of 1340, the Arab historian al–Umari notes that the Mansa commanded a force of 10,000 cavalry and was expending large sums in the importation from North Africa of Arabian mounts and Arab saddles. Saddle and stirrups greatly increased rider security, facilitating use of the slashing–sword and the massed charge with thrusting–spear, superior military tactics which would diffuse southward across the savanna (Hrbek 1977: 75–79, 90; Law 1980: 10, 28, 121).

But when Tuareg nomads seized Timbuktu in 1433, Mali lost control of the Sahel, and was superseded by Songhay which, according to later reports by Leo Africanus, imported Barb horses from northwest Africa. The very large Dongola horses were also procured from the Arab tribes invading the Darfur region. In the central Sudan, Kanem–Bornu and Hausa rose to prominence and constructed fortified towns in order to protect the gold trade. Cavalry, the striking force of their armies, appears early to have been composed of free men, whose commanders were rewarded by gifts and estates from the king; in later times cavalry forces would include both slave warriors and mercenaries. In the sixteenth century, as Ottoman imperium exerted control over the Middle East and North African littoral, the Turkish Pasha of Tripoli sent rich gifts of horses and firearms to the kingdom of Bornu, as did the Sultan of Morocco. There are also accounts of Arab merchants trading "infinite numbers" of horses from the Fezzan to Bornu (Law 1980: 28, 50, 147–48). Absolutely indispensable to the armies of the savanna, these horses, however, due to the extreme climate, were ever subject to serious disease, needing to be replaced as frequently as every two years. The price of one horse was 15 to 20 human slaves. Therefore, in order to replenish cavalry, regular raids were mounted by Muslims against the southern Negro populations in order to realize many thousands of captives each year. To conserve the element of surprise in night–time raids, mares were preferred, as opposed to stallions, since females were more manageable and less apt to cause a disturbance. Typically at dawn, a mobile force of horsemen and cameleers would surround and surprise an entire village, seizing fit individuals to be dispatched northward, massacring the weak and elderly (Heers 2003: 64–67). In the Sahel the war horse, vital for the control of gold, also drove the slave trade. With the build–up of military power came improvements in weaponry. Iron helmets with chin straps to withstand the shock of the spear, chainmail, and later musketry largely of Turkish and Egyptian origin, were introduced to complement spears and swords, thereby attaining a definitive edge over pagan adversaries (Adeleye 1976: 558, 565, 574). As shown in Figure 10, *lifidi* (quilted cloth armour stuffed with kapok), locally manufactured, was also used to protect both rider and mount. The principal cavalry weapons were the lance for thrusting and the hurled javelin; although less important, a broad–bladed slashing sword was carried for hand–to–hand combat, as were cudgels and battle–axes. Surprisingly, archery was only occasionally used by mounted troops. Ibn Battuta, for instance, mentions fourteenth–century Mali cavalry commanders as carrying bows. More commonly, archers were infantry troops (Law 1980: 119, 127–30) (Figure 10).

Late Pre–colonial Times

In sum, we see urbanization across the entire Sahel with three major routes along which larger horses were imported north to south: in the west Barb horses (14.5 hands) were traded from Morocco (Barbary) through Timbuktu; to the east prized Arabian horses (14–14.5 hands) travelled from Tunis across the Fezzan or from Cairo up the Nile valley. In the militaristic kingdoms of the Sahel, constantly vying with one another for control of the gold trade, these high–quality horses were destined specifically for warfare, as were the large Dongola horses (15 hands) brought from the east by Arab tribes migrating

Figure 10. Early nineteenth-century cavalryman and horse wear quilted cloth armor (*lifidi*); the horse head is also protected by a metal chamfron. From a sketch by Major Denham, engraved by E. Finden. (Denham and Clapperton 1826: 279).

Figure 11. Early nineteenth-century bodyguard of the Sheikh of Bornu. The horse's head is protected by plates of iron, brass, and silver, leaving room for the eyes of the animal. From a sketch by Major Denham, engraved by E. Finden. (Denham and Clapperton 1826: 64).

from Nubia across the Sudan (Figure 11). Maintenance of these large horses required specialist skills and slave officials were selectively assigned to tending and training activities. Despite this great influx of horses from Morocco, Tunis, Cairo, and Dongola, there is firm evidence that above the latitude 12°N the small, sturdy pony continued to be reared for war. It seems that both large and small varieties of horse coexisted and that different military strategies might have been associated with each type. On the savanna, large–horse cavalries engaging in the massed charge were long accustomed to ruthlessly mow down pedestrian combatants. But in mountainous regions, the massed charge was less effective. Specifically on the Jos Plateau, it is recorded that Muslim cavalry invasions were repeatedly repulsed by Sura tribesmen riding ponies, hurling javelins, and employing harrying guerrilla tactics. While the Arabian horse was superior in lowland battles, on rocky terrain the wiry Sura pony decisively outperformed the larger foreign horse. Use of geldings in war was uncommon, but apparently practiced in the hunt for ostriches (Blench 1993: 94; Law 1980: 76, 127).

The larger horse everywhere was the symbol of wealth and high political status. As Arab tribes continued their expansion across the Darfur region, horse rearing of the larger breeds began to be undertaken in restricted areas of the Sahel, particularly in the well–watered lands around Lake Chad. Considerable attention was accorded to securing stallions of quality for breeding purposes. Monopoly of mares was enforced, as only males were available for purchase. Artificial devices were utilized to discourage mountings by unauthorized stallions. Mares were thus fitted with spiked chastity belts or with metal rings obstructing the vulva in order to prevent impregnation.

In territorial expansion, equestrian power from the Sahel initially reached south to the fringes of the tropical forest, but not far beyond. Cavalry, a fearsome force in the deserts and savannas, could not operate effectively in the humidity of the tropics; warhorses could deploy for a charge only on open level ground. Dense tropical vegetation presented insurmountable obstacles to effective cavalry offensive, especially when bamboo traps were set along forest paths to trip advancing horses. From very early colonial reports, it is known that the Ijesa forested country of Yorubaland on multiple occasions had proven to be a graveyard for cavalry armies invading from the north. The high cost of transporting supplies of special grasses into the forest for foreign horse consumption was also prohibitive (Law 1980: 46, 139 f.). The pony in this regard held a certain advantage. Far longer inhabitant of the southerly regions, it had developed the ability to digest a wide range of plant foods. As we have seen, it had also achieved a limited degree of trypanotolerance. Not so the prized Arabian warhorse. The larger horses were especially prone to trypanosomiasis (Blench 1993: 91), transmitted by the tsetse fly and causing rapid fatality. Furthermore, periodic fluctuation of the tsetse belt northward at times even affected the kingdoms of the Sahel, resulting in the weakening of cavalry forces and consequent decline in military strength.

Acknowledgements

First, I wish to acknowledge my intellectual debt to the enduring scholarship of Mary Aiken Littauer and Joost H. Crouwel, which has probed numerous and diverse areas of equine archaeology. I also commend the early pioneering works of Henri Lhote and Jean Spruytte in documenting the ancient presence of charioteers and horsemen across the Sahara. To Alain Spruytte and Franck David, I must express my sincere gratitude for their graciously permitting the reprint of Saharan materials. For 15 years, David worked in close association with Jean Spruytte, in whose memory today he directs the conservation of the Saharan experimental models, chariots, and archives for future scholarly research. Franck David (2011) additionally continues to honor the tradition of Jean Spruytte by pursuing the study of early harnessing systems in Gallic and Roman cultures. I further wish to recognize another 'more recent pioneer' of equine archaeology, Sandra L. Olsen, whose groundbreaking field research in the French Paleolithic Saône Valley, Botai, Kazakstan, and Saudi Arabia has my utmost admiration and I applaud her spirited organization of the symposium *Equids in the Ancient Near East, Egypt, and Arabia* in Lexington, KY. Lastly, I greatly thank Bill Cooke, director of the International Museum of the Horse, for his extraordinary Kentucky hospitality during the 2010 symposium; my fellow participant Gail Brownrigg for her extensive equestrian experience and generous assistance in my completion of this article; finally Joost H. Crouwel, Katheryn M. Linduff, and Peter Raulwing for their sedulous editorship of this volume dedicated to the memory of Mary Aiken Littauer.

Abbreviations

EWAia	M. Mayrhofer, *Etymologisches Wörterbuch des Altindoarischen.* I.–III. Winter, Heidelberg, 1997–2001.
CTH	Emmanuel Laroche, *Catalogue des textes hittites*. Klincksieck, Paris.
Herodotus	*The Landmark Herodotus: The Histories*, edited by Robert B. Strassler and translated by Andrea L. Purvis. Pantheon Books, New York 2007.
Polybius 2011a	*The Histories*, Vol. 2, Loeb Classical Library. Translated by W.R. Paton. Revised by F. W. Walbank and C. Habicht. Harvard University Press, Cambridge.
Polybius 2011b	*The Histories*, Vol. 4, Loeb Classical Library. Translated by W.R. Paton. Revised by F. W. Walbank and C. Habicht. Harvard University Press, Cambridge.
Strabo	*Geography*, Vol. 8, Loeb Classical Library, edited by G.P. Goold, translated by H.L. Jones. Harvard University Press, Cambridge, 1996.

Bibliography

Adeleye, R. A. 1976. Hausaland and Borno 1600–1800 In *History of West Africa* Vol. 1, edited by J. F. A. Ajayi and M. Crowder: 556–601. 2nd ed. Columbia University Press, New York.

Amirsadeghi, H. (ed.). 1998. *The Arabian Horse. History, Mystery, and Magic*. Thames & Hudson, London.

Barth, H. 1857–59. *Travels and Discoveries in North and Central Africa. Being a Journal of an Expedition Undertaken Under the Auspices of H.B.M.'S Government in the Years 1849–1855.* Vol. 2. Harper & Brothers, New York.

Beachey, R. W. 1976. *The Slave Trade of Eastern Africa.* Barnes & Noble, New York.

Bin–Sultan al–Nahyan, H. H. Sheikh Zayed. 1998. Introduction. In *The Arabian Horse. History, Mystery, and Magic.* edited by H. Amirsadeghi, 7–9. Thames & Hudson, London.

Blench, R. M. 1993. Ethnographic and linguistic evidence for the prehistory of African ruminant livestock, horses and ponies. In *The Archaeology of Africa, Food, Metals, and Towns*, edited by T. Shaw et al.: 71–103. Series One World Archaeology 20. Routledge, London and New York.

Bulliet, R. W. 1990. *The Camel and the Wheel.* Columbia University Press, New York. [1st ed. Harvard University Press, Cambridge, MA and London, 1975].

Clutton–Brock, J. 1992. *Horse Power. A History of the Horse and the Donkey in Human Societies.* Natural History Museum Publications, London and Harvard University Press, Cambridge.

Cotterell, A. 2004. *Chariot. The Astounding Rise and Fall of the World's First War Machine.* Random House and Pimlico, London.

Culbertson, C. 2010. The Emergence and Spread of the Arabian Breed. In *A Gift from the Desert. The Art and Culture of the Arabian Horse* (Catalogue accompanying an exhibition held 29 May–15 October 2010 at the International Museum of the Horse at the Kentucky Horse Park, Lexington, Kentucky, USA), 62–91. The International Museum of the Horse, Lexington.

David, F. 2011. Les jouguets des attelages gallo–romains: Études expérimentales. *Histoire et Sociétés Rurales* 35: 7–58.

Denham, D. and H. Clapperton. 1826. *Narrative of Travels and Discoveries in Northern and Central Africa in the Years 1822, 1823, and 1824, Extending Across the Great Desert to the Tenth Degree of Northern Latitude, and from Kouka in Bornou, to Sackatoo, the Capital of the Fellatah Empire by Major Dixon Denham, Captain Hugh Clapperton and the Late Doctor Oudney.* Murray, London.

Donner, F. M. 1981. *The Early Islamic Conquests.* Princeton University Press, Princeton.

Epstein, H. 1971. *The Origin of the Domesticated Animals of Africa*, Vol. 2, African Publishing, New York.

Garlake, P. 1990. *The Kingdoms of Africa.* 2nd edition Bedrick, New York.

Gurney, O. R. 1990. *The Hittites*. Penguin Books, London. [1 1952, 6 2002].

Heers, J. 2003. *Les négriers en terres d'Islam: La première traite des noirs VIIe–XVIe siècle.* Perrin, Paris.

Hill, D. R. 1975. The role of the camel and the horse in early Arab conquests. In *War, Technology and Society in the Middle East*, edited by V. J. Parry and M. E. Yapp, 32–43. Oxford University Press, London and New York.

Horn, V. 1995. *Das Pferd im Alten Orient.Das Streitwagenpferd der Frühzeit in seiner Umwelt, im Training und im Vergleich zum neuzeitlichen Distanz–, Reit– und Fahrpferd.* Documenta hippologica. Olms–Presse, Hildesheim [etc.].

Hrbek, I. 1977. Egypt, Nubia and the Eastern Deserts. In *The Cambridge History of Africa* edited by J. D. Fage and R. Oliver. Vol. 3. *From c. 1050—c. 1600*, edited by R. Oliver: 10–97. Cambridge University Press, Cambridge.

Hrbek, I. and J. Devisse. 1988. The Almoravids. In *General History of Africa.* Vol. 3. *Africa from the Seventh to the Eleventh Century*, edited by M. Elfasi: 336–366. Heinemann, London [etc.], University of California Press, Berkeley and Unesco, Paris.

Hyland, A. 1990. *Equus. The Horse in the Roman World.* Batsford, London.

Hyland, A. 2003. *The Horse in the Ancient World.* Sutton, Stroud (UK).

Kelekna, P. 2009. *The Horse in Human History*. Cambridge University Press, New York.

Kennedy, H. 2007. *Great Arab Conquests: How the Spread of Islam Changed the World we Live in.* Da Capo Press, Philadelphia.

Köhler–Rollefson, I. 1996. The one–humped camel in Asia: Origin, utilization and mechanisms of dispersal. In *The Origins and Spread of Agriculture and Pastoralism in Eurasia*, edited by D.R. Harris: 282–294. University College London Press, London.

Law, R. 1980. *The Horse in West African History: The Role of the Horse in the Societies of Pre–Colonial West Africa.* Oxford University Press for the International African Institute, Oxford.

Levtzion, N. 1980. *Ancient Ghana and Mali.* In Africana Publishing Co., New York [u.a.]. [1st ed. Studies in African history 7. Methuen, London 1973].

Levtzion, N. 1999. Islam in Africa to 1800: Merchants, chiefs, and saints. In *The Oxford History of Islam*, edited by J. L. Esposito: 475–507. Oxford University Press, Oxford.

Lhote, H. 1953. Le cheval et le chameau dans les peintures et gravures rupestres du Sahara. *Bulletin de l'Institut Français d'Afrique Noir* 15(3): 1138–1228.

Littauer, M.A. 2002. The function of the yoke saddle in ancient harnessing. Reprinted in *Selected Writings on Chariots, other Early Vehicles, Riding and Harness*, by M.A. Littauer and J.H. Crouwel, 479–486. Culture and History of the Ancient Near East 6, edited by P. Raulwing. Brill, Leiden and Köln.

Littauer, M. A. and J.H. Crouwel. 1979. An Egyptian wheel in Brooklyn. *Journal of Egyptian Archaeology* 65: 107–120. Reprinted in *Selected Writings on Chariots and other Early Vehicles, Riding and Harness*, by M. A. Littauer and J. H. Crouwel, 296–313. Culture and History of the Ancient Near East 6, edited by P. Raulwing. Brill, Leiden and Köln.

Littauer, M. A. and J.H. Crouwel. 1985. *Chariots and Related Equipment from the Tomb of Tut'ankhamūn*, Tut'ankhamūn's Tomb Series 7. Griffith Institute, Oxford.

Mallory, J. P. 1996. *In Search of the Indo–Europeans: Language, Archaeology and Myth.* Thames & Hudson, London.

Mallory, J. P. and V. H. Mair. 2000. *The Tarim Mummies.* Thames & Hudson, London.

Nyland, A. 2009. *The Kikkuli Method of Horse Training, Revised Edition.* Maryannu Press, Sydney.

Oates, J. 2003. A note on the early evidence for the horse in Western Asia. In *Prehistoric Steppe Adaptation and the Horse*, edited by M. Levine, C. Renfrew and K. Boyle, 115–125. McDonald Institute Monographs. McDonald Institute for Archaeological Research, Cambridge.

Oliver, R. and B. M. Fagan. 1975. *Africa in the Iron Age, c. 500 BC to AD 1400.* Cambridge University Press, Cambridge.

Olsen, S. L. 2010. The Horse in the Ancient Near East, Egypt, and Arabia. In *A Gift from the Desert. The Art and Culture of the Arabian Horse* (Catalogue accompanying an exhibition held 29 May–15 October 2010 at the International Museum of the Horse at the Kentucky Horse Park, Lexington, Kentucky, USA), 22–61. The International Museum of the Horse, Lexington.

Raulwing, P. 2004. Indogermanen, Indoarier und maryannu in der Streitwagenforschung. Eine rezeptions– und wissenschaftsgeschichtliche Spurenlese. In *Rad und Wagen. Der Ursprung einer Innovation Wagen im Vorderen Orient und Europa*, edited by S. Burmeister and M. Fansa: 515–531. von Zabern, Mainz am Rhein.

Raulwing, P. 2006. The Kikkuli Text (CTH 284). Some Interdisciplinary Remarks on Hittite Training Texts for Chariot Horses in the Second Half of the 2nd Millennium B.C. In *Les Équidés dans le monde méditerranéen antique*, edited by A. Gardeisen: 61–75. Édition de l'Association pour le développement de l'archéologie en Languedoc–Rousillon, Lattes.

Raulwing, P. and J. Clutton–Brock. 2009. The Buhen Horse: Fifty Years After Its Discovery (1958–2008). *Journal of Egyptian History* 2(1–2): 1–106.

Raulwing, P. and R. Schmitt. 1998. Zur etymologischen Beurteilung der Berufsbezeichnung *aššuššanni* des

Pferdetrainers Kikkuli von Mittani. In *Man and the Animal World. Studies in Archaeozoology, Archaeology, Anthropology and Palaeolinguistics in Memorian Sandor Bökönyi*, edited by P. Anreiter, L. Bartosiewicz, E. Jerem, and W. Meid, 675–706. Archaeolingua; Alapítvány, Budapest.

Shaughnessy, E. 1988. Historical perspectives on the introduction of the chariot into China. *Harvard Journal of Asiatic Studies* 48: 189–237.

Spruytte, J. 1983. *Early Harness Systems. Experimental Studies. A Contribution to the History of the Horse.* [Translated by Mary Littauer]. Allen, London.

Spruytte, J. 1996. *Attelages antiques Libyens: Archéologie Saharienne expérimentale.* Collection Archéologie expérimentale et ethnographie des techniques 2. Éditions de la Maison des Sciences de l'Homme, Paris.

Thieme, P. 1960. The 'Aryan' gods of the Mitanni treaties. *Journal of the American Oriental Society* 80(4): 301–317.

Trimingham, J. Spencer. 1962. *A History of Islam in West Africa.* Glasgow University Publications. Published for the University of Glasgow by the Oxford University Press, London and New York.

10

Napatan Horses and the Horse Cemetery at El–Kurru, Sudan

Denise Doxey

In the late eighth and early seventh centuries BCE, four kings of Kush were buried at the site of el-Kurru in northern Sudan. The cemetery also featured the graves of the kings' prized horses, which were elaborately adorned for their journey to the afterlife. This paper discusses the horse burials in the wider context of the role of horses in ancient Sudan.

The kingdom of Kush, centered in the Sudanese Nile Valley between the Third and Fourth Cataracts and including the fertile Dongola Reach, witnessed periods of growth, influence and retraction over the course of its three–millennium history. These eras are now known by the names of the modern towns closest to the ancient capitals at the time. The Kerma kingdom flourished in the northern part of the region between about 2400 and 1550 BCE, dominating regional trade and agricultural output, and eventually threatening the security of southern Egypt (Morkot 2000: 62–8). A period of Egyptian domination followed, corresponding to Egypt's New Kingdom and lasting to the end of the second millennium BCE, when Egypt's central authority fractured. As Egyptian influence waned, a new Kushite kingdom in the southern Dongola Reach, known today as the Napatan dynasty, was emerging. The exact mechanisms driving Napata's rise remain the subject of scholarly debate (Morkot 2000: 140–44). However, by at least about 750 BCE, the Napatan state was unified under a single king, Kashta, whose power extended into southern Egypt. His daughter, Amenirdis, was adopted into the important religious and political role of God's Wife of Amun at Thebes and his successor, Piankhy, went on to conquer the remainder of Egypt. Five Kushite kings then ruled Egypt as the 25th Dynasty, expanding their territory into Libya and the Levant, before being driven out by an Assyrian invasion and retreating to Napata in about 671 BCE. Napatan kings continued to be buried near their capital at Gebel Barkal until about 270 BCE, when the royal cemetery moved to Meroe, from which the powerful Meroitic dynasty reigned until about CE 320.

The cemetery at el–Kurru, located on the west bank of the Nile 35 km south of the Fourth Cataract, was the resting place of the earliest Napatan rulers and other members of the royal family (Dunham 1950; Kendall 1999: 421–23). Those interred there included kings Piankhy (743–712 BCE), Shabaka (712–698 BCE), Shebitka (698–690 BCE) and Tanwetamani (664–653 BCE). Shebitka's successor. Taharqa (690–665 BCE) was buried at Nuri, seven km to the north, as were the later kings of the Napatan dynasty. The royal tombs at El–Kurru, following Egyptian burial traditions, took the form of pyramids with adjacent offering chapels and subterranean burial chambers. El–Kurru was excavated in 1919 by an expedition sponsored by Harvard University and the Boston Museum of Fine Arts (now Museum of Fine Arts, Boston [MFA]), under the direction of George Andrew Reisner, the Museum of Fine Arts' Curator of Egyptian Art and Professor of Egyptology at Harvard. The results were published in 1950 by Reisner's successor at the MFA, Dows Dunham (Dunham 1950). All of the tombs had been looted and little remained of their superstructures, but significant amounts of funerary material survived, now divided primarily between the Sudan National Museum in Khartoum and the Museum of Fine Arts, Boston. The following discussion will focus on one segment of the El–Kurru cemetery, namely the burial place of the royal horses.

Horses in Ancient Sudan

It remains uncertain exactly when and by what mechanism the horse first arrived in Sudan. Although the dating has been questioned, the skeleton of a horse discovered at the fortress of Buhen in 1958 may date as early as 1675 BCE (Raulwing and Clutton–Brock 2009). Another horse, found buried in a well shaft in the cemetery at Soleb, has been dated to the reign of Amenhotep III, but this date, too, is not universally accepted (Mallory–Greenough 2005: 105). In 2011, excavations led by Stuart Tyson Smith and Michele Buzon at Tombos, near the 3rd cataract, discovered the intact burial of a horse dating to the early Napatan period (Schrader, Smith, Buzon and Olsen 2018). The earliest textual and artistic references to horses in Sudan, discussed in greater detail below, derive from the time of Piankhy's conquest of Egypt. However, evidence from contemporary Assyrian texts demonstrates that Kushites were already well known as horse breeders and trainers by the mid–eighth century BCE (Dalley 1985: 43–7; Heidorn 1996: 105–10). The Nimrud Wine List number nine, dating to the reign of Tiglath–Pileser III, a contemporary of Piankhy, mentions Kushites present at the royal court, and Stephanie Dalley has argued persuasively that they were responsible for importing horses into Assyria from Kush via Egypt (Dalley 1985). The Horse Lists from

Nimrud, dating to the reign of Sargon II (744–727 BCE), and the Horse Reports from Nineveh refer specifically to Kushite horses as a superior breed for pulling chariots (Dalley 1985: 43). When the Assyrians under Esarhaddon (680–669 BCE) invaded Egypt and defeated the Napatan king Taharqa, he took "large horses" as booty (Heidorn 1996: 107). Hopefully, further excavation in Sudan will one day elucidate the Kushite end of this exchange.

Beginning in the reign of Piankhy both archaeological and textual sources provide abundant evidence of the value early Kushite rulers placed on their horses. The best–preserved historical evidence for the Kushite conquest of Egypt is the "Victory Stele" of Piankhy, discovered in 1862 at Gebel Barkal. Here, as well as on the walls of his Gebel Barkal temple, the king recounts his defeat of the four rulers of a divided Egypt, Nimlot of Hermopolis (ancient Khemenu), Peftjawybast of Herakleopolis (ancient Nen–nesut), Tefnakht of Sais and Osorkon of Bubastis (ancient Per–Bastet). Although the text presents Piankhy as a stereotypical example of a just and pious ruler and cannot therefore be accepted at face value as evidence of his true personality, the prominence given to horses is exceptional and must reflect a degree of historical accuracy. At Hermopolis, Nimlot and his wife surrendered to Piankhy by offering him gifts including a horse, which is portrayed on the victory stele (Morkot 1999: 168). When Piankhy entered the royal palace, Nimlot is said to have offered him his wives and daughters, but Piankhy "did not turn his gaze upon them" (Morkot 2000: 188). He was, however, deeply moved when he reached the stables, where Nimlot's horses had become malnourished during the siege of Hermopolis. According to the stele, he said to Nimlot, "[the fact] that my horses were made to hunger pains me more than any other crime you committed in your recklessness" (Morkot 2000:188). Relief decoration on the north wall of Gebel Barkal temple B501 also shows the victorious king receiving a procession of horses, a position in which human prisoners would be more expected in traditional Egyptian representations (Dunham 1070: pl. 50). The evidence from Kush itself therefore dovetails with the surviving historical documentation from Assyria to illustrate the significance of the horse in Napatan times.

The El–Kurru Horse Cemetery

On March 23, 1919, the Harvard–MFA Expedition discovered the first of a series of graves, called Cemetery 200, about 120 meters southeast of El–Kurru 51–55, the tombs of Napatan royal women, including the wives of Piankhy. The excavators went on to uncover a series of 24 horse burials and two small, circular graves of dogs (Dunham 1950: 110–17). The horse graves were arranged in four rows, from southwest to northeast, with four burials each in the first (SW) and fourth (NE) row and eight burials each in the second and third (Figure 1). Inscribed objects dated the second and third row to the reigns of Shabaka and Shebitka, respectively, and Reisner concluded that that the two outermost and shorter rows therefore most likely dated to the reigns of the other two

Figure 1. The horse cemetery at el-Kurru (A2820). Harvard University–Boston Museum of Fine Arts Expedition. Photograph © Museum of Fine Arts, Boston.

kings buried at El–Kurru, Piankhy and Tanwetamani. All the graves had been plundered, especially those in the outer two rows, and some were entirely empty. Others, however, were far better preserved. Dunham discussed the best–preserved burials in his publication of the El–Kurru cemetery (Dunham 1950: 111–17). This paper is intended to review Reisner and Dunham's work, to supplement it with unpublished material from the graves and to add thoughts on the possible configuration and function of the horse trappings.

In all cases where evidence survives, the horses were buried standing up with their heads facing toward the northeast (Dunham 1950: 111), the direction of the rising sun. Their surviving skeletal remains were divided between the Museum of Comparative Zoology at Harvard University, which received one of the best–preserved skeletons, and the Sudan National Museum, which houses all the remaining skeletons and bones. Because they found no skulls, Reisner and Dunham postulated that the horses had been decapitated. However, Dunham himself lists teeth among the finds from grave Ku 201, and Sándor Bökönyi's more recent study of the horse remains at Harvard University confirmed the presence of teeth as well as skull fragments (Bökönyi 1993: 302–05). Furthermore, one of the horses of Tanwetamani, from horse grave Ku 219, was buried with a gilded silver plume holder now in Boston (MFA 23.877) that must originally have been attached to a bridle or halter. It therefore seems evident that the horses were not decapitated but that their heads, projecting above ground and presumably covered by mounds of earth, were particularly subject to damage.

There has been no firm agreement regarding the breed of horse interred at El–Kurru and further study of the skeletal remains is warranted. Reisner described the horses as "a short, rather small breed, not unlike the Arab" (Reisner 1919: 253). A study by George M. Allen, the Curator of Mammals at Harvard's Museum of Comparative Zoology also remarked on the relatively small stature of the horses, stating that "their skeletal character was similar to that of animals now living in Europe and America, except that their lines were a little more delicate, amounting to a few millimeters in length of shaft of the longer limb bones" (Dunham 1950: 111). Bökönyi reached a very different conclusion, stating that the horses "were large animals [over fifteen hands], of much greater size than the average oriental horses of their time" and that they had very slender limbs and wide, flat hooves (Bökönyi 1993: 307). Bökönyi's description accords well with the Assyrian description of Nubian horses as particularly large animals, as well as with the characteristics of the Dongolawi horses that became well–known in Sudan in medieval times and later (Heidorn 1996: 107). For now, however, the question remains unresolved. Likewise, the sex of most of the horses is uncertain. Based on the presence of a canine tooth among the remains published by Bökönyi, Mallory–Greenough has demonstrated that at least one and probably another of Shebitka's horses (buried in Ku 211 and 212) was a stallion (Mallory–Greenough 2005: 108).

The sacrifice of animals to accompany elite burials has a long history in ancient Sudan. Kerma period graves routinely contained the bodies of dogs and sheep (Kendall 1997: 59). A horse grave discovered by Reisner in the South Cemetery at Meroe contained the skeleton of a young horse without associated artifacts (Dunham 1963: 441). The tomb has not been securely dated but may be of the mid–seventh century BCE (Heidorn 1996: 107, n. 7). Horse burials and trappings accompanied several later tombs in the cemeteries at Meroe, dating from the first century BCE to the third century CE (Dunham 1957). In these cases, the horses, wearing functional tack, were buried at the bottom of the stairways leading to the burial chambers. Royal tombs at Ballana and Qustul from the fourth to sixth centuries CE featured the burial of multiple horses in open courtyards or the entrance ramps to the tombs (Brooklyn Museum of Art 1978: 112–15), often with very elaborate trappings. The El–Kurru horse cemetery is therefore part of a long tradition of horse burials. What distinguishes it from other instances is that it is the only known case in which a distinct cemetery was set aside for horses.

The Horse Trappings

Although the horse burials at El–Kurru were consistently plundered and some were entirely devoid of artifacts, certain patterns can be established based on the better–preserved graves. Virtually all the horse trappings were found at the head end of the graves, where they had fallen from the heads and necks of the horses (Reisner 1919: 253; Dunham 1950: 111). There is no evidence for bits or other functional tack, or of chariots or yokes. The trappings are therefore "horse jewelry" designed specifically for adornment or for ritual purposes. Within each row, the preserved trappings are nearly identical, indicating that the horses of each king were distinctively attired. The following sections will treat each king's horses and their graves in chronological order.

The Horses of Piankhy

The earliest graves in the horse cemetery at El–Kurru, numbered Ku 221 through Ku 224 by Reisner, almost certainly belong to the horses of king Piankhy (743–712 BCE) and were located in the southwestern row of burials (Table 1). Of the four rows of graves, those dating to Piankhy's reign are the most complex in type, rectangular in plan with deep holes for the fore and rear legs, central supports for the belly, and a shelf at the head end to support the neck; grave Ku 221 had separate holes for each of the forelegs (Dunham 1950: 116 f.). In grave Ku 222 remains of the ribs and spinal column of the horse remained in place on the central support and leg bones in the pits (Figure 2). These graves were the most severely plundered in the cemetery. Grave Ku 221 was completely empty and graves Ku 223 and Ku 224 contained only a few bones each. Little remains of the trappings from grave Ku 222, but fragments of cord and matting with looped edges and traces of fine cloth suggest that the horse was covered in a shroud. Strings of small faience ring beads, oval gilt silver beads and another gilt silver amulet in the form of the sacred eye of Horus (*wedjat*) are all that remain of the horse trappings. This last amulet features a motif long popular in Egypt and later adopted by the Kushites. According to myth, Horus, the god of kinship,

Table 1. Horses of Piye: remains recorded in the HU-MFA Object Register*

No.	Horse remains	Trappings
221	None	None
222	Partial skeleton disturbed, head missing, ribs and spine in place on central support	*19-4-122 (MFA 19.2889): fragments of cord.* 19-4-123: Fragment of matting, with threads crossing at right angles, looped edging, and traces of fine clothe. *19-4-124 (MFA 19.2890): strings of small faience ring beads.* *19-4-125 (MFA 19.2891): 2 hollow, oval, silver gilt beads, decayed.* *19-4-126 (MFA 19.2892): decayed silver fragments of hollow wedjat eye (gilt), embossed.* *19-4-127 (MFA 19.2893): bit of pierced, light blue faience object.*
223	Disturbed bones	None
224	Disturbed bones (19-4-128)	None

*Artifacts in italics are currently housed at the Museum of Fine Arts, Boston. Those not listed in italics are now in the Sudan National Museum.

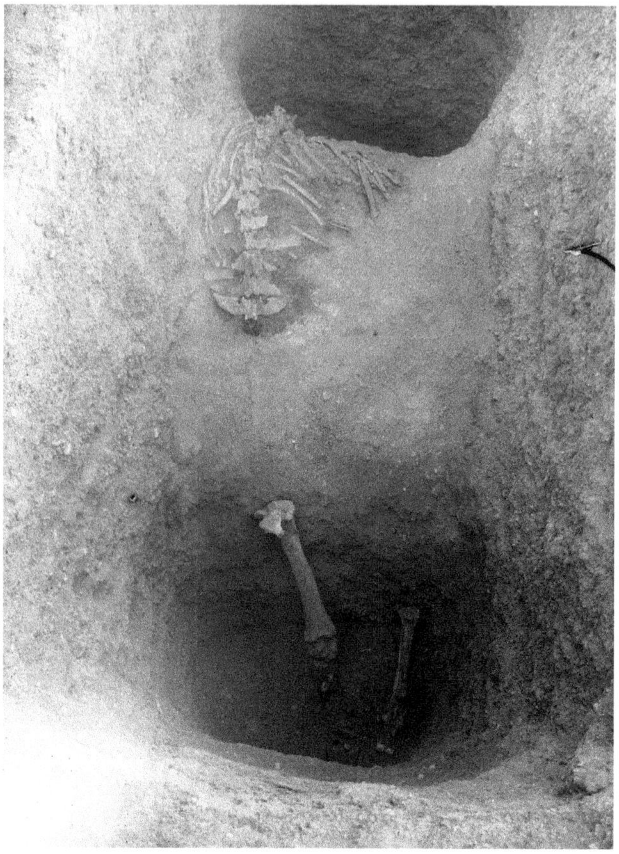

Figure 2. Horse remains in grave Ku 222 (C8804). Harvard University–Boston Museum of Fine Arts Expedition. Photograph © Museum of Fine Arts, Boston.

lost his eye in conflict with his murderous uncle Seth, the god of chaos, only to have it miraculously restored. Amulets featuring the restored eye were therefore believed to offer protection both to the living and the dead.

The Horses of Shabaka

Shabaka's eight horses were interred in graves—numbered Ku 201–Ku 208 (Table 2)—of a type similar to those of Piankhy, but with a somewhat simpler design, rectangular in plan with a central support for the belly and holes for the fore and rear legs (Figure 3). Only one, Ku 203, featured a low support for the horse's neck (Dunham 1950: 111 f.). Very few bones from Shabaka's horses survived, but grave Ku 202 contained both rear hooves, enabling the excavators to determine the original orientation of the animal. Although most of the graves were severely looted, the surviving remains of horse trappings strongly suggest that Shabaka's horses were identically adorned for burial. Based on the best–preserved set, from grave Ku 201, a tentative idea of at least some of the original trappings can be proposed.

Grave Ku 201 contained a very large quantity of blue faience ring beads, along with a smaller quantity of ring beads in red, white and yellow faience. Beads of this type were often used for the bead nets that covered contemporary human mummies. Many of the ring beads in Ku 201 were found in association with blue faience beads in the form of the head of Hathor, the Egyptian goddess of love, fertility, and rebirth in the afterlife. Some of the Hathor head beads featured two horizontal holes for stringing and others were pierced with a single, vertical hole, indicating that they were part of a network of beads. Also found alongside these beads were blue, yellow and red faience ball beads, blue and yellow pendant beads shaped like hanging flower blossoms, and a blue faience cartouche bearing Shabaka's throne name, Neferkara. The arrangement of the beads allowed the excavators to reconstruct a partial network of beads (MFA 21.10560) like that which must originally have adorned the neck and chest of the horse (Figure 4). Dunham noted that the arrangement was neither complete nor definitive, as it does not account for all the beads found in the burial (Dunham 1950: 112). The original network must have been considerably larger, probably continuing around the horse's shoulders. In addition, among the beads not incorporated into the reconstruction are similar Hathor–head beads in silver, along with fragmentary silver pendants, suggesting that the nets originally incorporated precious metals. Other beads that must have been part of additional trappings, that cannot be reconstructed, include faience and silver barrel beads, a star–shaped bead in blue faience, a bi–conical blue faience bead, a blue faience bead in the shape of a cowrie shell (associated in antiquity with fertility), and a large, elaborate bead with a floral motif, also of blue faience (MFA 19.2470). This last bead is inconsistent in both size and shape with the other beads in the grave and is unlikely to have been part of a bead net. It may have adorned a bridle.

The finds from Ku 203 and Ku 204, though less well–preserved, are virtually identical to those from Ku 201, including faience and silver Hathor beads, some designed for vertical stringing and others for horizontal double stringing, abundant faience ring beads, faience ball beads and faience floral pendants. Although they are yet to be reconstructed, these beads must have formed part of a network like that from Ku 201 (Figure 5).

The Horses of Shebitka

The graves of Shekitka's horses, numbered Ku 209 through Ku 216, are the best–preserved group in the cemetery (Table 3). The style of the graves themselves continues a progression from a more complex to a simpler design. Rectangular in plan, the graves lack the central belly supports seen in the earlier horse burials, and none has a neck support (Figure 6). While two burials (Ku 214 and Ku 215) were completely plundered, and three (Ku 212, 213 and 216) were heavily looted, five of the skeletons (Ku 209–213) were found virtually intact aside from the missing heads (Figure 7). The three best–preserved graves (Ku 209–211) also retained considerable remains of horse trappings. The trappings from Ku 209 are now housed in Khartoum, while the rest are in Boston.

The trappings of Shebitka's horses differ significantly from those of his predecessor. Based on the evidence from the three relatively well–preserved graves, as well as remnants

Table 2. Horses of Shabaka: remains recorded in the HU-MFA Object Register*

No.	Horse remains	Trappings
201	A few bones and teeth	*19-3-687 (MFA 21.10560): Hathor head beads and ring beads.* *19-3-688 (MFA 21.10560): 15 blue floral pendants.* *19-3-689 (MFA 19.2458): 6 yellow floral pendants.* *19-3-690 (MFA 19.2459): fragments of 4 silver pendants.* *19-3-691 (MFA 21.10560, 19.2460): 42 large blue and yellow ball beads.* *19-3-692 (MFA 19.2461): 2 hollow silver Hathor head pendants.* *19-3-693 (MFA 19.2460): 2 blue and one white faience barrel bead.* *19-3-705: (MFA 19.2470) decayed faience object with floral motif.* *19-3-706 (MFA 21.10560): 4 blue faience ring beads.* *19-3-707 (MFA 21.10560): blue faience cartouche, pierced vertically.* *19-3-708 (MFA 19.2472): blue faience cartouche, pierced vertically.* *19-3-709 (MFA 21.10560): 10 blue faience Hathor pendants.* *19-3-710 (MFA 21.10560): 8 blue faience Hathor pendants.* *19-3-711 (MFA 21.10560): 6 blue faience Hathor pendants.*
		19-3-712 (MFA 21.10560): blue faience Hathor pendant. *19-3-713 (MFA 19.2460): red-brown faience ball bead.* *19-3-714 (MFA 19.2460): 3 yellow and 1 blue faience ball beads.* *19-3-715 (MFA 19.2460): faience star-shaped ring bead.* *19-3-716 (MFA 19.2476): remains of 4 decayed silver Hathor pendants.* *19-3-717 (MFA 21.10560): quantity of faience ring beads.* *19-3-718 (MFA 19.2478): 2 fragments of large, hollow silver barrel beads.* *19-3-719 (MFA 19.2479): silver floral pendant.* *19-4-31 (19.2841): 2 blue faience beads (1 biconical, 1 cowrie).* *19-4-32 (MFA 21.10560): red, blue and white faience ring beads.* *19-4-33 (MFA 19.2843): fragment of alabaster instrument?* *19-4-34: 2 baskets of horse bones and teeth.* *19-4-35 (MFA 19.2844): 3 small fragments of iron.*
202	Broken bones, including both rear hooves	*19-4-36 (19.2845): a few red, yellow and blue faience ring beads.*
203	Few bone fragments	*19-4-37 (21.10561-3): 100 ½ blue faience Hathor heads, some with double horizontal stringing and some with single vertical.* 19-4-38: Horse bones. *19-4-39 (21.10563): Red, white and blue faience ring beads.* *19-4-40 (19.2847): 14 decayed silver Hathor pendants, decayed.* *19-4-41 (19.2848): 3 silver imitation cowrie shells.* *19-4-42 (19.2849): 7 blue faience ball beads.* *19-4-43 (19.2850): Fragment of a blue faience floral pendant.*
204	Disturbed bones	*19-4-44 (19.2851): 2 Hathor-head beads and 1 star-shaped ring bead.* *19-4-45 (19.2852): Large quantity of blue and white ring beads.* 19-4-46: a few horse bones. *19-4-47 (19.2853): ½ a silver barrel bead.* *19-4-48 (19.2854): 3 fragments of a blue faience floral pendant.*
205	Disturbed bones	19-4-49: a few horse bones. 19-4-50: blue faience ball bead
206	Disturbed bones	19-4-51: basketful of horse bones.
207	Disturbed bones	19-4-52: 2 basketsful of horse bones. 19-4-53 (19.2855): a few blue faience ring beads.
208	Disturbed bones	19-4-54: basketful of horse bones. *19-4-55 (19.2856): a few blue and white faience ring beads.* *19-6-56 (19.2857): blue faience, cowrie-shaped bead.*

*Artifacts in italics are currently housed at the Museum of Fine Arts, Boston. Those not listed in italics are now in the Sudan National Museum.

in the partially plundered graves, the trappings of all eight horses were identical. Rather than networks of beads and pendants, the horses were buried wearing numerous individual strands of beads, many of which were found either in their original position or still on their original strings, allowing for a more accurate reconstruction of their composition (Figure 8).

Two burials (Ku 209 and 211) retained fragments of wide, silver bands, pierced at the edge for attachment to leather or fabric backings. These may have been parts of nape straps similar to those that appear in some early first millennium Assyrian reliefs (Littauer and Crouwel 1979: 127, figs. 53 ff.). Four of the horses (Ku 209–212) wore a series of twelve large, heavy, solid bronze ball beads

Table 3. Horses of Shebitka: remains recorded in the HU-MFA Object Register*

No.	Horse remains	Trappings
209	Skeleton nearly intact except for head	19-4-71: 12 yellow and blue cartouches, strung as found in situ, plus 2 more blue found loose. 19-4-72: string of wedjat eye and cowrie shells, ½ strung in the original order as found. The other ½ is restored. Order: 1 large faience pendant eye flanked by 2 small eyes of faience, followed by 7 cowries and then 3 small faience eyes. Total of 22 eyes and 42 shells. *19-4-73 (19.2866): 28 blue faience lotus pendants. Original order of stringing with small blue faience ring beads.* 19-4-74: 36 blue faience small eyes strung in original order with 2 blue faience ball beads as ends. 19-4-75: a large quantity of blue (and?) ring beads, strung and loose. 19-4-76: some horse hoof bones. 19-4-77: Large, hollow, silver gilt wedjat eye amulet. 19-4-78: 6 pieces of wide silver band collar, pierced at edges. 19-4-79: blue openwork eye bead. 19-4-80: 2 large blue faience ball beads with wedjat eyes and ankhs in black. 19-4-81: blue faience ball bead with black circles. 19-4-82: green faience ball bead with black circles. 19-4-83: 2 large gold (hollow) ball beads. 19-4-84: several loose cowrie beads. 19-4-85: 12 large solid bronze ball beads with some of the original cord. 19-4-86: some decayed cord and matting.
210	Skeleton nearly intact except for head	*19-4-87 (21.10572): strings (in original order?) of 36 small wedjat eyes in blue faience.* *19-4-88 (19.2867): string of wedjat eyes and cowries similar to 19-4-72.* *19-4-89 (19.2868): strings of small faience ring beads.* *19-4-90 (19.2869): bronze balls as 19-4-85, some with small ring beads and their original stringing attached.* *19-4-91 (19.2870): masses of small faience ring beads on strings.* *19-4-92 (21.10573): 1 blue faience cartouche.* *19-4-93 (19.2871): 1 blue faience floral pendant.* *19-4-94 (19.2872): 2 gold ball beads as 19-4-83.* *19-4-95 (19.2873): several loose cowrie shell beads.* *19-4-96 (19.2874): 6 blue faience beads (as in grave 209).* 19-4-97: Horse bones.
211	Skeleton intact except for head	19-4-98: Horse bones (Peabody). *19-4-99 (21.11729): 2 large, hollow gold ball beads.* *19-4-100 (21.10564): string of 35 small wedjat eyes in blue faience.* *19-4-101 (21.10565 and 19.2875): 11 ½ yellow and blue cartouches strung in original order, Djedkara. In the middle of the string is a cartouche of Menkheperra.* *19-4-102 (19.2876): 15 small faience eyes and one large eye pendant on string of 35 cowries. Eyes incised on both sides.* *19-4-103 (19.2877): 2 ball beads, blue with black circles.* *19-4-104 (21.10566): 1 wedjat bead, 2 ball beads with ankhs and wedjats in black paint, and 1 openwork barrel bead with incised wedjats* *19-4-105 (19.2878): Several cowrie beads.* *19-4-106 (21.10567): 27 jasmine flower pendants and ring beads strung in original order.* *19-4-107 (19.2879): a few frags of silver sheet collar.* *19-4-108 (21.10568): 12 bronze ball beads (as 19-4-85).* *19-4-109 (19.2880): quantity of faience ring beads, strung.*
212	Skeleton intact except for head	*19-4-57 (21.10569a-b): String of 29 jasmine flower pendants, originally strung with a double row of ring beads in order in which they were found.* *19-4-58 (21.10570): 5 openwork blue faience wedjat eye beads and fragments of 3 more, containing remains of original string.* *19-4-59 (21.10571): 12 large, solid bronze ball beads with original stringing in place.* 19-4-60: 12 cowrie shell beads, copper stained. *19-4-61: a quantity of horse bones.* *19-4-62 (MFA 19.2858): bag of contents from burial (brown) and faience ring beads.*
213	Skeleton intact except for head	*19-4-110 (19.2881): Shawabty of Tanwetamani (intrusive).* *19-4-111 (19.2882): blue faience barrel bead.* 19-4-112: horse bones. *19-4-113 (19.2883): Several cowries.* *19-4-114 (19.2884): strings of faience ring beads.*
214	Completely plundered	Nothing.
215	Completely plundered	Nothing.
216	Part of a disturbed skeleton	19-4-136: horse bones. *19-4-137 (19.2900): 2 cowries (beads).* *19-4-138 (19.2901): 1 faded faience bead in the shape of a cowrie shell.*

*Artifacts in italics are currently housed at the Museum of Fine Arts, Boston. Those not listed in italics are now in the Sudan National Museum.

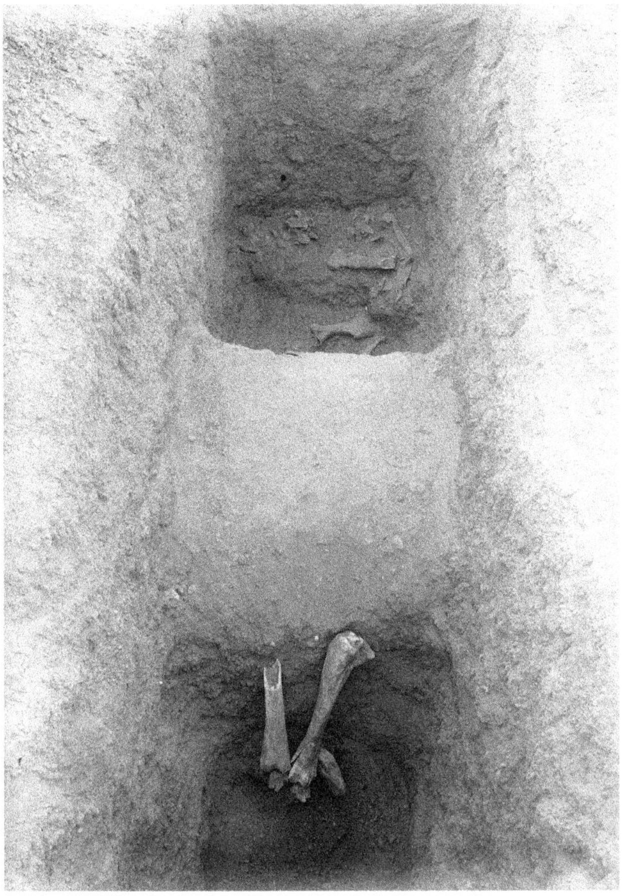

Figure 3. Grave Ku 207, with horse bones in situ (B3667). Harvard University–Boston Museum of Fine Arts Expedition. Photograph © Museum of Fine Arts, Boston.

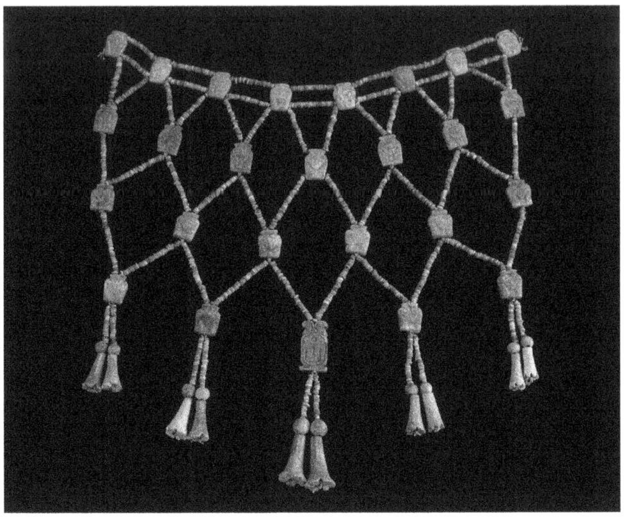

Figure 4. Bead net from the horse buried in grave Ku 201, MFA 21.10560. Harvard University–Boston Museum of Fine Arts Expedition. Photograph © Museum of Fine Arts, Boston.

strung at regular intervals on thick, twisted cords (Figure 9). Reisner surmised that these strings were fastened to a bridle and passed down the horses' chests and through the front legs like a martingale (Reisner 1919: 253). Assyrian representations of horses, however, show necklaces of similar beads worn halfway up the neck (Littauer and

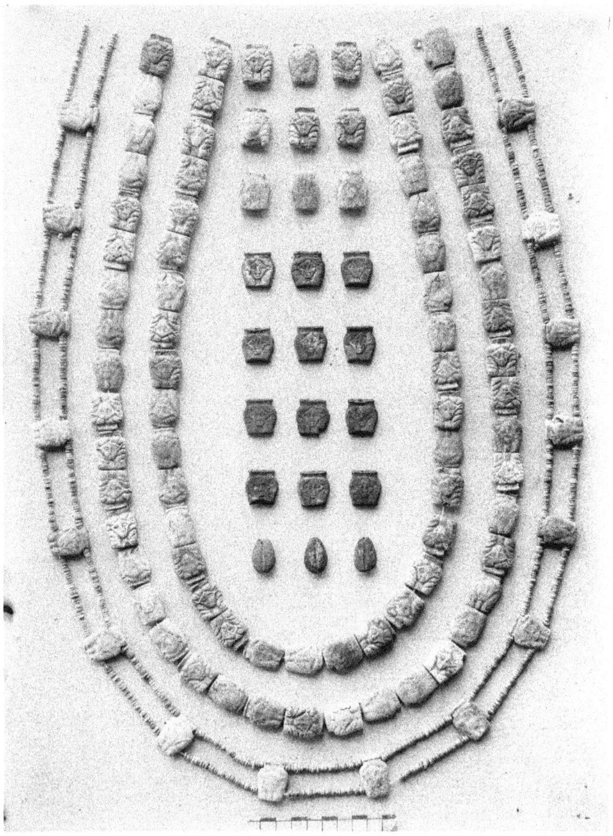

Figure 5. Horse trappings from grave Ku 203 (A2810). Harvard University–Boston Museum of Fine Arts Expedition. Photograph © Museum of Fine Arts, Boston.

Figure 6. Grave Ku 213 (B3668). Harvard University–Boston Museum of Fine Arts Expedition. Photograph © Museum of Fine Arts, Boston.

Figure 7. Horse skeleton in grave Ku 209 (B3666). Harvard University–Boston Museum of Fine Arts Expedition. Photograph © Museum of Fine Arts, Boston.

Figure 8. Grave Ku 209 with beads in situ (B3677). Harvard University–Boston Museum of Fine Arts Expedition. Photograph © Museum of Fine Arts, Boston.

Crouwel 1979: 127, figs. 53, 76), and the length of the strands suggests that the beads found with Shebitka's horses may have served the same function. Graves Ku 209, 210 and 211 each also contained a pair of large, hollow gold ball beads, the original placement and function of which are unclear.

Excavation photographs indicate that the horses' necks were also adorned with a series of "necklaces" of faience and shell beads. Immediately below the bronze ball beads, they wore a strand of fourteen alternating blue and yellow faience beads in the shape of cartouches topped by double plumes and bearing Shebitka's throne names, Djedkara and Menkheperra (Figure 10). Lower on the neck was a string of approximately thirty floral pendants of blue and yellow faience strung with multicolored faience ring beads (Figure 11) and a strand of red, yellow, blue and white faience *wedjat* eyes. Longer strings of beads draped around the neck and chest of the horses were composed of a large, central blue faience *wedjat* eye bead strung with alternating groups of three smaller *wedjat* eyes and seven cowrie shells, totaling twenty–two *wedjats* and 42 shells (Figure 12). Very large numbers of small, multi–colored faience ring beads, some still threaded on the original string, were found throughout the head ends of the graves, apparently from multiple bead strands draped randomly around the forepart of the horses. In some cases the beads were fused together, and in Ku 210 they were also fused

Figure 9. String of bronze ball beads, MFA 21.10568. Harvard University–Boston Museum of Fine Arts Expedition. Photograph © Museum of Fine Arts, Boston.

to the large, bronze ball beads. The abundant presence of these bead strands, along with cartouche beads and floral pendants nearly identical to those that adorned Shabaka's horses, suggests that the beads were originally intended for nets that were never completed. Hathor–head beads, so prevalent in the graves of Shabaka's horses, appear to be absent among the horse trappings of his successor.

Several other types of distinctive beads were found in smaller quantities and their placement on the horses' bodies remains uncertain. These include large, oblong,

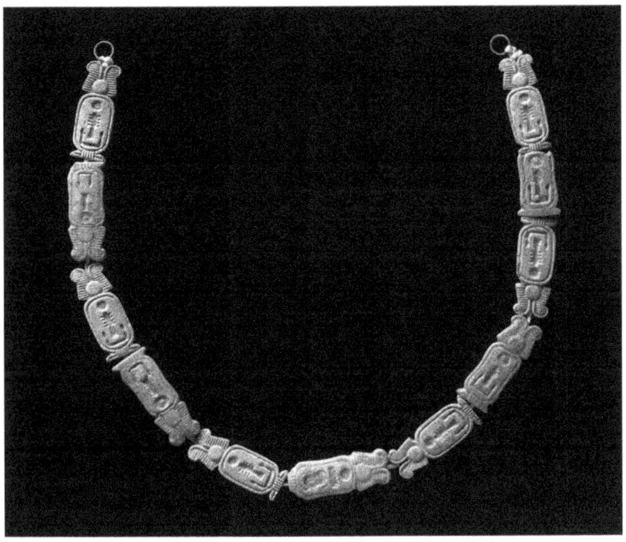

Figure 10. String of faience cartouche beads, MFA 21.10565. Harvard University–Boston Museum of Fine Arts Expedition. Photograph © Museum of Fine Arts, Boston.

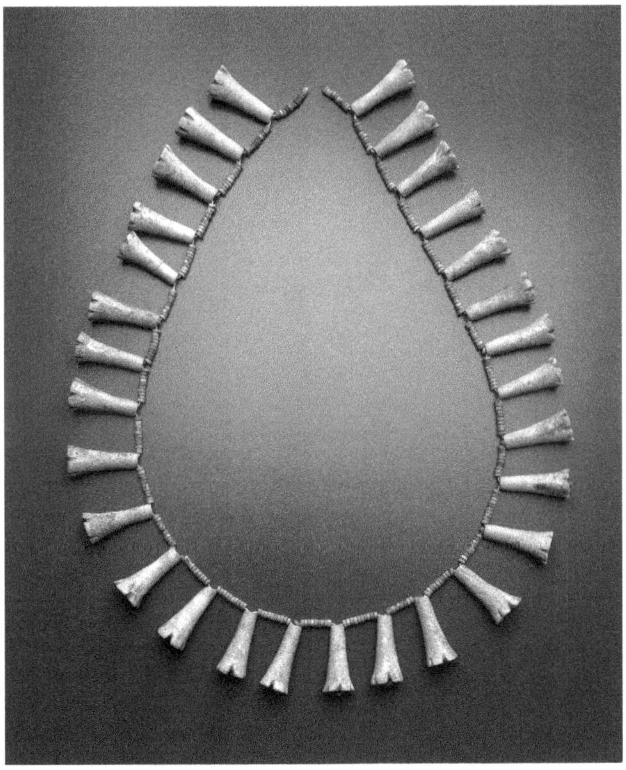

Figure 11. String of faience floral pendants, MFA 21.10567. Harvard University–Boston Museum of Fine Arts Expedition. Photograph © Museum of Fine Arts, Boston.

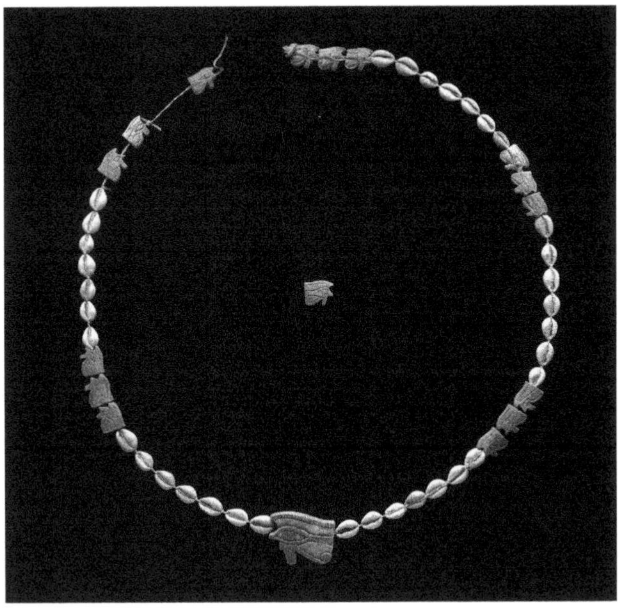

Figure 12. String of *wedjat* eye and cowrie shell beads, MFA 19.2876. Harvard University–Boston Museum of Fine Arts Expedition. Photograph © Museum of Fine Arts, Boston.

Figure 13. Beads with *ankh* and *wedjat* designs from Ku 210, MFA 19.2874 Harvard University–Boston Museum of Fine Arts Expedition. Photograph © Museum of Fine Arts, Boston.

blue–green faience ball beads approximately 2.5 cm in diameter, decorated with either concentric circles or alternating *ankh* and *wedjat* signs in black, of which no more than two were found in a given burial (Figure 13). Accompanying them in two cases were smaller, ridged, bi–conical beads in blue faience. Four graves contained large beads of blue faience bearing openwork *wedjat* eyes approximately 2.9 cm long, the largest group, from grave 212, including five complete beads and fragments of three more on their original string.

The Horses of Tanwetamani

The four graves of Tanwetamani's horses (Ku 217–220, Table 4), in the northeastern row, are, like the graves of Piankhy's horses, very poorly preserved. The graves are simple, oval pits without neck or belly supports (Figure 14). All four were greatly disturbed, with graves Ku 217, Ku 218, and Ku 220 containing only a small quantity of disturbed bones and scattered beads. A substantial portion of the skeleton remained only in Ku 219, but very few other artifacts survived making it impossible to speculate on the nature of the horse trappings. Despite

Table 4. Horses of Tanwetamani: remains recorded in the HU-MFA Object Register*

No.	Horse remains	Trappings
217	A few bones	19-4-115: horse bones. *19-4-116 (19.2885): a few small faience ring beads.* *19-4-134 (19.2898): a few small faience ring beads.* *19-4-135 (19.2899): natural pebble with hole.*
218	A few bones	*19-4-117 (19.2886): a few small faience ring beads.* 19-4-118: horse bones.
219	Skeleton disturbed, head missing	*19-4-119a (19.2887.1): 1 sm. sacred eye as 19-4-100, but slightly smaller.* *19-4-119b (19.2887.2): 1 sm. ball bead.* 19-4-120: horse bones. Unregistered (23.877): Plume holder for a horse's bridle.
220	A few scattered bones at head end, head missing	*19-4-121 (19.2888): 1 sacred eye, light blue faience, design inscribed on one side only.* *19-4-131(Boston?): horse bones.* *19-4-132 (19.2896): a few ring beads and ½ a cylinder bead.* *19-4-133 (19.2897): 1 cowrie bead.*

*Artifacts in italics are currently housed at the Museum of Fine Arts, Boston. Those not listed in italics are now in the Sudan National Museum.

Figure 14. Grave Ku 219 (C8815). Harvard University–Boston Museum of Fine Arts Expedition. Photograph © Museum of Fine Arts, Boston.

the plundering, Ku 219 produced possibly the most important find from the horse cemetery, a gilded silver plume holder that must once have adorned a bridle or halter (Figure 15). It takes the form of a papyrus blossom on a curved stem, atop which sits the falcon–headed sun god Ra as a falcon, wearing the sun disk and royal *uraeus* cobra. A hollow tube at the back of the sun disc enabled the insertion of the plume, most likely made of ostrich feathers.

Figure 15. Plume holder for a horse, MFA 23.877. Harvard University–Boston Museum of Fine Arts Expedition. Photograph © Museum of Fine Arts, Boston.

Comments and Conclusions

Cemetery 200 at El–Kurru is exceptional in a number of ways. First, it is the only Nubian cemetery yet known devoted exclusively to animals. While other horse burials are attested both in ancient Sudan and elsewhere in the ancient world, they are typically found in or near the tombs of their owners. Cemetery 200 was clearly planned and constructed specifically for the royal horses. The design of the graves, in which the horses were buried standing, also remains unparalleled to date. A better understanding of the role horses played in early Napatan royal funerary ideology and practice requires additional evidence and further study. The horse trappings include no trace of functional tack of any kind, instead featuring exclusively ornamental and amuletic elements. Thus, although the arrangement of the animals in groups of four suggests that they may have served as chariot teams, the lack of harnesses or other equipment does not necessarily support a role in providing transport to the kings in the afterlife.

Based on the apparently uniform nature of the horse trappings within each reign, it seems clear that each group of horses was buried simultaneously, perhaps at the time of the king's internment. It is also possible that they were killed at another point in the reign. Herodotus recounts a Meroitic king's sacrifice of four horses to the sun god following a victory in battle (Mallory–Greenough 2005: 106). The El–Kurru horses were almost certainly sacrificed as well, although it is unclear how they were killed. In Meroitic cemeteries where the horses' skulls have survived, the cause of death can be seen to have been a blow to the head, possibly from a pole–ax (Mallory–Greenough 2005: 107). Reisner's original suggestion that the El–Kurru horses were beheaded has conclusively been proven to be incorrect. There is no evidence that they were mummified.

The nature of the horse trappings suggests that their function was not unlike that of human funerary accoutrements. Egyptian style amulets such as *wedjat* eyes, *ankh* signs, heads of the goddess Hathor, and cowrie shells, common in the graves of Shabaka and Shebitka's horses, served to offer the wearer protection en route to the afterlife and to facilitate rebirth after death. All appear as grave goods of elite people, as well. Human mummies in Late Period Egypt and Napatan Nubia frequently wore nets of faience, as did the horses of Shabaka. One group of distinctive faience ball beads adorned with *ankh* and *wedjat* symbols, found in the graves of Shebitka's horses, which have a close parallel in Ku 53, the tomb of Piankhy's wife Tabiri (MFA 19.2617), although the queen's beads, at approximately 1.2 cm in diameter, are about half the size of the beads found with Shebitka's horses.

Clearly, additional research into the nature of Napatan horses and horsemanship is warranted, and further excavation and study will no doubt illuminate the importance of horses to the ancient Kushite civilization. At present, cemetery 200 at El–Kurru offers unique testimony regarding the significance of horses in royal ideology and demonstrates the strong bond between Napatan rulers and their horses, supplementing the evidence of textual and visual sources.

Addendum

Since this article went to press, articles by Gerard P. F. Broekman and Frédéric Payraudeau have convincingly argued that Shebitka's reign preceded that of Shabaka. This has significant impact on the interpretation of Cemetery 200. Because the two outermost rows of graves lacked inscriptions, it was assumed that the horses nearest Shabaka's were those of Piankhy, while those nearest Shebitka's belonged to Tanwetamani. In light of the new evidence it now seems more likely that the order should be reversed, so that graves 217-220 should be assigned to Piankhy's horses and graves 221-224 to Tanwetamanis's.

Bibliography

Albenda, P. 1982. Observations on Egyptians in Assyrian art. *Bulletin of the Egyptological Seminar* 4: 5–32.

Bökönyi, S. 1993. Two horse skeletons from the cemetery of Kurru, Northern Sudan. *Acta Archaeologica Academiae Scientiarum Hungaricae* 45: 301–316.

Brooklyn Museum of Art. 1978. *Africa in Antiquity. The Arts of Ancient Nubia and the Sudan.* Brooklyn Museum, Brooklyn.

Dalley, S. 1985. Foreign chariotry and cavalry in the armies of Tiglath–Pileser III and Sargon II. *Iraq* 47: 31–48.

Dunham, D. 1950. *The Royal Cemeteries of Kush.* Vol. I. *El Kurru.* Museum of Fine Arts, Boston and Harvard University Press, Cambridge/MA.

Dunham, D. 1957. *The Royal Cemeteries of Kush.* Vol. 4. *Royal Tombs at Meroe and Barkal.* Museum of Fine Arts, Boston.

Dunham, D. 1963. *The Royal Cemeteries of Kush.* Vol.5. *The West and South Cemeteries at Meroë.* Museum of Fine Arts, Boston.

Dunham, D. 1970. *The Barkal Temples.* Museum of Fine Arts, Boston.

Heidorn, L. 1994. Historical implications of the pottery from the earliest Tombs at El Kurru. *Journal of the American Research Center in Egypt* 31: 115–131.

Heidorn, L. 1997. The horses of Kush. *Journal of Near Eastern Studies* 56: 105–114.

Kendall, T. 1997. *Kerma and the Kingdom of Kush 2500–1500 B.C.* National Museum of African Art, Washington, DC.

Littauer, M. A. and J. H. Crouwel. 1979. *Wheeled Vehicles and Ridden Animals in the Ancient Near East.* Handbuch der Orientalistik Siebente Abteilung. Kunst und Archäologie. Erster Band. *Der Alte Vordere Orient.*

Zweiter Abschnitt. *Die Denkmäler*. Brill, Leiden and Köln.

Mallory–Greenough, L. 2005. The horse burials of Nubia. *Journal of the Society for the Study of Egyptian Antiquities* 32: 105–119.

Morkot, R. G. 2000. *The Black Pharaohs. Egypt's Nubian Rulers*. The Rubicon Press, London.

Raulwing, P. and J. Clutton–Brock. 2009. The Buhen Horse: Fifty years after its discovery (1958–2008). *Journal of Egyptian History* 2: 1–106.

Reisner, G. A. 1919. Discovery of the tombs of the Egyptian XXVth Dynasty at El–Kurruw in Dongola Province. *Sudan Notes and Records* 2: 237–254.

Rommelaere, C. 1991. *Les chevaux du nouvel empire égyptien: Origines, races, harnachement*. Connaissance de l'Égyptien Ancienne 3. Connaissance de l'Égyptien Ancienne, Bruxelles.

Schulman, A. R. 1957. Representations of horsemen and riding in the New Kingdom. *Journal of Near Eastern Studies* 16: 263–271.

Schrader, S. A., S. T. Smith, S. Olsen and M. Buzon. 2018. Symbolic equids and Kushite state formation: a horse burial at Tombos. *Antiquity* 92 362: 383–397.

Seele, K. C. 1974. University of Chicago Oriental Institute Nubian Expedition: Excavations between Abu Simbel and the Sudan Border. *Journal of Near Eastern Studies* 33:1–43.

11

Horses, Asses, Hybrids, and their Use as Revealed in the Ancient Rock Art of the Syro-Arabian Desert

Michael C.A. Macdonald

The desert rock art of Syria, Jordan and northern Saudi Arabia shows a variety of equids which are often identified in accompanying inscriptions. The conventions used, which emphasize the distinguishing characteristics of the subjects, enable us to identify horses, domestic donkeys, hemiones, and, probably, hybrids (mules and hinnies) and provide an insight into the wild and domesticated equids of these deserts in the last half millennium BCE and the first half millennium CE.

The rock drawings which I shall be discussing in this paper were carved by nomads in the deserts of what is today southern Syria, Jordan, and northern Saudi Arabia, in the second half of the first millennium BCE and the first four centuries CE. Much of this area consists of broken-up lava flows which cover the desert floor with millions of basalt stones and boulders. As a result of the chemicals in the basalt interacting with those in the atmosphere the exposed parts of these rocks are covered with a thin patina (or "desert varnish") which over millions of years has produced a shiny black surface (Figure 1). If this surface is pierced, the resulting mark shows the natural pumice-grey colour of the rock beneath, which looks almost white against the surrounding black. Over the millennia, this mark gradually patinates back to the black of its surroundings, at a speed dependant on the depth and width of the mark and the degree of exposure it suffers.

People in this area had, of course, been carving rock-drawings since time immemorial. The difference at this period—the last few centuries BCE and the first few centuries CE—was that, for the only time in their history, the nomads of this region were able to read and write (Macdonald 2009a: I; 2010), a skill which they practised with great enthusiasm on the rocks with which the desert was so conveniently supplied. This means that for the first time, they were able to "sign" their drawings and often identify the subjects. Although there are many which simply say "By so-and-so is the drawing", the majority say "By so-and-so is the oryx, the ibex, the gazelle, the horse, the donkey, etc." The script they used is called "Safaitic" and the language they spoke and wrote was a dialect of pre-Islamic Arabic (Al-Jallad 2015: 10–12).

There are a number of conventions in the rock art of this area and period (Macdonald 2005: 336–38; 2009b: 159, 161). For our purposes, the most important of these is the habit of emphasizing the most obvious distinguishing characteristics of the animals they portray. Thus, in the case of horses the relatively small head as compared to those of

Figure 1. The basalt desert in southern Syria near al-Namārah.

asses, the flowing mane, and the tail made up of individual hairs starting at the root (Figure 2a),[1] as opposed to asses which are shown with a heavier head, stiff upright manes and stalklike tails with a tassel at the end (Figure 2b–c).[2] At least some of these features tend to be accentuated even in otherwise crude, schematic, or frankly bizarre drawings (Figure 2e).

[1] Figure 2a is an excellent example of another very widespread convention in Arabian rock art associated with Ancient North Arabian inscriptions. This is the attempt to show even those elements which, in reality, would be invisible to an observer (see Macdonald 2009b: 159, 161; 2016). Thus both ears of the horse are shown, and both the rider's legs, as are the parts of the oryx's horns which are embedded in the horse's body. Note also that the three identifying features of the oryx—its long straight horns, the hump at the base of its neck, and its short tasselled tail—are present here as in almost all drawings of the animal, however crude. In Figure 2c, both the donkey's eyes are shown on the visible side of its head for the same reason.

[2] Note that, for clarity, on Figures 2a–c and on some other photographs, I have removed extraneous marks—and sometimes also the inscriptions—in order to show the drawings more clearly. In these cases, when the drawing is already published the reader can check the original in the publication. When the drawing is published here for the first time (as with Figure 2c), I have added a photograph of the original for comparison (Figure 2d). Readings of all the inscriptions on the previously unpublished stones can be found in the Appendix.

Figure 2a. A horseman (identified in the inscription as *h-frs*[1]) spearing a charging oryx, with the inscription (KRS 331) and blemishes removed to show the drawing more clearly. (Original photograph by Geraldine King, see KRS 331 in OCIANA).

Figure 2b. A drawing of a female domestic donkey referred to as *h-ʾtn* in the accompanying inscription (Winnett and Harding 1978: no. 3423+3424, which has been removed to show the drawing more clearly).

Figure 2c. An unpublished drawing of a wild, feral, or domestic ass, not mentioned in the accompanying inscription (removed for greater clarity). Note that both eyes are shown on the visible side of the face. (Photograph by William and Fidelity Lancaster).

Figure 2d. The original of the drawing shown in 2c.

Figure 2e. A bizarre unpublished drawing (not accompanied by an inscription) showing two equids which display the diagnostic features of small head, short ears and flowing tail suggesting (despite the legs) that they were intended to represent horses. (Photograph by Alison Betts).

After the camel, the horse and other equids used as mounts are by far the commonest subject in these rock drawings. Moreover, in contrast to the camel, the mounted equid is almost always shown in what might be described as "heroic" contexts, i.e. hunting large and/or dangerous beasts, raiding camels, single combat, or a full-scale battle (Figure 3a–c). This accords with the place the horse enjoyed a few centuries later in the pre-Islamic Arabic poetry of the sixth and early seventh centuries CE, and in the nomadic society of the Syro-

Figure 3a. Littmann 1943:no. 325 with the inscriptions (which do not identify the animals) and blemishes removed to show the drawing more clearly. It depicts a rider on an equid raiding a Bactrian camel and a dromedary (see Macdonald et al. 1996:468, fig. 18, 471–472 for the original). Note the plaited mane. For the use of the spear in raiding see Macdonald 1990. (Photograph by the author).

Figure 3b. A drawing showing a horseman (identified in the inscription, KRS 3288, as *h-frs*ʾ) hunting an oryx. On his costume see Crone 2008: 5–6. Note the flowing mane represented by dashes on the horse's neck and the apparently plaited tail. (Photograph by Geraldine King).

Figure 3c. An unpublished drawing showing two horsemen fighting two men on foot. The inscription (KWQ 19) mentions "the two horsemen" (*h-frs*ʾ*n*). Note that no attempt has been made to represent the mane or the hairs of the tail. (Photograph by Geraldine King).

Arabian desert recorded by travellers and ethnographers of the nineteenth and early twentieth centuries, where all these pursuits, if carried out successfully, were considered to bring honour to the individual and his tribe. The camel was the nomad's capital: the measure of his wealth and the source of his ability to be generous, one of the essential means by which honour is achieved in Bedouin society. It was also, of course, his primary beast of burden, his everyday mount, the source of the milk products which formed a major part of his diet, and the animal which allowed him to inhabit regions which were inaccessible to non-nomadic peoples, thus providing him with the other fundamental element of Bedouin society, individual independence. The camel was therefore clearly a symbol of the Bedouin's way-of-life and identity. While, it cannot be proved that the nomads at the turn of the era had the same relationship to the camel as the Bedouin, their rock art and inscriptions certainly give this impression.

The role of the horse in Bedouin life is very different. Although the accounts of the travellers and ethnographers reveal that Bedouin horses often had an unexpectedly good tolerance of poor grazing, long periods on dry fodder and limited amounts of water, and were capable of swift travel over surprisingly long distances, the horse was not a *requirement* of Bedouin life but a status symbol and a relatively pampered pet. While the camel represented the Bedouin's fundamental identity, the horse, because of its relative rarity, the fact that it was kept individually near the tent rather than in herds, and because it was the mount used in hunting, raiding, and battle, provided him with a heroic self-image. Once again, this is also the impression we get from the inscriptions and rock drawings of the ancient nomads of this area.

In the rock drawings, horses appear in different shapes and sizes, though they all share at least some of the characteristics of smallish heads, shortish ears, and tails made up of long, individual hairs starting from the root. Some appear to have very "deep" chests, a feature which may recur in later representations of horses in Iran, if this is not simply artistic licence (Figures 2a, 4a–d). Could this represent a recognizable physical feature of certain breeds of horse? In other cases, the chest appears to be shallower.

The mane is sometimes shown as flowing free (Figure 2a), but more often as trimmed or plaited (Figure 3a, 5a), so that, unless the artist has identified the animal as a horse, it can occasionally be difficult to tell whether he intended to show a plaited horse's mane or the short erect mane of an ass or mule. The forelock is also often shown (Figure 5a–b), and occasionally what looks like an ostrich feather is attached to the horse's forehead (Figures 2a, 3a, 5a), and sometimes the head of its rider (Figures 4a, 5b). A practice which nowadays has been transferred to the front of cars and trucks (Figure 5c)!

Figure 4a. A previously unpublished drawing, showing a "deep-chested" equid which is not identified in the inscriptions. (Photograph by William and Fidelity Lancaster).

Figure 4b. A drawing of a "deep-chested horse" beside KRS 2560 which identifies it as *h-frs*[1]. (Photograph by Geraldine King).

Of the equipment shown on the equids used for riding, bridles are rare, though see Figure 6a.[3] However, one (Figure 6b),[4] or usually two, reins are common (Figures 6c–d). The difference in number is almost certainly the result of inconsistent adherence to the convention of showing parts of the composition which would normally be invisible (as with both the rider's legs in Figures 2a, 6c, 6d, 7c), rather than a variation in riding practice.[5] Occasionally, a saddle seems to be represented (Figures 5a, 7a, 7b), though when there is a rider it is often difficult to tell. Sometimes there may be a saddle cloth (Figures 6c, 7c, 7d ?), though this could be simply be a gap in the fanciful decorative markings with which both horse and rider can be adorned (Figures 2e, 7d). Naturally, there are no stirrups.[6] Unlike drawings of camels, where girths are

Figure 4c. A tenth-century CE bronze model of an apparently "deep-chested" horse from Iran (The State Hermitage Museum, St. Petersburg. Photograph © The State Hermitage Museum. Photo by Leonard Kheifets).

Figure 4d. A ninth- to tenth-century CE bowl from Nishapur, Iran, showing an apparently "deep-chested" horse. (Museum für Islamische Kunst, Berlin, inv. no. I. 11/62).

[3] This, however, does not have an accompanying inscription and so cannot be dated securely to any particular period.

[4] Note that the rein is the zigzag line. It is not clear to me what the straight lines or the crossed lines represent but cf. those on Figure 3a.

[5] Note that in Figure 6b where only one rein is shown, only one of the rider's legs is shown. However, on Figure 6c one leg but two reins are shown, and 6d there are two legs and two reins.

[6] Although the Scythians may have had a rudimentary stirrup as early as the fourth century BCE (Clutton-Brock 1992: 76) and the Kushans in Central Asia seem to have had them in CE 100 (Azzaroli 1985: 158),

Figure 5a. A drawing of an equid with a plaited mane beside the Safaitic inscription KRS 3297, which does not identify the equid, (removed for greater clarity). Note also the forelock, the ostrich feather, and the saddle. (Photograph by Geraldine King).

Figure 5b. A drawing of a horse with a forelock (beside KRS 1865 which identifies it as *h-frs*[1], and has been removed for greater clarity). (Photograph by Geraldine King).

Figure 5c. A modern drawing of a truck with an ostrich feather on the front. (Photograph by William and Fidelity Lancaster).

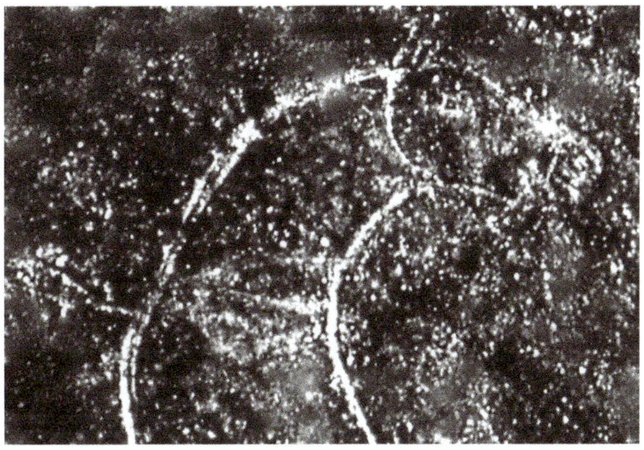

Figure 6a. Part of an unpublished drawing of an equid which may have a bridle. There is no inscription. From the surroundings of the site of Jawa in north-eastern Jordan. (Photograph by Geraldine King).

Figure 6b. A drawing showing an equid on which only one rein (the wavy line) is shown. The equid is not identified in the inscription. The whole face of the stone can be seen in Macdonald 2012:279, Fig. 9c. (Photograph by Alison Betts).

quite often included, I know of no examples of girths being shown in the drawings of mounted equids discovered so far.

There is one extremely interesting drawing (Figure 8), unfortunately known only from a nineteenth-century hand copy, which shows the breaking in or training of a filly. The accompanying inscription (*CIS* v 1186) states that the drawing is by the author and that he owns the filly (*mhrt*) that he is training.[7]

stirrups do not seem to have penetrated to the Middle East until the sixth or seventh centuries CE (Azzaroli 1985: 157–61).

[7] I am most grateful to Gail Brownrigg for telling me at the symposium that this drawing represents quite accurately the process by which young equids are still broken in today.

Figure 6c. Part of a drawing showing a horseman using two reins, beside KRS 177 which does not mention the equid and has been removed for greater clarity. (Photograph by Geraldine King).

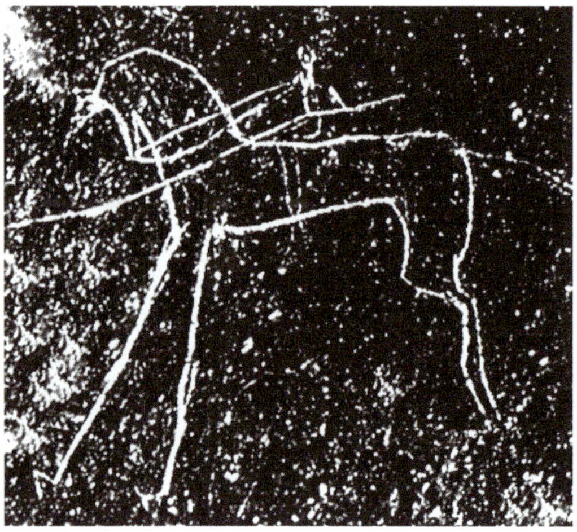

Figure 6d. Part of a drawing showing a horseman using two reins, beside KRS 2989 which does not mention the equid and has been removed for greater clarity. (Photograph by Geraldine King).

Figure 7a. Part of a drawing showing a saddled equid beside KRS 993 which does not mention the equid and has been removed for greater clarity. (Photograph by Geraldine King).

Figure 7b. A possible hinny, not identified in the accompanying inscriptions (KRS 1005–1006, removed for greater clarity). Note the saddle. (Photograph by Geraldine King).

Figure 7c. A drawing showing a hunter mounted on an equid identified in the inscription as *h-ʿr*, possibly on a saddle-cloth. Compare the plaited tail with that of the horse in Figure 3b. Clark 1984–85: no. B 1 (Amman Museum J.14537) with the inscription removed for greater clarity. (Photograph by the author).

Figure 7d. A possible hinny, not identified in the accompanying inscription (KRS 253, removed for greater clarity). (Photograph by Geraldine King).

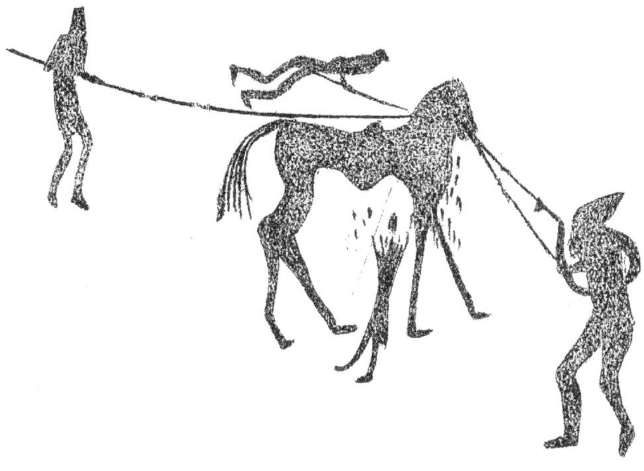

Figure 8. A 19th-century copy of a rock drawing from southern Syria showing the breaking-in of a foal. The drawing is accompanied by five Safaitic inscriptions (*CIS* v 1185–1189) and a drawing of a camel by another artist, all of which have been removed for greater clarity. One of the inscriptions (*CIS* v 1186) says that the author drew the filly (*mhrt*) which he was training.

Figure 9a. A drawing (Harding 1969) showing a horseman, and a woman with a lyre riding on a donkey or mule. The inscriptions, which do not identify the equids, have been removed for greater clarity.

Figure 9b. A drawing showing a woman with a lyre on an equid identified in the inscription as *h-ʿr* (*CIS* v 2839–2840, by permission of the Académie des Inscriptions et Belles-Lettres, Paris).

As might be expected, representations of the poor relation in the equid family, the domestic donkey, are much rarer than those of the horse. Indeed, it is remarkable that we have as many as we do. The terms used in Safaitic are *ḥmr* (cf. Arabic *ḥimār*) for the jackass and *ʾtn* (cf. Arabic *ʾatn*) for the jenny, terms which are common to most Semitic languages. In contrast to the drawings of the horse, those of asses usually show the animal in isolation rather than as part of a heroic scene, such as a hunt or a fight. But, as with horses, the distinguishing features such as ears, mane and tail, are often emphasized (Figures 2b, 2c).

A particularly fine and interesting drawing (Figure 9a) shows a man riding a horse[8] and a woman with a *kithara* riding a domestic donkey or a mule. The long ears, the apparently stalk-like tasselled tail, and possibly the position of the rider[9] might suggest that the animal is a donkey, even though it is shown as bigger than the horse. It is possible, however, that it could be a mule and that the tail (which should be like that of a horse) has been plaited but left loose at the end.[10] This would explain the size of the animal in relation to the horse, though relative proportions are rarely observed in these drawings. Another drawing, known only from a rough hand copy (Figure 9b), appears to show a similar scene and this time the animal is described as a *ʿr*, on which see below.

To the best of my knowledge, the only species of wild equid which was native to any part of the Arabian Peninsula in historical times was the Syrian wild ass,[11] *Equus hemippus*[12] (Figures 10a–c),[13] which was hunted to extinction in the early twentieth century.[14] Bones thought to be of *Equus africanus* have been found at some prehistoric sites on the Arabian side of the Persian Gulf (Uerpmann 1991: 20, 29 f.; Uerpmann *et al.* 2000: 230 f.),[15] but I know of no other evidence for its early presence in the Peninsula. *Equus hemippus* appears to have been relatively common in the northern areas of the Peninsula, in what is now southern Syria, north-eastern Jordan, and Israel, but, to the best of

[8] The horseman is actually wielding a spear represented by a faint line running diagonally to the feet of the standing woman. A disproportionately large sword was added later and has been removed in Figure 9a.

[9] The male rider in the drawing sits forward, as close as possible to the withers of the horse. The female rider sits much further back, as one would on a donkey. However, since her legs are not shown she may be in some form of litter, in which case that would account for her position.

[10] See Harding 1969 and Macdonald 2012: 278–80 for a description of the drawing as a whole.

[11] I use this term rather than the more common "onager" following Mary Littauer and Joost Crouwel's statement that "the more usual 'onager' is correctly applicable only to the Iranian subspecies" (2002: 396).

[12] Formerly, *Equus hemionus hemippus*, but see the reclassification in Groves and Grubb 2011: 15. I am most grateful to Gertrud and Helmut Denzau for this reference.

[13] See also the painting published by M. Milne-Edwards in 1869 which is reproduced in Denzau and Denzau 1999: 137.

[14] I understand that the wild asses in conservation areas in Israel are not *Equus hemippus* but Iranian Onagers (*Equus hemionus onager* Boddaert, 1785) and Turkmenistani kulans (*Equus hemionus kulan* Groves & Mazák, 1967) see Groves and Grubb 2011: 15. I am most grateful to Gertrud and Helmut Denzau for this information.

[15] Uerpmann (1991: 16 f., §2.6) reports bones of *Equus africanus* at the Natufian site at Ra's al-Naqab in southern Jordan, "although the onager seems to be present as well". "Dating from the Natufian, they must belong to wild equids" (*ibid.* p. 18).

Abb. 1. Syrischer Halbesel in einer Galoppwendung. Das Bild zeigt gut die kleinen Ohren, großen Nüstern und die prachtvolle Rücken- und Schultermuskulatur.
Keller phot. 1915

Figure 10a. What is thought to be the last surviving specimen of *Equus hemippus*, which died in the Tiergarten Wien-Schönbrunn in 1929. Photographed by Keller in 1915. (From Antonius 1929).

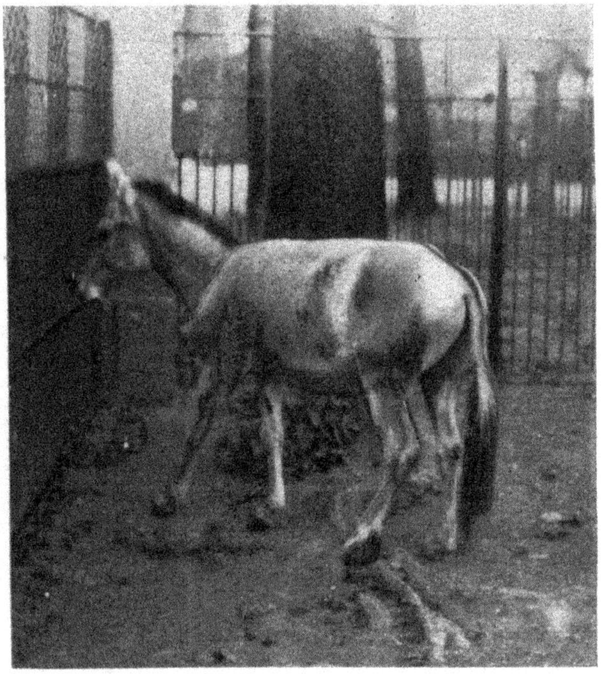

Abb. 2. Syrischer Halbesel. Spiegel und Aalstrich deutlich sichtbar.
Kaiser phot. 1924

Figure 10b. The same animal photographed by Kaiser in 1924. (From Antonius 1929).

my knowledge, neither clear rock-drawings nor excavated bones of this animal have been discovered further south, in central and southern Arabia. On the other hand, there are mediaeval and early modern reports of it in Nejd and the Ḥijāz (Denzau and Denzau 1999: 135 f.), and the wild ass

Figure 10c. Stampeding wild asses (from a late seventh-century CE wall painting at Quṣayr ʿAmrah, north-eastern Jordan. Photograph by M.C.A. Macdonald).

is used as an image in the pre-Islamic Arabic poetry, and is discussed by writers of the Islamic period, which suggests that it was known in other areas of the Peninsula (Denzau and Denzau 1999: 194 f.).

Equus hemippus is shown relatively seldom in these drawings and is called a *ʿrd* (feminine *ʿrdt*, plural *ʾʿrd*)[16] in the accompanying inscriptions. There are several drawings of individual wild asses and one which shows a group of five adults and one foal being hunted by men on foot with bows (Figure 11a).[17] The heads are relatively large, the manes are erect, but the tails are simply represented by straight lines.[18] Here the wild asses are identified as *ʾʿrd*.

Another scene (Figure 11b–c), which alas is only labelled "the picture", may show wild asses being attacked or driven by men with lances mounted on equids. However, the riderless animals are similar in size to those carrying riders and the drawing is sufficiently crude to make it difficult to be sure what was intended in the representation of manes and tails. It may therefore be that the scene was intended to show riders driving off horses, or hybrids, in a raid, rather than hunting wild asses.[19] On the other hand, Figure 11d clearly shows a horseman, with a helmet and

[16] For *ʿrd* see Ababneh 2005: nos 327, 944, 1032; for *ʿrdt* see *ibid.* no. 689. The plural *ʾʿrd* occurs in the inscription accompanying the previously unpublished photograph on Figure 11a.

[17] I am most grateful to Gertrud and Helmut Denzau for identifying the foal which I had previously mistaken for a hound! There is possibly another drawing, though the animals are not identified in the accompanying inscription. This is in Harding 1953: no. 73, face B. See Macdonald 2005: 335–37.

[18] As can be seen from Figure 10b, the first 30 or 40% of the Syrian wild ass's tail seems to have been "stalk-like", with the rest being much more like that of a horse. This is also suggested in the wall-painting shown in Figure 10c. It is of course impossible to know whether the much more donkey-like tail shown in Milne-Edwards' drawing (Denzau and Denzau 1999: 137) is an accurate representation of a variation. Be that as it may, the tails in the rock drawing shown on Figure 11a are unfortunately simply straight lines with no attempt at accurate representation. The one exception is the animal on the far left which has the same mane and ears as the others, and is identified as a foal by Gertrud and Helmut Denzau (see note 18), but bizarrely appears to have a curling tail! I have no explanation for this.

[19] Compare the clear example of hunting wild asses—though this time by men on foot with nets—in the late seventh-century CE wall-paintings

Figure 11a. A previously unpublished inscription and drawing from north-eastern Jordan showing men on foot with bows and arrows hunting wild asses (called *ʾʿrd* in the inscription). (Photograph by Geraldine King, with blemishes removed).

Figure 11b. A previously unpublished drawing showing men mounted on equids driving, hunting or raiding other equids none of which are identified in the inscription which has been removed for clarity. (Photograph by Alison Betts).

Figure 11c. The original of Figure 11b.

Figure 11d. A rock drawing from southern Syria showing a horseman wielding a short throwing spear (Arabic *ramḥ*) having thrown another spear at a wild ass. The decoration on both the horse and the wild ass is purely the fantasy of the artist and is not intended to represent realistic features. (Photograph by the author).

a short throwing spear (Arabic *ramḥ*) hunting what is presumably a wild ass, having already thrown one spear at it.

So far, we have examined drawings of horses (Safaitic *frs*^{1},[20] feminine *frs*^{1}*t*), domestic jack-asses (Safaitic *ḥmr*) and jennies (Safaitic *ʾtn*) and the Syrian wild ass (Safaitic *ʿrd*, feminine *ʿrdt*, plural *ʾʿrd*). But, there are also drawings of equids which are identified as *ʿr* or *ʿyr* in the accompanying inscriptions.[21]

This word presents something of a problem. In later Arabic, *ʿayr*, seems to have been used for the wild ass in the pre-Islamic poetry.[22] However, the Classical Arabic lexicographers say that it could be used for both the domestic and the wild ass. On the other hand, in these drawings—which are of course several centuries earlier—most of the animals which are called *ʿr* appear to have a mixture of horse and ass features, which would suggest that they are hybrids.

The whole subject of the types of hybrids and the extent of their use in Arabia in late antiquity and the Islamic period is very under-researched, as is the origin and exact significance of the words for them in the different Semitic languages.[23] As a working hypothesis, I would tentatively

at Quṣayr ʿAmrah in north-eastern Jordan (Figure 10c), not far from the location of these rock-drawings.

[20] The word for horse *frs*^{1} (cf. Arabic *faras*^{1}) and horseman *frs*^{1} (cf. Arabic *fāris*) are indistinguishable in the orthography of Safaitic.

[21] So far, we have no examples of either a feminine or a plural of *ʿ(y)r*, even though the sexes are clearly indicated in the drawings some of which show more than one *ʿ(y)r*.

[22] See for example the discussion in Wagner 1987: 106, 108 f.

[23] Although, in the Islamic period, there are Hadith (Traditions of the Prophet) forbidding the production of hybrids, these were generally regarded as forged and were widely ignored. Indeed it is reported that Prophet himself possessed a mule called *Dildulah*, a gift he had received from Egypt. It is interesting, however, that although hybrids aroused a certain amount of interest in academic circles in the Islamic period, there was only a very limited vocabulary in Arabic for describing them, with apparently no distinction being made between mules and hinnies, both being called *baġl* (feminine *baġlah*, plural *ʾabġāl* and *biġāl*). Since

suggest that the breeding of hybrids among the Bedouin might have died out some time before the surviving pre-Islamic Arabic poetry was composed, or at least by the time it was written down and analysed in the eighth and ninth centuries CE. But while the breeding of these animals had died out, what may have been the word for them, *ʿayr*, remained in the vocabulary of the nomads but came to be applied to one of the parents, the ass, whether wild or domestic. This process of retaining a word for something which has ceased to exist and applying it to something else, is a common feature of historical linguistics.[24]

It is certainly possible that the nomads who produced these inscriptions and drawings bred hybrids from horses and asses, though it is not clear whether the asses were wild, or domestic, or both. Apart from the possible interpretation of Figure 11b mentioned above, we have no drawings as yet showing the capture of wild asses, and it may be that they were only hunted for their meat, while hybrids were bred from horses and domestic asses.[25]

It is well-known that hybrids in any species while having the disadvantage that they are usually sterile, have the great advantage that they tend to be larger, stronger, and often capable of greater endurance, than either parent (Macdonald 2009b: n. 21 and references there). These advantages in an equid would clearly be very attractive to nomads, producing mounts with the speed and agility of the horse but with more power and endurance. In these drawings, the animals which I have tentatively identified as hybrids are shown in the same contexts as horses which suggests that the two were considered to be of equal status.

Hybrids combine the features of both parents, but in different ways. A mule (Figure 12a) has a jackass as sire and a horse mare as dam. Its head tends to be larger than that of a horse and more like that of a donkey, its ears are longer than a horse's but usually slightly shorter than a donkey's and it has the horse's long thick tail, though the hairs may not start right from the root. Its height can vary and it can be taller than either parent.

Figure 12a. "Man manipulates harness on mule". (Photograph by Ned Richards http://animalphotos.info/a/2008/01/05/man-manipulates-harness-on-mule/).

Figure 12b. A hinny. (Photograph by B. Hutchins, published in Clutton-Brock 1999: 45, and reproduced by kind permission of Natural History Museum Publications).

Figure 12c. A hinny with its jenny dam behind it. (Photograph from http://www.maulesel.info/what_is_a_hinny.html).

Hinnies (Figure 12b–c) have a horse stallion as sire and a jenny as dam and seem to vary in appearance much more than mules, some indeed can look very like horses.[26] The size and shape of the ears vary considerably and the mane is something between that of a horse and that of a donkey.

the nineteenth century this word, which does not occur in the Ancient North Arabian inscriptions, but is found once in the Qurʾān (16: 8), has been regarded as a loan-word from Ethiopic. However, there are certain linguistic problems with this derivation and more recently some scholars have suggested that the loan went the other way. The only thing that seems reasonably certain is that by the fifth century CE hybrids seem no longer to have been a feature of nomadic life, at least in those parts of Arabia which produced the pre-Islamic poetry, though they may have been used in the towns.

[24] Indeed a very similar process had occurred in Akkadian where, before the introduction of the horse in the mid-third millennium BCE, wild asses were captured to produce donkey x wild ass hybrids for which the name was anše-BAR.AN. With the introduction of the horse, the use of wild asses was phased out and the name was transferred to horse x donkey hybrids. Postgate 1986: 194–206, cited in Clutton-Brock 1992: 89 f.

[25] Note that in those drawings showing animals labelled *ʿrd* (feminine *ʿrdt*, plural *ʾʿrd*) the humans are armed with bows and arrows, rather than lassos or nets. See also the discussion in relation to the late Assyrian reliefs which show wild asses being shot in one context and captured (for purposes unknown) in another (e.g. Clutton-Brock 1992: 87–90; Macdonald 2009b: n.20).

[26] See www.lovelongears.com/about_mules.html and http: //www.maulesel.info/what_is_a_hinny.html both consulted 1st August 2012.

As with the mule, the tail is normally horse-like[27] but with a short length of "stalk" before the hairs begin.

The animals described as *'r* are shown in several different contexts (Figures 13–17). In Figures 13a–d we see the *'r* used as a riding animal with riders armed with spears and swords in contexts in which, in other drawings, we find the horse (*frs'*). The animal shown on Figure 13a looks to all intents and purposes like a horse, with small head, erect ears, what could be a flowing mane,[28] and a tail in which the hairs start from just below the root. The rider appears to be positioned over the withers which would again suggest that the animal is a horse. Thus we seem to have a thoroughly horse-like animal described as an *'r*.[29]

In Figure 13b, on the other hand, we find an animal which is very similar, except that it has an erect mane. The head is

Figure 13a. A horse-like *'r*. KRS 3199–3201. (Photograph by Geraldine King).

Figure 13b. A previously unpublished drawing of a rider with a long lance mounted on an equid described as a *'r* in the accompanying inscription (KWQ 161). (Photograph by Geraldine King).

Figure 13c. An equid described as a *'r* in the accompanying inscription (KRS 3134). (Photograph by Geraldine King).

small, the hairs of the tail, though represented differently, start from just below the root, and the rider is seated over the withers. What then are we to make of the mane?

Figure 13c shows an animal with a tiny head, two long ears, an erect mane,[30] and what looks like a stalk-like tail with a tassel starting half-way down its length. The rider is sitting back from the withers, though not really in the position one would adopt on an ass, and appears to be wielding a spear.

Another combination of features is found in Figure 13d, where the animal has a largish head, longish erect ears, possibly an erect mane, and a tail on which the hairs seem to start at the root. As in Figures 13a and b the rider is over the withers. Although the animal is clearly a male, there is what may be intended to represent a foal below it, though it looks more like a small dog. This has erect ears, no mane, and a stalk-like tail without a tassel.[31]

Animals described as *'r* also occur as mounts in scenes of hunting, raiding, and fighting (Figure 14a-e). Thus, in Figure 14a, the rider holds two reins in his right hand and two hunting spears in his left, and appears to be pursing an oryx[32] and what look like two equids.[33] He sits over the withers of an equid with a strangely shaped head which is possibly more horse-like than ass-like, with no apparent

[27] I am most grateful to Sandi Olsen for pointing this out to me. See the photograph on http: //www.maulesel.info/what_is_a_hinny.html consulted 1st August 2012.

[28] This is indicated by closely spaced vertical scratches over the cross-hatching on the neck.

[29] Although it appears to have fairly long legs, if the height of the rider's feet above "the ground" are anything to go by, the relative proportions of the horse and rider might suggest that it is not very tall. But on this see below.

[30] This can be seen when the photographed is enlarged.

[31] There are other drawings within this category but either they are not sufficiently detailed to help in this discussion (e.g. Ryckmans 1955: 8, no. 1, Clark 1979: no. 59) or they are only known from copies (e.g., Ababneh 2005: nos 47, 48, 427; *CIS* v: nos 2839–2840 (Dunand 294a–b)).

[32] The animal on the far left. Note the hump behind the base of the neck and the long straight horns.

[33] The one below the rider, which is more skillfully drawn, has a small, horse-like, head and ears pointing forward. The tail is represented by a single line which is of no help in identifying the species.

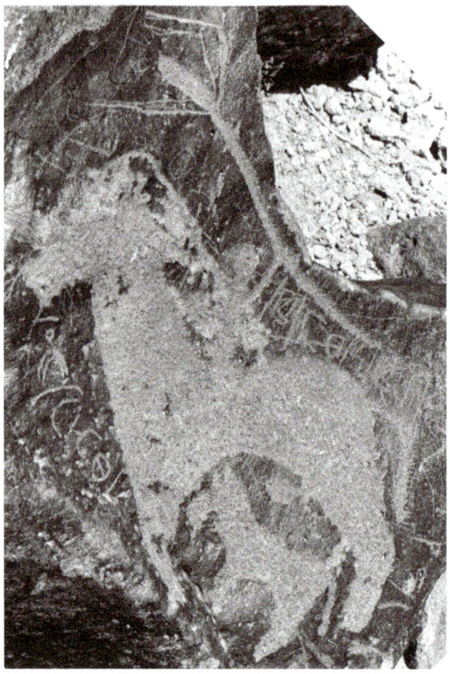

Figure 13d. An equid described as a *'r* in the accompanying inscription (Winnett and Harding 1978: no. 3767).

Figure 14a. A hunting scene with a rider on what is described as a *'r* in the accompanying inscription (KRS 3204) pursuing an oryx and possibly two wild asses. (Photograph by Geraldine King).

ears or mane, a tail in which the hairs start from the root, and long legs.

In Figure 14b, the rider is sitting well back, holding in his right hand a rein (or reins) represented by a zigzag line, and in his left a long lance raised above his head, and is pursing an oryx. The equid's head consists of two parallel vertical lines which do not seem to be joined at the bottom and is useless for the purposes of identification. Unfortunately, the upper part of the tail is badly damaged and only the long hairs of the lower half remain. The ears,

Figure 14b. A rider on what is described as a *'r* in the accompanying inscription (KRS 3429) pursuing an oryx. (Photograph by Geraldine King).

Figure 14c Two riders on what are described as *'r* pursuing a camel in what is possibly a camel raid. (British Museum ME 122182, see Ryckmans 1951: 84 f. Photograph by courtesy of the Trustees of the British Museum).

if they existed, would be under a patch of damage and no mane is shown. It is therefore only the position of the rider which *may* give a clue as to the identity of the animal.

A third hunting scene (Figure 7c) again shows a rider hunting an oryx using a long lance. Here, the equid's head is fairly non-descript but a magnificent erect mane is shown.[34] Unfortunately, the lower half of the tail is damaged and it is difficult to tell whether the top half is stalk-like or plaited. The legs are shown as very long in relation to those of the rider, who is sitting back from the withers.

Figure 14c may show a raiding scene.[35] The centre of the picture is dominated by a rider on an exaggeratedly long-legged equid. He is seated over the withers, holding a rein represented by a zigzag line in his right hand, and a stick or

[34] It is difficult to believe that a plaited mane could have this shape.

[35] It is not absolutely certain that this is a raid, since usually the rider is shown touching the camel with a long lance (see Macdonald 1990).

Figure 14d. A previously unpublished drawing of rider on what is described as a *ʿr* with bow arrow and quiver, fighting a man on foot who is similarly armed with the addition of a shield. (Photograph by Geraldine King).

Figure 14e. A previously unpublished drawing showing an animal described as a horse (*frs*^{1}, top left) and one described as a *ʿr* (bottom left). (Photograph by Geraldine King).

short spear in his left.[36] The head of the animal is perhaps larger than one would expect in a horse and the mane is either erect or plaited. The tail is covered with long hairs starting a short way below the root. The small rider below is on a similar equid, though here the ears are shown and the mane is clearly erect, while the tail is probably intended to show hairs starting just below the root.[37]

Scenes of fighting are represented in Figures 14d and 14e. In the former, a mounted archer with bow, arrow and quiver confronts a similarly armed archer who is on foot and has the small round shield familiar in these drawings and still used by the Bedouin in the early twentieth century. The equid has an ass-like head, but with relatively small ears, an erect mane, and a tail in which the long hairs start after a short stalk. The rider is seated over the withers and is very small in relation to his mount.

Figure 14e is particularly interesting because it shows two equids by two different artists one of whom (at the top) identifies his drawing as *h-frs*^{1} ("the horse") and the other (at the bottom) calls his *h-ʿr*. Unfortunately, neither artist was very skilful and the animal identified as a horse looks more like a bovid. The drawing identified as *h-ʿr*, has a head which looks ass-like (though this is entirely subjective), two longish ears, and a tail represented unhelpfully by a single line.[38]

There are also representations of *ʿr* by themselves, and these seem to have predominantly ass-like features. Thus, on 15a and 15b, the animal has a large head, long ears, an erect mane and stalk-like tail with a tassel. On 15c, it has the large head, long ears, and an erect mane, but there does not seem to be a tassel on the end of the tail, while on 15d it has the ears, the mane[39] and the stalk-like tasselled tail, and on 15e it has the head, the ears and the tail. On other

Figure 15a. A previously unpublished drawing of an equid described as a *ʿr*. (Photograph by the author).

Figure 15b. A previously unpublished drawing of an equid described as a *ʿr*. (Photograph by the author).

[36] This assumes that the rider is looking out of the picture, as is normal in these drawings, rather than with his back to the viewer.

[37] Although at first sight it looks as though the artist intended to show a stalk-like tail, if this were so, one would have expected him to show the tassel in a drawing as detailed as this.

[38] This line sticks out horizontally and then descends diagonally crossing the head of an ostrich. The drawing of a camel being attacked from behind by a hyena (?) is by yet another artist who simply calls it *h-ḥyt* "the animals". It is not clear who was responsible for the drawings of the six ostriches.

[39] This can be seen when the photograph is enlarged.

Figure 15c. A previously unpublished drawing of an equid described as a *ʿr*. (Photograph reproduced by kind permission of by Gertrud and Helmut Denzau).

Figure 15d. A previously unpublished drawing of an equid described as a *ʿr*. (Photograph by Geraldine King).

Figure 15e. A previously unpublished drawing of an equid described as a *ʿyr*. (Photograph by the author).

drawings there are only two distinguishing features such as the ears and the tail (Figure 15f and 15g) or head and ears (Figure 15h).

Figure 15f. A previously unpublished drawing of an equid described as a *ʿr*. (Photograph by Alison Betts).

Figure 15g. A previously unpublished drawing of an equid described as a *ʿr*. (Photograph by the author).

Figure 15h. A previously unpublished photograph of Littmann 1943: nos 513–514, showing a drawing of an ass-like equid described as a *ʿr*. (Photograph by the author).

There are also drawings of animals which are referred to as *ʿyr* (Figures 15e and 16a–d). One might assume that this represents a plural (cf. later Arabic *ʿiyār* and *ʿuyūr*) but while Figures 16a and c show more than one animal, there is only one on Figures 15e and 16d. *ʿyr* could represent a diminutive (cf. later Arabic *ʿuyayr*/*ʿiyayr*), but none of these equids seems significantly different from those referred to as *ʿr* and, on present evidence, one can only suppose that the variations in spelling represent either dialectal differences, or slightly different pronunciations, one as a pure diphthong (* *ʿayr*, or a monophthong e.g. * *ʿēr*, which would not show up in Safaitic orthography) and the other where the /y/ has become vocalized (e.g. *ʿayər).

Figure 16a. A previously unpublished drawing of men fighting and equids described as *ʿyr*. Blemishes and the inscription have been removed for greater clarity. (Photograph reproduced by kind permission of Gertrud and Helmut Denzau).

Figure 16b. The original of Figure 16a.

Figure 16c. A copy of a drawing of equids described as *ʿyr* in the accompanying inscription. (Ababneh 2006:no. 96, by permission of Shaker Verlag GmbH, Aachen).

Figure 16d. A previously unpublished drawing of an equid described as *ʿyr*.

There are also representations of animals called *ʿr* apparently being hunted. This should be distinguished from the hunting of the Syrian wild ass (*Equus hemippus*, see Figure 11a) which is called *ʿrd* in the texts (see above). In Figure 17a, an archer on foot is about to shoot an equid with long erect ears, an erect mane, and a tail on which the long hairs start shortly below the root. On 17b–c an archer on foot is shooting at a *ʿr* with very long ears and a stalk-like tail with a small tassel.

Figure 17d is particularly interesting since it shows an animal with long ears which is described as a *ʿr* being attacked by a hound (?) and two men with bows and arrows, within an enclosure which looks remarkably like a "desert kite".[40] These enclosures, most if not all of which are prehistoric, are generally thought to have been built originally for hunting herds of gazelle, but later to have been used for other purposes. Occasional rock-drawings accompanied by Safaitic inscriptions show them with a variety of animals including ostrich.[41] This would seem to make it clear that animals called *ʿr* were hunted as well as being used as mounts.

[40] See Macdonald 2005 and references there.

[41] See Macdonald 2005: 333 and Figure 1.

Figure 17a. A previously unpublished drawing of an archer on foot shooting at an equid described as a *ʿr*. (Photograph reproduced by kind permission of Gertrud and Helmut Denzau).

Figure 17b. A previously unpublished drawing of an archer on foot confronting an equid described as a *ʿr*. The blemishes and the inscription have been removed for greater clarity. (Photograph by the author).

Figure 17c. The original of Figure 17b.

What are we to make of all this? From the drawings, it would seem that the Safaitic word *ʿr* (and occasionally *ʿyr*) can refer to both a horse-like and an ass-like equid, and

Figure 17d. An unpublished drawing showing an animal described as a *ʿr* being attacked within an enclosure which may be a "desert kite".

animals with features of both. Moreover, they are found in the service of man as well as being attacked by him.

The simplest explanation would be to suggest that those who carved these inscriptions and drawings used the word(s) *ʿr*/*ʿyr* for any kind of equid and were careless in the features they gave them in their drawings. This seems to me most unlikely. The nomads who carved these drawings had separate words for horse stallion and mare, for jack-ass and jenny, and for male and female wild asses. They also show a detailed knowledge of their domestic animals and a familiarity with the features of the commoner large wildlife, such as gazelle, oryx, ibex, etc., though slightly less expertise in representing lions and hyenas. It is surely inherently unlikely that they would make mistakes about the basic distinguishing features of the equids they were drawing and use a catch-all term such as "equid" to describe them.

If I am correct in this assumption, then we need to explain the features of the animals shown in the drawings labelled *h-ʿ(y)r*. As a working hypothesis, I would suggest that in Safaitic *ʿ(y)r* was used of mules and hinnies of either sex.[42] Thus, if we compare the *ʿr* shown on Figure 13a which is remarkably horse-like with the photographs of hinnies on Figures 12b and c one can see that, apart from the exaggeratedly long legs, the drawing could well represent a hinny. As noted above, the relative size of the rider and the animal suggest that, despite the length of its legs, it was closer in size to the hinny in the photograph. The horse-like head, ears and tail all tally with this identification.

[42] So far, we have no examples of either a feminine or a plural of *ʿ(y)r*, even though the sexes are clearly indicated in the drawings.

Figures 13b and c might be taken to represent a mule with erect manes and horse-like tails with the hairs starting a little below the root. Note that in 13c the relative proportions of the rider and his mount suggest that the latter is smaller than it looks. The drawings shown on Figures 14a–c also show this mixture of horse and ass features which suggest a hybrid.

On the other hand, the *'(y)r* shown unmounted in Figures 15e and 16a–d look distinctly more ass-like, as do those which are being hunted shown on Figure 17a–d. Certainly, the most detailed of these (17a) displays the long ears and erect mane but horse-like tail of a mule and I would suggest that they all represent hybrids of both types and both sexes (though males predominate). The reason for the difference between those shown mounted and unmounted may be due to the "heroic" contexts in which mounted equids are shown: hunting, raiding, and fighting. These contexts may have influenced the artists unconsciously to increase the size of the animals and give them more horse-like proportions. As I said earlier, there does not seem to have been any difference in status between the horse and the *'(y)r* and the artists seem to have been happy to identify their mounts as the latter and to give them the features of a hybrid. The context encouraged them to make the animals look imposing, something generally achieved by a disproportionate increase in the length of the legs, something which they also did in drawings of horses (e.g. Figures 3a, 5a and b, etc.).

My suggestion that *'(y)r* refers to a "hybrid" (i.e. either mule or hinny) is, of course, purely a working hypothesis and I should be delighted if it can be replaced by a better one! If I am correct, one has to ask why do we find drawings of people apparently attacking them. Is it possible that these may be hybrids which have escaped and that the scenes are intended to show their recapture, or that they are hybrids belonging to an enemy? Alas, it is impossible to say.

Thus, while these drawings provide a unique record of the equids used by Arabian nomads at the turn of the era, much of the evidence they provide is ambiguous and is capable of multiple interpretations. A thorough study of all the available material by someone with a good knowledge of equids together with someone with expertise on the inscriptions could be extremely useful. If this brief paper should chance to stimulate interest in such an endeavour, it will have fulfilled its purpose.

Acknowledgements

I am most grateful to Gertrud and Helmut Denzau for reading an earlier draft of this paper and teaching me a great deal about living equids. Their comments, corrections, and suggestions have greatly enriched this work, but, naturally, any errors that remain are entirely my responsibility. I would also like to thank Sandi Olsen and Gail Brownrigg for their very helpful comments which have saved me from many errors. Once again, however, any that remain are, of course, entirely my fault.

Abbreviations

CIS v Safaitic inscriptions published in the *Corpus Inscriptionum Semiticarum. Pars V. Inscriptiones Saracenicas continens, Tomus 1. Inscriptiones Safaiticae*. Imprimerie nationale, Paris, 1950–51.

KRS Safaitic inscriptions recorded by Geraldine King on the Basalt Desert Rescue Survey. Published in the Online Corpus of the Inscriptions of Ancient North Arabia (OCIANA) database at http://krc.orient.ox.ac.uk/ociana/index.php.

KWQ Unpublished Safaitic inscriptions and drawings recorded by the late Geraldine King in Wādī Qaṭṭafī, north-eastern Jordan, and Published in the Online Corpus of the Inscriptions of Ancient North Arabia (OCIANA) database at http://krc.orient.ox.ac.uk/ociana/index.php.

Ms Safaitic inscriptions and drawings photographed by a number of different people and edited by the author. Published in the Online Corpus of the Inscriptions of Ancient North Arabia (OCIANA) database at http://krc.orient.ox.ac.uk/ociana/index.php.

Bibliography

Ababneh, M. I. 2005. *Neue safaitische Inschriften und deren bildliche Darstellung*. Semitica et Semitohamitica Berolinensia 6, Shaker, Aachen.

Alexander, D. (ed.) 1996. *Furusiyya*. 2 vols. King Abdulaziz Public Library, Riyadh.

Azzaroli, A. 1985. *An early history of horsemanship*. Brill, Leiden.

Antonius, O. 1928. Beobachtungen an Einhufern in Wien-Schönbrunn. I. Der Syrische Halbesel. *Der Zoologische Garten* N.F. 1: 19–25.

Clark, V. A. 1979[83]. *A Study of New Safaitic Inscriptions from Jordan*. [Doctoral thesis, Department of Middle Eastern Studies, University of Melbourne]. University Microfilms International, Ann Arbor, MI.

Clark, V. A. 1984–85. New Safaitic Inscriptions from Sakaka and Azraq. *Abr-Nahrain* 23: 14–21.

Clutton-Brock, J. 1992. *Horse Power. A History of the Horse and the Donkey in Human Societies*. Natural History Museum Publications, London.

Crone, P. 2008. "Barefoot and Naked": What did the Bedouin of the Arab conquests look like? *Muqarnas* 25: 1–10.

Denzau, G. and Denzau, H. 1999. *Wildesel*. Thorbecke SPECIES, 3. Stuttgart: Thorbecke.

Groves, C. and Grubb, P. 2011. *Ungulate Taxonomy*. Johns Hopkins University Press, Baltimore.

Harding, G. L. 1969. A Safaitic drawing and text. *Levant* 1: 68–72.

Al-Jallad, A. M. 2015. *An Outline of the Grammar of the Safaitic Inscriptions*. Leiden: Brill.

Littmann, E. 1943. *Safaïtic Inscriptions*. SYRIA Publications of the Princeton University Archaeological Expeditions to Syria in 1904–1905 and 1909. Division IV. Section C. Brill, Leiden.

Littauer, M. A. and J. H. Crouwel. 2002. Terracotta models as evidence for wheeled vehicles with tilts in the ancient Near East. In *Selected Writings on Chariots, other Early Vehicles, Riding and Harness*, by M. A. Littauer and J. H. Crouwel, edited by P. Raulwing, 380–402. Culture and History of the Ancient Near East 6. Brill, Leiden.

Macdonald, M. C. A. 1989. Cursive Safaitic inscriptions? A preliminary investigation. In *Arabian Studies in Honour of Mahmoud Ghul: Symposium at Yarmouk University December 8–11, 1984*, edited by M. M. Ibrahim, 62–81. Yarmouk University Publications: Institute of Archaeology and Anthropology Series 2. Harrassowitz, Wiesbaden.

Macdonald, M. C. A. 1990. Camel hunting or camel raiding? *Arabian archaeology and epigraphy* 1: 24–28.

Macdonald, M. C. A. 2005. Of rock-art, "desert kites" and *meṣāyid*. In *Arabia Vitalis: Arabskiĭ Vostok, islam, drevnyaya Araviya: Sbornik Naychnykh stateĭ, posvyashchennyĭ 60-letiyu V. V. Naumkina*, edited by A. V. Sedov and I. M. Smulyanskaya, 332–345. Rossiĭskaya Akademiya Nauk, Moscow.

Macdonald, M. C. A. 2009a. *Literacy and Identity in Pre-Islamic Arabia*. Variorum Collected Studies 906, Ashgate, Farnham.

Macdonald, M. C. A. 2009b. Wheels in a land of camels: another look at the chariot in Arabia. *Arabian archaeology and epigraphy* 20: 156–184.

Macdonald, M. C. A. 2010. Ancient Arabia and the written word. In *The development of Arabic as a written language*, edited by M. C. A. Macdonald, 5–28. Supplement to the Proceedings of the Seminar for Arabian Studies volume, 40, Archaeopress, Oxford.

Macdonald, M. C. A. 2012. Goddesses, dancing girls or cheerleaders? Perceptions of the divine and the female form in the rock art of pre-Islamic North Arabia. In *Dieux et déesses d'Arabie : Images et représentations*, edited by I. Sachet and Ch. J. Robin, 261–297. Orient et Méditerranée, 7, De Boccard, Paris.

Macdonald, M. C. A. 2016. Three dimensions in two: Conventions and experiment in the rock art of ancient North Arabia. In *The Archaeology of North Arabia: Oases and landscapes*, edited by M. Luciani, 317–335. Oriental and European Archaeology, 4, Austrian Academy of Sciences Press, Vienna.

Macdonald, M. C. A., al-Muʾazzin, M. and Nehmé, L. 1996. Les inscriptions safaïtiques de Syrie, 140 ans après leur découverte. *Comptes Rendus de l'Académie des Inscriptions et Belles-Lettres*, 435–492.

Postgate, J. N. 1986. The equids of Sumer again. In *Equids in the Ancient World*, volume II, edited by R. H. Meadow and H.-P. Uerpmann, 194–206. Reichert, Wiesbaden.

Ryckmans, G. 1951. Inscriptions ṣafaïtiques au British Museum et au Musée de Damas. *Le Muséon* 64: 83–91.

Ryckmans, G. 1955. Ṣafaitica. Inscriptions ṣafaïtiques relevées par Henry Field. *Bibliotheca Orientalis* 12: 8–9.

Uerpmann, H.-P. 1991. *Equus africanus* in Arabia. In *Equids in the Ancient World*, Vol. II, edited by R. H. Meadow and H.-P. Uerpmann, 12–33. Tübinger Atlas des Vorderen Orients 19/1. Reichert, Wiesbaden.

Uerpmann, M., Uerpmann, H-P. and Jasim, S.A. 2000. Stone Age nomadism in SE–Arabia—palaeo-economic considerations on the neolithic site of Al-Buhais 18 in the Emirate of Sharjah, U. A. E. *Proceedings of the Seminar for Arabian Studies* 30: 229–234.

Wagner, E. 1987. *Grundzüge der klassischen Arabischen Dichtung.* I *Die altarabische Dichtung.* Wissenschaftliche Buchgesellschaft, Darmstadt.

Winnett, F. V. & Harding, G. L. 1978. *Inscriptions from Fifty Safaitic Cairns*. Near and Middle East Series 9. University of Toronto Press, Toronto.

Appendix: Inscriptions Accompanying the Previously Unpublished Drawings Shown Here

Editorial conventions in these readings of inscriptions are as follows: { } enclose a doubtful letter in the reading and the word or name which includes a doubtful letter in the translation; [ ] enclose additions in the translations; ---- indicate parts of the inscription which have been lost.

Figures 2c and 2d:

(1) running around the drawing, *l ḫlf bn ḥrb bn ʾs*^{1}*n h-ḫṭṭ h yṯʿ ʿwr ḏ y ʿwrn h-* ---- "By Ḫlf son of Ḥrb son of ʾS^{1}n is the carving: O Yṯʿ blind those who would scratch out the ----".

(2) running vertically above the ass's back and into its body, *l yḥmʾl {b}{n} ms*^{2}*kr* "By Yḥmʾl {son of} Ms^{2}kr".

Figure 3c

KWQ 19: *l s*^{2}*mt bn ḥrs*^{1} *h-frs*^{1}*n* "By S^{2}mt son of Ḥrs^1 is [the drawing of] the two horses/horsemen".

Figure 4a

Ms 50: in large letters running around the drawings: *l khl bn qnʾl bn s*^{1}*ʿr bn nẓr w h ʾlt rwḥ m s*^{1}*qm w h lt nqʾt l ḏ y ʿwr h-ḫṭṭ* "By Khl son of Qnʾl son of S^{1}ʿr son of Nẓr and O ʾlt [grant] relief from illness and O Lt [inflict] ejection from the grave on whoever scratches out the carving".

Ms 51: in tiny letters starting to the right of the oryx: *l ys^{1}lm bn grmʾl bn qn w* ---- "By Ys^{1}lm son of Grmʾl son of Qn and ----".

Ms 52: in tiny letters below the equid: ----*s^{2}rt bn ṣbḥ* "---- S^{1}rt son of Ṣbḥ".

Ms 53: along the bottom edge of the face: *l {t}m bn khl bn {q}ḥf* "By {Tm} son of Khl son of {Qḥf}".

Figure 11a

l glḥn bn mqmʾl h-ʾʿrd "By Glḥn son of Mqmʾl is [the drawing of] the wild asses".

Figures 11b and c

l bs^{1}r bn fny bn nẓmt h-dmyt w h yṯʿ mḥlt l ḏ yʿwr h-ḫṭṭ "By Bs^{1}r son of Fny son of Nẓmt is the picture. And O Yṯʿ [inflict] dearth on whoever scratches out this carving."

Figure 11 d

Mr.A 6: *l s^{1}ḫr bn ʿḏ bn ʿḏ bn ġṯ w s^{2}ry m-ʿ{ḏ} bn ʾḫ-h h-dmyt w ʿ{w}r l-ḏ {y}ʿ{w}r* "By S^{1}ḫr son of ʿḏ son of ʿḏ son of Ġṯ and he bought from {ʿḏ} his nephew the drawing and {blindness} to whoever {scratches [it] out".

Mr.A 7: *l {ʾ}nʿ{m} bn ʾnʿm bn rbʾl {b}{n} ʾnʿm* "By {ʾnʿm} son of ʾnʿm son of Rbʾl {son of} ʾnʿm".

Figure 13b

KWQ 161: *l mlṯt bn ʿly bn bdl h-ʿr* "By Mlṯt son of ʿly son of Bdl is [the drawing of] the ʿ*r*".

Figure 14d

KWQ 10: *l ʿk bn ṯʿt bn frk h-ʿr* "By ʿk son of Ṯʿt son of Frk is [the drawing of] the ʿ*r*".

Figure 14e

KWQ 122 (at the top): *l ʿzz bn kfrl h-frs^1* "By ʿzz son of Kfrl is [the drawing of] the horse(man)".

KWQ 123 (diagonally immediately above the camel): *l ntz bn ʾs^{2}yb* "By Ntz son of ʾs^{2}yb".

KWQ 124 (diagonally between the camel and the horse): *l grmh bn rb h-ʿr* "By Grmh son of Rb is [the drawing of] the ʿ*r*".

KWQ 125 (vertically between the ʿ*r* and the camel): *l ʾṯʿ bn ḫft* "By ʾṯʿ son of Ḫft".

KWQ 126 (diagonally behind the camel): *l ẓrr bn ʾbyn h-ḥyt* "By Ẓrr son of ʾByn is [the drawing of] the animals".

Figure 15a

Above the equid's tail: *l ḥbb h-ʿr* "By Ḥbb is [the drawing of] the ʿ*r*".

The other inscriptions are too damaged to allow a reading of more than one or two letters.

Figure 15b

l ms^{1}k bn mrṭn hn-ʿr "By Ms^{1}k son of Mrṭn is [the drawing of] the ʿ*r*".[43]

Figure 15c

l ḏky bn ʿty h{n}-{ʿ}r "By Ḏky son of ʿty is [the drawing of] the ʿ*r*".[44]

Figure 15d

KWQ 21 at the top: *l {s^2}ẖt bn qdm{ʾ}l {h-} ʿr* "By {S^{2}ẖt} son of {Qdmʾl} is [the drawing of] {the} ʿ*r*".[45]

KWQ 22 scratched in tiny letters above the equid's tail: *l ʿ{l}y bn ṯbr{n} h-ʿr* "By {ʿly} son of {Ṯbrn} is [the drawing of] the ʿ*r*".

Figure 15e

l nʿm bn ms^1{k} h-ʿyr "By Nʿm son of {Ms^{1}k} is [the drawing of] the ʿ*yr*".

Figure 15f

Ms 79: *l gd{y} bn ʿnq h-ʿr* "By {Gdy} son of ʿnq is [the drawing of] the ʿ*r*".

Figure 15g

l s^{2}f bn ġs^{1}l h ʿr "By S^{2}f son of Ġs^{1}l is [the drawing of] the ʿ*r*".

Figure 16a and b

l {h}ḥrm bn {ʾ}byʿ bn {h}{m}y h-ʿyr. "By {Hḥrm} son of {ʾbyʿ} son of {Hmy) is [the drawing of] the ʿ*yr*".

Figure 16d

l ʿkdt h{n} ʿyr "By ʿkdt is [the drawing of] {the} ʿ*yr*".[46]

[43] If the reading of the definite article as *hn-* is correct this would be one of very few examples in Safaitic of this form before pharyngal consonants. This feature is normally associated with the Dadanitic dialect (see Macdonald 2009a III: 41 f.).

[44] Some of the letters have been playfully joined together (see Macdonald 1989 and 2009a II: 386 f.). Note, if the reading is correct, the definite article here would be *hn-* as in Figure 15b and 16d.

[45] The letters in this inscription too have been playfully joined.

[46] Once again the definite article appears to *hn-*.

Figure 17a

l tmn bn fhrn h-ʿr "By Tmn son of Fhrn is [the drawing of] the *ʿr*".

Figure 17b and c

(1) In thinly scratched letters in the top right-hand corner: *l s¹wd bn mḫlm bn rbʾl bn ʾnʿm w ḥl{l} h-{d}{r} f h lt s¹lm w ġnmt m-s²nʾ w nqʿt l-ḏ yʿwr h-s¹fr* "By S¹wd son of Mḫlm son of Rbʾl son of ʾnʿm and {he camped} {here}. So, O Lt [grant] security and booty from enemies and [inflict] ejection from the grave on whoever scratches out the inscription".

(2) In thick hammered letters immediately behind the equid: *l ʿzhm bn gḥfl bn ẓnn w m mḫbl h-s¹fr nqʿt l-ʾḫ -h* "By ʿzhm son of Gḥfl son of Ẓnn: and whoever is a destroyer of this inscription [may there be] ejection from the grave for his brother".

(3) In large thinly incised letters starting to the right of the drawing (not visible on the photograph): *l wdm bn ʿmr bn s¹ry bn ḥmyn h-ʿr* "By Wdm son of ʿmr son of S¹ry son of Ḥmyn is [the drawing of] the *ʿr*".

(4) Below the equid: *l ʾtm bn mʿyr bn ʾlhn* "By ʾtm son of Mʿyr son of ʾlhn".

(5) Below (4): *l ʿzhm bn gḥfl bn ẓnn* "By ʿzhm son of Gḥfl son of Ẓnn".

Figure 17d

KWQ 94: *l ṭrq bn ḥl h-ʿr bn khl* "By Ṭrq son of Ḥl son of Khl is [the drawing of] the *ʿr*."

Part IV

The Horse in Ancient Asia

12

The Heavenly Horses Visualized in Han China (220 BCE–220 CE)

Katheryn M. Linduff

Although well known in Western Asia and the Near East, multiple horse types (breeds?) played a less significant role in China until the Han (汉朝) period. Based on visual records of horses, the earliest evidence documenting what must be blood–sweating horses, likely related to the "great" Nisean horse prized in Persia (Brownrigg, personal communication, 11/12/2017), was in the Western Han Dynasty, or during the 2nd century BCE. This is the time when the Han Emperor sought to trade gold coins for the remarkable steeds from Kokand (eastern Uzbekistan, the main transportation junction in the Ferghana Valley) and bring them to his court. Impressive horses came into China only after Emperor Wudi (汉武帝140–87 BCE) commanded an army of 60,000 men, 30,000 horses, 100,000 head of cattle, and thousands of donkeys and camels and marched on Ferghana. After he reached the capital and besieged it, he returned to China with a great haul of horses. The Han period bronze statue called the 'Flying Horse of Gansu' (2nd c. CE) is perhaps the best–known equine statue in China and probably represents one of these special gaited horses. This wingless steed was probably meant to be understood metaphorically rather than literally; the image was certainly idealized. It is the significance of this and its visual models in Chinese culture that are the focus of this paper.

Introduction

The excavation in 1969 of a large tomb built of brick and stone dating from the later years of the Eastern Han Dynasty, or from between 186–219, at Wuwei (武威) in Gansu Province in present–day western China yielded the spectacular find known as the "Flying Horse of Gansu" (Figure 1). It was among 231 relics in the tomb, including 99 bronze statues of mounted warriors and horses. The bronze horse is 34.5 cm high, 45 cm long and weighs 7.15 kg. The horse is balanced on a single hoof that rests on the back of a wingspread bird with three hoofs and its head and tail raised high in the air. The animal represented is a gaited horse, one frozen in time in a recognizable gait that is called a singlefoot, or amble, where one foot touches the ground at a time. Here we see lateral pacing wherein the front and hind feet on the same side move in sequence. Not all horses can perform an ambling gait; however, today many breeds can be trained to produce it, and there are several breeds of horses that inherit the ability to perform this gait either naturally from birth or with a minimal amount of training. This depiction confirms, nevertheless, an animal of training and ability.

Figure 1. Flying horse of Gansu, bronze, from tomb in Wuwei, Gansu. Height 34.5cm, length 45cm, weight. 7.15kg, Eastern Han Dynasty, c. 186–219. From: Zhongguo Meishu Quanji Bianji Weituanhui 1990: p. 202, pl. 238.

In China, the conformation and countenance of this 'Flying Horse' create a model for special horses from this time forward. With all its attributes—high spirits, straight neck, broad, slightly shovel–shaped forehead, flaring nostrils, and high raised tail—this statue seemingly depicts the attributes of those eventually assigned to the Arabian breed (Figure 2). If that is an accurate identification, as I believe it is, how can we explain its appearance in the late 2nd century CE in Han China, when the earliest noted textual references to Arabians in China date to the Tang (唐朝), or the 8th century (Cooke 2000: 48)? I shall hopefully make a convincing case that it does in fact make its debut in western China as early as the Han Dynasty when the texts tell us that the Chinese emperors became interested in expanding their stock of horses. Moreover, the likeness lived on into the Tang, not in its original form, but only in representations of the head to suggest the powerful spirit of the untamed

Figure 2. Arabian horse. From: Edwards 1977: 36.

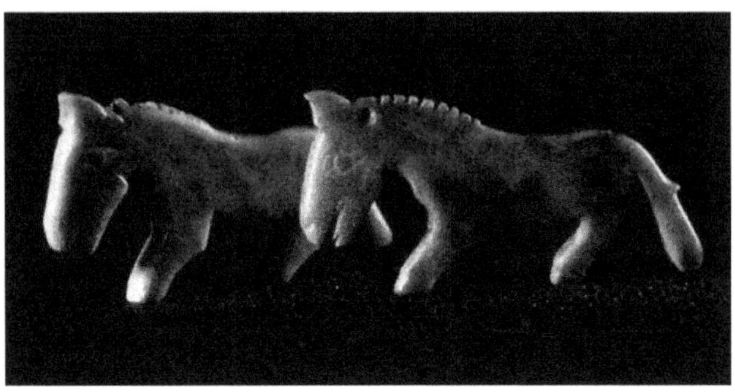

Figure 3. Jade horse from the tomb of Lady Hao, Yinxu, Anyang, Henan Province. Shang Dynasty, c. 1250 BCE. From: Institute of Archaeology 1980, pl. 30:2.

animal. The full–chested, long–legged animal of the Tang is probably a mixed breed.

Background—Depictions of Horses in China Prior to the Han

We must begin with the assumption that Chinese sculptors and painters presented images that were observationally accurate or at least recognizably modeled on the actual animal and that the depiction of horses was very frequently charged with special connotations—meanings that are based on an actual prototype that determine its manner of visual interpretation.

The early representations of equines from the late second millennium BCE, such as the little jade equids from the Fu Hao Tomb (妇好墓 M5) at Anyang that probably depict the tenacious, wild *Equus przewalskii* (Figure 3) (Institute of Archaeology, Beijing. 1980, pl. 30:2) and the inscription in the shape of a long–eared wild ass on the base of the bronze ritual vessel *gui* (簋) from Jingjiecun (旌介村), in Lingshi (灵石), Shanxi Province (Shanxi 1986: 14), are rare (Figure 4). Even so, their portrayal follows the late Shang (商朝) artistic predilection for depiction of wild animals on artifacts used in ritual. Their features—stocky builds, erect manes, and shaggy tails—are not comparable, however, to the longer legged and elegantly proportioned bodies of the horses that were buried in sacrificial pits and tombs at the site, or from the last capital of the Shang Dynasty at Yinxu (殷墟), Anyang (安阳), dating from about 1250 BCE (Linduff 2006: 304) (Figure 5). Clearly

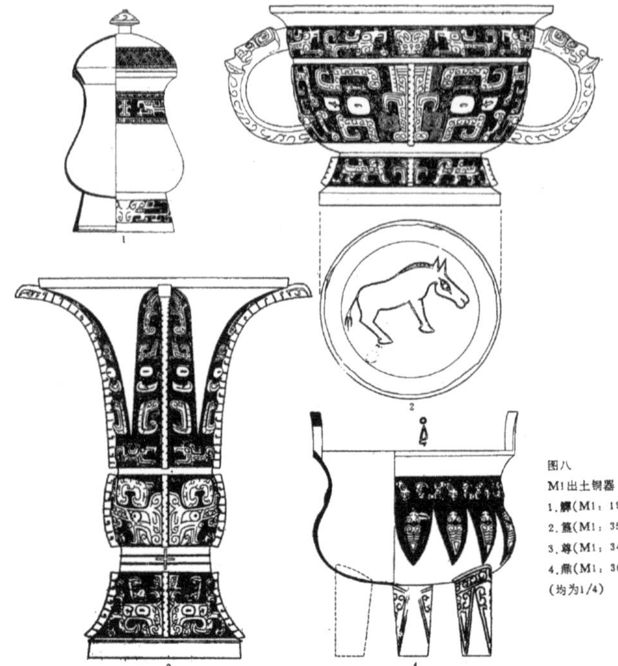

Figure 4. Cast inscription of wild ass and bronzes from M1, Jingjiecun, Shanxi Province, Shang Dynasty, c. 1250 BCE. From: Shanxi Institute of Archaeology and Cultural Center of Lingshi 1986: p. 5, fig. 8.

the Shang artisans made a distinction between the wild and the tamed.

Evidence of highly trained horses and associated equipment, such as those buried in these tombs, has not been documented elsewhere in ancient China before about 1250 BCE. The practice appears full–blown at Anyang as part of a well–established set of rituals—of burial, hunt, and war (Creel 1970; Yetts 1934). The simultaneous appeal of representations of wild horses and the sacrifice of tamed ones in ritual use in late Shang society raises questions about the attraction of this dichotomous manner of display, as well as the availability of both tamed and untamed animals. This bimodal way of thinking about horses appears to have continued throughout early Chinese history, and is preserved in the standards by which horses were illustrated. On the one hand, the trained horses were considered useful and important, both actually and figuratively in battle and as a marker of courage and strength, but on the other, some were conceived of as metaphorical and behaved in supernatural, or even wild or untamed ways (Linduff 2003: 139).

Taken together all the evidence (visual, archaeological, and inscriptional) points to limited human control of horses primarily for dietary and burial purposes in northwestern present–day China before they were introduced for ritual purposes for the Shang dynastic elite at Anyang. The Shang were the first for whom we have evidence of the use of horses for parade and not primarily for food. They probably maintained their stables only through frequent contact, both cordial and hostile, with non–Shang groups. Intensive breeding and training are not indicated in dynastic lands before that time, which further supports the

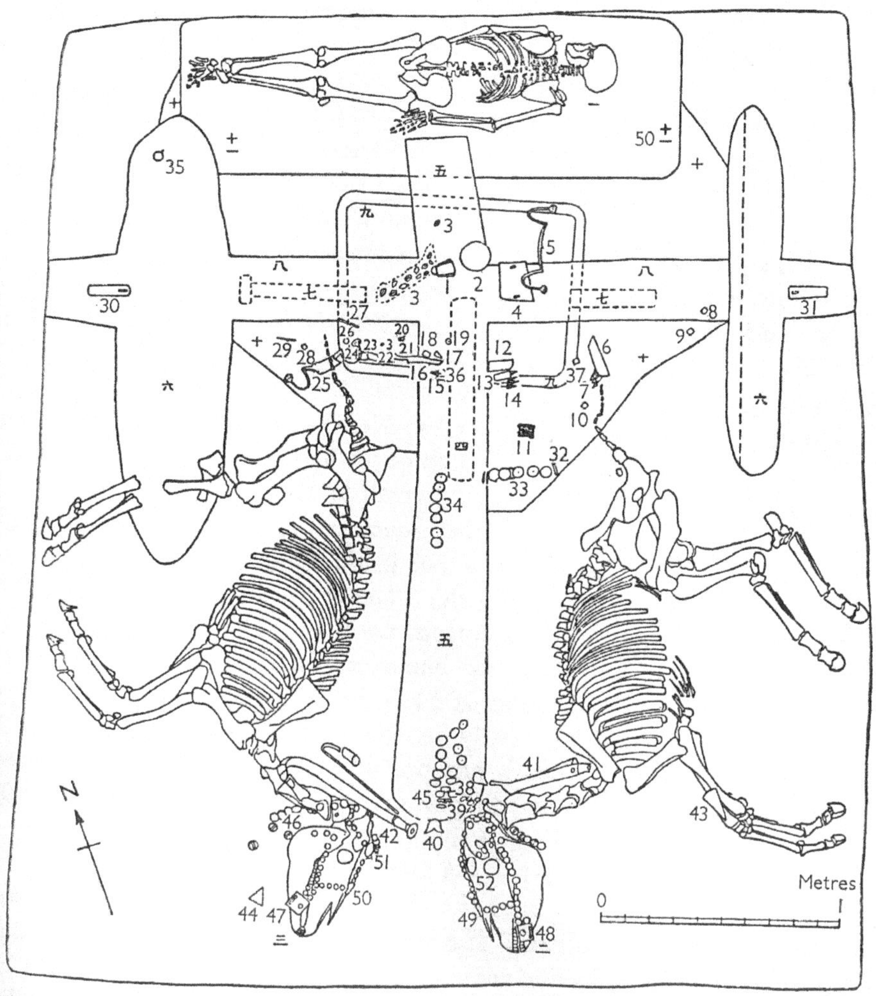

Figure 5. Chariot burial, Tomb 175, Dasikong, Anyang, Henan, Shang Dynasty, c. 1250 BCE. Chinese numerals: 1 Driver, 2–3 Horses, 4–9 Impressions of various parts of the chariot, 10 Traces of red lacquer, 11 Dark-colored material. From: Cheng 1960: 71.

notion that alliances with groups to the west and north of Anyang provided the best sources for horses, but also for breeders, trainers and equipment (Linduff 2003: 157).

The possession of horses must have given all of their owners an advantage—in transporting their leaders, as well as goods—whether the late Shang themselves or those from whom they got horses, either through trade or military effort. Burial evidence from Anyang underscores the notion that horses counted as well when sacrificing to the ancestors or other spiritual beings. Inscriptional evidence expands our understanding of the high regard in which horses were held by the Shang—their features, including color, size, number, and conformation, were recorded. The horses were also associated with special aspects of character (speed, strength, power), like those accorded to their owners. Burial practice and the contents of the tombs, however, indicate that the individuals associated directly with trading, breeding, training, and managing horses were 'outsiders' (Linduff 2003: 158).

In the Shang, association with the horse and horse gear could also indicate cultural identity or affiliation. As standard goods associated with royal tombs, the horse and the chariot were part of the complex apparatus of power, both political (chariot) and spiritual (horse sacrifice). The wild counterparts of the great parade animals, on the other hand, were memorialized in jade (Figure 3), or were described in pictographs and were thereby transformed into static symbols signifying 'outsiders.' This was the case, for instance, with the burial of Lady Hao, the consort of the third king at Anyang. Her little jade horses and a set of frontier items buried with her probably identified her with her natal family to the northwest of Anyang (Linduff 1996; 2000: 19–22).

Most agree that warfare and sacrifice were integral and essential parts of the religious system of the following Zhou dynasty (周朝), and that the state and social order were dependent on them (Yates 1999: 8 f.). Throughout the realm in the early Zhou by about 1000 BCE, nearly all burials of high–level officials included chariot and horse sacrifices, providing a unifying image of the power of the Zhou (c. 1050–900 BCE) and affiliation with them (Figure 6). Highly trained horses and drivers were available to, and perhaps necessary for, the political elite, and the mysteries or significance of the horse and chariot as exotic booty was presumably forgotten (Linduff 2003: 158). By this time,

Figure 6. Chariot burial, Early Zhou Dynasty, c. 1000 BCE, Liulihe, near Beijing, Yan state capital. From: Joint Archaeological Team, IA, CASS and Municipal Archaeological Team of Beijing 1984: 410.

the horse and chariot conveyed the idea of military power under the centralized political authority of the all–powerful Zhou, and supported the notion of a ranked society. One's kinship and position in the hierarchy could be documented by the number of certain artefacts and the size of the burial (Hsu and Linduff 1988: 163–77).

For instance, during the reign of the third King, Kang Wang (周康王, 1004–967 BCE), the Zhou continued to secure their hold and expand their territories. They engaged in at least two major wars with the pastoral people living to their north. An account of their triumph was inscribed on a ceremonial bronze vessel from this period (Hsu and Linduff 1988: 142). Upon return from battle, the Zhou troops presented the court with 4800 enemy heads, 13,000 captives, 300 cattle, and numerous horses, sheep, and chariots. With a second victory, the troops brought back 140 horses, as well as enemy heads, captives, and chariots (Shirakawa 1962–77: 648–714; Hsu and Linduff 1988: 141 f.). These accounts record the striking display of the power and perceived status of the victorious Zhou in relation to their adversaries. The debilitated state of the defeated enemy was marked by the capture of precisely those pastured animals (sheep and horses) that were at the center of their livelihood and without which they could not function (Linduff 2003). The balance of power between the Zhou state and groups of their pastoral neighbors was delicate and essential to the perceived if not actual hegemony of the Zhou.

During the reign of the fifth monarch, King Mu (周穆王, 947–928 BCE), his fame was tied to his travels and horses and had become the stuff of legends. In the *Zuo Zhuan* (左传, c. 551–479 BCE), he is described by his desire to leave his chariot tracks everywhere (Legge 1935: 641), and to have a chariot pulled by eight semi–divine stallions. As the hero of a later fictional account, he even traveled in his chariot to a sacred mountain to visit the Queen Mother of the West (西王母 *xiwangmu*) to seek immortality.

These legends of King Mu's wanderings and his eight wonder horses probably reflect some historical truth, for he was thought to have owned especially good horses (Hsu and Linduff 1988). Evidence of that is documented in an inscription on a wine vessel in the shape of a colt (盠驹尊 *lijuzun*) excavated in Lijiacun (李家村), Meixian (眉县), Shaanxi, and dated from his reign (Figure 7). The inscription on the body of the vessel (尊, *zun*) says

Figure 7. Bronze ritual vessel (*zun*) in the shape of a colt, Mei County, Shaanxi, China, Western Zhou Reign of King Mu, c. 947–928 BCE. Height 32.4 cm; length 34 cm. From: Zhongguo Meishu Quanji Bianji Weituanhui 1990: vol. 4, p. 220, pl. 221.

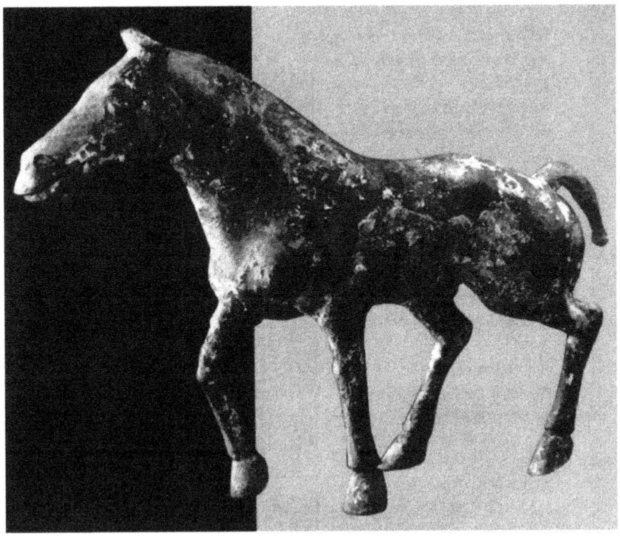

Figure 8. Bronze horse sculptures in the tomb of the king of the Zhao State, reign of King Mu, c. 947–928 BCE, Mei County, Shaanxi, China. Height 32.4 cm; length 34 cm. From Hao 2003: 7.

that before the king bestowed two young colts on the owner of the bronze, he handled a foal. In doing so, the king himself presumably initiated the taming, or the process of 'civilizing and sinicizing' the untamed foal. Two more inscriptions appear on the lid, each of which gives specifics about these two colts, even facts related to their short stature, large heads, and thick neck (Shirakawa 1962–77: 342 f.; Guo 1961: 312–19). Such horses were probably bred and pastured outside of the dynastic center since, as many argue, grassland suitable to pasturing horses was not readily available close by (Sinor 1972). Another inscription on a bronze vessel (鼎, *ding*) from the reign of King Xiao (周孝王, 897–888 BCE) records a gift of 32 horses given in pairs and in fours to be harnessed to chariots, not for use as sacrificial offerings. Such evidence points out the importance of having good stock and the high regard in which horses were held at court. But it was not their propagation, nutrition, or training that counted in these equations as it might have with breeders and handlers, it was their total numbers, potency, and outer appearance (color and size) that mattered. Even so, these portrayals visualized the actual physical characteristics of range ponies in the pre–Qin period.

In this same period, only the bronze images of horses from the tomb of the King Wuling of Zhao (赵武灵王) at Handan (邯郸) are depicted in motion (Figure 8). Wuling was the King who demanded that his army adopt cavalry and riding gear like the 'barbarians' in 307 BCE. When Official Cheng claimed, "[…] if you adopt nomadic clothing and cavalry in the battles with them, you are going to lose your Chinese identity!" King Wuling replied, "There is no one way to rule the world, nor is there any need to copy the old to benefit my country."—(战国策, *Zhanguo Ce*, "Stratagems of the Warring States" [475–221 BCE], Ch. 18; Crump 1996). Here, the King claimed that

Figure 9. A pair of cavalrymen, earthenware with painted surfaces. Height 23.5 cm, length 18 cm; height 23.5 cm, length 17.5 cm, Eastern Zhou Dynasty, Warring States Period (475–221 BCE) excavated in 1995 from Tomb 2, Xianyang City, Shaanxi Province, Xi'an Institute of Cultural Relics and Archaeology. From Cooke, 2000: p. 119, pl. 96.

his state could only survive by becoming more like the enemy, and his decision to adopt 'barbarian' ways changed the Chinese military forever, and perhaps provoked the freely moving depictions of these horses! This episode calls attention to what the Chinese regarded as the essential 'foreign' nature of horses and riding gear, most especially pants. Even so, horses for use as cavalry, such as the pair dating from the Warring States Period, were necessary for the stability of the state and were depicted in clay in stasis for burial display (Figure 9).

Throughout Chinese history, the peoples of the northern steppes and the sedentary farmers in the Central Plains had come into conflict over land and resources. The key to military success, and to fortifying the borders to the

Figure 10. Terracotta figures of a saddle horse and warrior, terracotta. Horse: height 174 cm, length 213 cm, weight 340 kg; warrior: height 185 cm, weight 184 kg. Qin Dynasty (221–207 BCE), Excavated from Pit 2, necropolis of Qin Emperor, Lintong County, Shaanxi Province, Museum of the Qin Shi Huang Terracotta Army. From Cooke 2000: p.125, pl.103.

west and north of the Chinese state confederation, was the number and quality of horses that could be used to repel the invaders. In times of peace, the horse was an indicator of rank; in times of instability and war, it provided the foundation of the late Zhou military (Yang 2008: 107). The Qin created a state agency that specialized in managing and breeding horses, and depictions of horses in the tomb of the first emperor bore the same physical attributes as those in the Zhou (Figure 10). The terracotta draft and riding horses in the Tomb are stocky, have well developed musculature, trimmed manes, and tails knotted into bobs that show they had been carefully prepared, were trained for parade and display and stand about 13.2 hands (Cooke 2000: 40).

In Search of the Han 'Celestial' Horses

The Han period (256 BCE–220 CE) was the first Chinese golden age of horses. The evolution of draft and cavalry mounts finally culminated during the Han in a vitality that is best described in Chinese in the language of dreams: *tianma* (天马, heavenly horses), *feima* (飞马,flying horses), *longma* (龙马,dragon horses), *shenma* (神马, divine horses), and *hanxuema* (汗血马, blood–sweating horses). These particular creatures were peerless and alone in the universe (Yang 2008: 18) and were represented as such! Their special regard is documented early in the dynasty in the reign of Emperor Wudi (汉武帝, r. 140–87 BCE), as are the extremes to which he quested for such animals. One result of his leadership was the introduction of new stock, perhaps new strains (Cooke 2000: 41), which also must have invigorated imagery of horses.

Emperor Wudi, the 'Martial Emperor' was the sixth emperor of Han and devoted himself to military conquests and territorial expansion besides introducing many domestic reforms. While Marius and Sulla were invading the Mediterranean world in the West, Wudi built a powerful empire that stretched from North Korea into Central Asia. Throughout the Qin (秦朝) and Han period, China's most formidable opponent was the Xiongnu (匈奴), a confederation of peoples with a mounted leadership and warrior class that, at its apex of power early in the second century BCE, held sway over a territory that extended all the way from Eastern Mongolia to the Aral Sea. When Wudi took over the Chinese empire, he had two things in mind—defense and trade (Cooke 2000: 41–43).

After losing nearly 80 percent of his cavalry horses in the initial campaigns against the Xiongnu, Emperor Wudi adopted a multi–pronged approach—he improved horse husbandry within the empire, and sought to expand the equine breeds (Figure 12). In his quest to bring the Xiongnu under control, he planned to form an alliance with the Xiongnu's old enemies, the Yuezhi (月氏), and to attack the Xiongnu from the east and west simultaneously. In 138 BCE Zhang Qian (张骞, d. 114 BCE), the commander of the guards at the imperial palace gates, volunteered to undertake the hazardous journey to persuade the Yuezhi to become allies. The Xiongnu captured and retained his men and him in the Hexi region in Gansu Province for more than 10 years. He escaped and continued to travel westward south of the Tianshan, through Kashgar and the Pamirs, finally arriving in Kokand (Qo'qon), Ferghana (in modern day Uzbekistan) where he became friends with the King. He resumed his journey to the west, met up with the Yuezhi, then made his way back to Chang'an (长安) (西安, Xi'an), the capital of the Han dynasty, after a trip of 13 years.

Between 119 and 115 BCE, Wudi sent him to Central Asia again, this time to seek an alliance with another pastoral group, the Wusun (乌孙), a people who lived in the Ili river region of Xinjiang (新疆, شىنجاڭ). Wudi also intended to create friendly ties with the western kingdoms by offering gifts. From then on, diplomatic missions were dispatched regularly along with commercial trading; thus the byproduct of the Xiongnu campaigns and the exploration by Zhang Qian was the opening of the eastern routes of the Silk Road. The Chinese imported horses, cattle, and furs and hides from Central Asia, while silk was sent out from China.

While Zhang Qian was in Ferghana, he was particularly impressed with the fine horses of Kokand (大宛) that 'sweated blood' (Figure 11). From him, Wudi learned that these great steeds stood very tall (estimated from Chinese measure to sixteen hands, Cooke 2000: 42), and were bred two thousand miles away in Ferghana. Since the campaigns against the Xiongnu demanded vast numbers of warhorses with the qualities of size, stamina, and muscle not possessed by indigenous breeds, Wudi hoped to bring Ferghana–bred horses to his court.

Wudi's initial attempt to trade gold coins for Celestial Horses was rejected by the king of Ferghana, and the Han

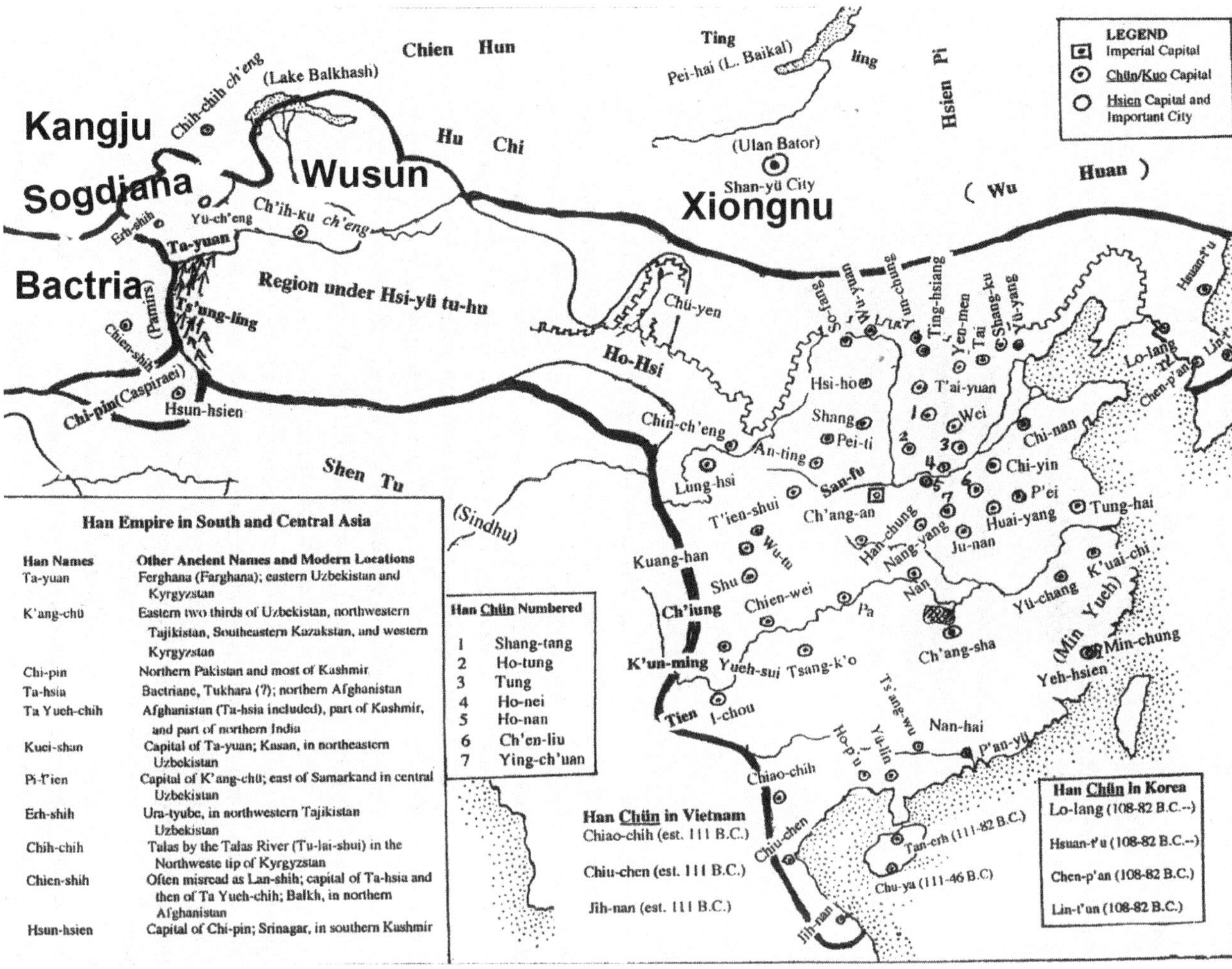

Figure 11. Map indicating the location of the Han Dynasty Western Borderlands. Adapted from Chang 2007: 2.

envoy sent for the negotiation was killed. In 102 BCE, the Emperor himself embarked on a military campaign with an army of 60,000 men, 30,000 horses, 100,000 head of cattle, and thousands of donkeys and camels marching toward Ferghana. They reached the capital, successfully besieged it, and returned to China with a great haul of the famous Ferghana steeds. The story continues and claims that Ferghana continued to supply the Han court with the best of these horses, including 2000 'ordinary stallions and mares', and *two celestial horses* every year.

Recently discovered inscriptions on bamboo strips found at Xuanquan (悬泉), in Dunhuang (敦煌), Gansu Province, that are dated to the late Western Han Dynasty (2nd c. BCE), record an embassy from Kangju (康居) that brought a tribute of nine horses, 31 donkeys, 25 camels, and one cow (Bi 2010) (Figure 12). The Kingdom of the Kangju was a powerful nomadic state or confederacy in Central Asia and was the first state in that region to send an ambassador to the Western Han court (2nd c. BCE). Following the import from several locations in addition to Kangju, Ferghana, and Wusun, the horses were renamed. Those from the Wusun were called *xiji* (西极, the Westernmost horse) and ones from Ferghana were called *tianma* (天马, heavenly/celestial horses). Such attributes are the first to be distinguished, other than color and size, and to record discernable strains.

Moreover, the physical attributes of equine sculptures were also transformed in tandem with the importation of different stock types. Not only was the conformation of the horses more varied, but also suddenly the beasts broke into action. One such example is the little jade statuette discovered in 1966 near Pingling (平陵), Shaanxi Province, in the mausoleum of Emperor Zhao (汉昭帝 r. 87–74 BCE) (Xianyangshi bowuguan 1973.3: 45 f.) (Figure 13). The limbs of the horse are bent forward as though galloping, and beneath his legs is a cloud–shaped platform, clearly a symbol of his flight through the air! The horse has a large head, open mouth, flaring nostrils, a thick neck, and short legs. Although lacking in grace—this celestial beast has features of the Arabian.

Fine examples dating from the Eastern Han have been found in Fanglingcun (防陵村) in Xushui (徐水), in Hebei Province (northeast), that depict a horse with a high head and tail carriage, but the most celebrated examples were unearthed in a general's tomb of the second century CE, near Wuwei (武威), in Gansu Province. Among them is the riderless steed known as the 'Flying Horse of Gansu,' meant to be understood metaphorically rather than literally, in that it represents a horse that moves so fast that it seems to outrun the wind (Cooke 2000: 44). Its head and tail are raised high as he proudly gallops with his right rear hoof resting on the back of a swift–flying

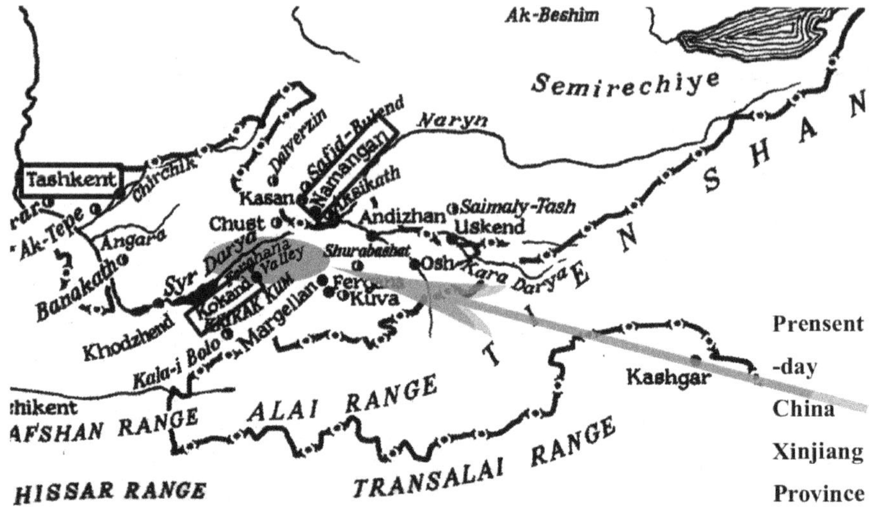

Figure 12. Map showing the Ferghana Valley. Adapted from Knobloch 2001: 156.

Figure 13. Winged horse, jade, Western Han (2nd c. BCE), from Weiling in Xianyang City, Shaanxi Province. From Cooke 2000: p.137. pl. 120.

hawk, probably a Eurasian Sparrowhawk, or goshawk (*Accipiter nisus*), a bird of prey with short, broad wings, and a long tail (Figure 1). The complete image documents the interest in importation of special breeds as a result of Wudi's interest in expanding and supplying his stables. With its distinguishing attributes—high spirits, straight neck, broad, slightly shovel–shaped forehead, flaring nostrils, and raised tail—this image apparently documents Arabian–type horses in western China no later than the 2nd century CE, and perhaps earlier. This is no ordinary horse, but regardless of its land of origin, the Flying Horse of Gansu is not as tall as the blood–sweating breed (16.2 hands) from Ferghana, so much admired and recorded in historical documents and still bred today in Turkmenistan. Does it represent one of the heavenly creatures of which only two were sent each year from Ferghana, or possibly the *xiji* (西极Westernmost horse) type gifted by the Wusun? That the Gansu horse is paired with a Eurasian bird of prey (and not a swallow as it has been so often identified), further acknowledges its foreignness, its admired breeding, and perhaps even its training and spirited airborne capacity!

The Flying Horse of Gansu is a very, very special horse, maybe like the two supplied yearly from the Kingdom of Ferghana. This specialness continues to be appreciated elsewhere in East Asia at later times, for example in Korea and Japan in mortuary settings in the 5th and 6th centuries, where the celestial aspects of this special animal were revered. In addition, the type was admired well into the Tang Dynasty in the 8th century in China. This is exemplified in the raised head of *Night–Shining White* (照夜白 *zhaoyebai)*, the favorite, tethered charger of Emperor Xuanzong (唐玄宗, r. 712–56), which may have represented both aspects of horses discussed earlier, the untamed and the tamed or civilizing aspect. So, too, were the untamed spirits and demeanors of the horses and their trainers pictured as a constant reminder of their status as outsiders. With unbridled spirit suggested in the head of this animal, it was likened to its wild ancestors as well as admired for its potentially civilized conduct (Linduff 2004).

Concluding Remarks

Use of the horse in East Asia begins early in the second millennium BCE with its exploitation in the north and west of the Central Plain as a source of food (Linduff 2003: 142–45; Yuan and Flad 2004). During the Shang, the evidence clearly documents intentional breeding and training for royal service for draft, for sacrificial symbols of power, and, in their wild form, as representatives designating 'outsider' groups. By the late Zhou, during the period that saw the rise of Confucian humanism in the late sixth and early fifth centuries BCE, horses were depicted in ritual narratives, their importance as a mounted parade animal emerged, and their symbolic uses increased to include funerary and mythological journeys. In the Han, wild and tamed aspects were defined and strains were distinguished. Although none of these attributions is certain, it is clear that as the Chinese became interested in expanding their use of horses for cavalry and parade, they simultaneously began to seek out the variety of types available to them. These new categories were probably initially secured from several locations far outside dynastic borders in what the Chinese called the Western Territories (西域). These locations included the Ferghana valley, as well as the regions of the Wusun, the Kangju, and probably others. In this context, it is possible to distinguish between the blood–sweating horses of Ferghana as well as others. The Flying Horse is testimony that Gansu was probably a prime location for the eventual breeding and handling of multiple breeds inside dynastic territory.

Acknowledgements

I acknowledge the encouragement of Sandra Olsen, who insisted that I take a look at the Chinese material on horse types. I am glad I did, as I think the inquiring mind of Mary Littauer would have enjoyed the attempt! Also the assistance of Dong Lihui in the preparation of the figures is gratefully appreciated. My thanks to Lauren Tomsic, my student who made the suggestion that the horses were mixed breeds in her undergraduate paper on Tang horses, Spring 2010.

Bibliography

Bi Bo. 2010. New Evidence from Dunhuang, China and Central Asia for the Kangju (康居): An Enigmatic Power on the Silk Road, Lecture given at the Institute for the Study of the Ancient World, New York University, April 27.

Chang, C. S. 2007. *The Rise of the Chinese Empire*, vol. 2. University of Michigan Press, Ann Arbor.

Cooke, W. 2000. The Horse in Chinese History. In *Imperial China. The Art of the Horse in Chinese History*, edited by W. Cooke, 27–62. International Museum of the Horse, Kentucky Horse Park, Lexington/KY.

Creel, H. G. 1970. The Role of the Horse in Chinese History. *What is Daoism?* University of Chicago Press, Chicago.

Cutter, R. J. And W. G. Crowell. 1999. *Empress and Consorts: Selections from Chen Shou's Records of the Three States with Pei Songzhi's Commentary.* University of Hawaii Press, Honolulu.

Duiker, W. and J. Spielvogel, 2002. *World History.* Wadsworth/Thomson Learning, Belmont/CA.

Guo M. (郭沫若). 1961. Collected Work on Cultural History (*Wenshi lunji* 文史论集). People's Publishing House, Beijing.

Hao, L. 2003. Research on Several Issues of Tomb of the King of the Zhao State and the Excavated Bronze Horses (Zhaoguo Wangling Jiqi Chutu Qingtongma de Ruogan Wenti Tanwei 赵国王陵及其出土青铜马的若干问题探微), *Wenwu Chunqiu* 2003(3): 5–12.

Hsu, C. Y. And K. M. Linduff. 1988. *Western Chou Civilization.* Yale University Press, New Haven.

Institute of Archaeology, Chinese Academy of Social Science (Zhongguo shehui kexueyuan kaogu yanjiusuo 中国社会科学院考古研究所). 1980. *Fuhao Tomb in Yin Dynasty Ruins (Yinxu fuhaomu 殷墟妇好墓).* Wenwu Press, Beijing.

Knobloch, E. 2001. *Monuments of Central Asia. A Guide to the Archaeology, Art and Architecture of Turkestan.* Tauris, London.

Legge, J. 1935. The Ch'un Ts'ew, with the Tso Chuen. In *The Chinese Classics*, edited by J. Legge, 1–837. vol. 5. Oxford: Oxford University Press.

Linduff, K. M. 1996. Art and Identity: the Chinese and Their 'Significant Others' in the Shang. In *Cultural Contact, History and Ethnicity in Inner Asia*, edited by M. Gervers and W. Schlepp, 12–48. Toronto Studies in Central and Inner Asia, Joint Centre for Asia Pacific Studies, Toronto.

Linduff, K. M. 2003. A walk on the wild side: Late Shang appropriation of horses in China. In *Prehistoric Steppe Adaptation and the Horse*, edited by M. Levine, C. Renfrew and K. Boyle, 139–162. McDonald Institute Monographs. McDonald Institute for Archaeological Research, Cambridge.

Linduff, K. M. 2006. Imaging the Horse in Early China: From the Table to the Stable, In *Horses and Humans. The Evolution of Human–Equine Relationships,* edited by S.L. Olsen, S. Grant, A. Choyke, and L. Bartosiewicz, 303–322. BAR International Series 1560, BAR Publishing, Oxford.

Shanxi Institute Of Archaeology And Cultural Center Of Lingshi (Shanxisheng kaogu yanjiusuo, Lingshixian wenhuaju 山西省考古研究所、灵石县文化局). 1986. The Shang Dynasty Tombs of Jingjiecun (Shanxi Lingshi Jingjiecun Shangmu 山西灵石旌介村商墓). *Wenwu* 11: 1–26.

Shirakawa, S. (キンブン ツウシャク白川靜). 1966. Kinbun Tsushaku (金文通釋). *Hakutsuru Bijutsukan shi* (白鶴美術館誌), vol. 12: 666.

Sinor, D. 1972. Horse and Pasture in Inner Asian History. *Oriens Extremus* 19: 171–183.

Xianyang City Museum (Xianyangshi bowuguan 咸阳市博物馆). 1973. Recent Finds of the Qin and Han Periods at the City of Xianyang (Xianyangshi jinnian faxian de yipi qin han yiwu 咸阳市近年发现的一批秦汉遗物), *Kaogu,* 3: 167–170.

Yang, H. (杨泓). 2008. *Heavenly Horse. The Horse in Chinese Art and Culture (Tian ma shen jun: Zhongguo ma de yishu he wenhua 天马神骏：中国马的艺术和文化).* Sunshine Press, Hong Kong.

Yates, R. D. S. 1999. Early China. In *War and Society in the Ancient and Medieval Worlds*, edited by K. A. Raaflaub and N. S. Rosenstein, 7–45. Center for Hellenic Studies, Harvard University Press, Cambridge/MA.

Yetts, W. P. 1934. The horse: a factor in early Chinese history. *Eurasia Septrionalis Antiqua* 9: 231–255.

Yuan J. and R.K. Flad. 2003. Two Issues Concerning Ancient Domesticated Horses in China. *Bulletin of the Museum of Far Eastern Antiquities* 75: 111–127.

Zhanguo Ce (战国策, *Stratagems of the Warring States*), [475–221 BCE]. Transl. by J. Crump, Center for Chinese Studies, University of Michigan, Ann Arbor, MI, 1996.

Zhongguo Meishu Quanji Bianji Weiyuanhui (中国美术全集编辑委员会). 1990. Editorial Board of The Complete Works of Art in China. *The Complete Works of Art in China (Zhongguo meishu quanji 中国美术全集).* Wenwu Press, Beijing.

13

Emperor Tang Taizong's Six Stone Horse Reliefs and Sasanian Art

Xiuqin Zhou

The Emperor Taizong 唐太宗 (599–649) of the Tang dynasty (618–907) in China commissioned six large relief sculptures of horses to stand in his mausoleum. These pieces have been known since then as unusual within the traditional Chinese mortuary sculpted tradition, and also because they depict horses in relief and in compositions that recall those known to have been made far to the west in Sasanian dynastic period (224–651) in present–day Iran. Sasanian rulers also focused on commissioned relief sculptures of these noble beasts. Examples were made not only in stone, but also in silver in the form of plates. This paper reconstructs the historical connections between China and the region of Sasanian Iran in order to show that these western Asian examples were possible sources of inspiration for the Tang Chinese reliefs.

Introduction

Horses played an important role in the development of early human history, reshaping global political matrices, as well as changing cultural practices. Ancient representations of the horse in various art forms and media have preserved traces of cultural interaction associated with Eurasian grasslands, the focus of this study. The set of six stone horse reliefs, an integral component of the Chinese Emperor Tang Taizong's 唐太宗 (599–649) burial complex, is an exceptional example. The six stone horses, known as *Zhaoling Liujun* 昭陵六駿 [Six Stone Horses of Zhao Mausoleum] are masterpieces historically and artistically. They are associated with one of the greatest Chinese sovereigns, Emperor Tang Taizong (r. 627–649). After assisting his father in founding the Tang dynasty in 618, Li Shimin 李世民, then the Prince of Qin 秦王, continued battling against tough contenders to secure the borders of the newly founded country. Several years later, he eliminated major rivals and returned triumphantly, but found himself to be a political rival to his elder brother, the crown prince. The political struggle ended in the death of the crown prince and abdication of the first Tang emperor. In 626, Li Shimin became the second Tang ruler,[1] Emperor Taizong posthumously, and ruled for 23 years, inaugurating one of the most glorious reigns in Chinese history and laying a foundation for a prosperous Tang Empire lasting almost three hundred years. Emperor Taizong selected Mount Jiuzong 九嵕山, 90 km northwest of the capital Chang'an 長安 (now Xi'an 西安) in 636, to bury his empress and to attend to his own necropolis (later named Zhaoling 昭陵 [Zhao Mausoleum]). He ordered the images of his six favorite horses, selected from many he had ridden when defeating the contenders to the throne, to be carved in stone to accompany him in his necropolis. He named and composed a laudatory poem for each of them, all of which have survived in literature.

Carved in relief, the six horses are rendered in different postures: three are galloping, two are treading, and one is standing still. The latter is accompanied by a general who is pulling an arrow from the horse's chest. The general is clad in a war robe and armor, wearing a felt hat and carrying a long knife, bow, and arrows. All the horses are adorned with full trappings, and most of them are depicted pierced with arrows still visible today. Their sculptural form in relief and exceptionally realistic depictions earn them a unique place in the history of Chinese art. The term, *qianwuguren* 前無古人 [having no predecessors] (Lu 1973, 300), has often been used in praising these reliefs, inviting interpretations that they were inspired by visual traditions outside China.

The study of particular features of these horse reliefs, including their sculptural form and presentation of the mane, tail, stirrups, saddle, war trappings, and weaponry has yielded rich evidence demonstrating interaction among Chinese and non–Chinese cultures. The study reveals that Chinese sculptors borrowed from various ancient groups, including Scythians, Sasanians, and Turkics.

This article frames the discussion about the interaction between China and ancient Iran, particularly the Sasanian Empire (224–651), by examining the sculptural form of reliefs and the presentation of crenellated manes. The sculptural form adopted for the Emperor's horses resembles that of the Sasanian rock reliefs; the crenellated manes depicted on the Tang horses bear close resemblance to those depicted on the Sasanian silver plates.

[1] Tang Taizong became the second emperor of the Tang dynasty on the day of *guihai* 癸亥 of the eighth moon (month) of 626 (September 2, 626), but his reign, Zhenguan 贞观, did not officially start until 627.

Sculptural Forms of the Six Stone Reliefs and Sasanian Rock Reliefs

The vivid representations of the horses in relief and their placement on the tomb site depart from the Chinese traditions of using stereotyped mortuary sculpture.[2] The famous stone sculpture group at the burial of Han General Huo Qubing 霍去病 (d. 117 BCE) outside of Xi'an is an exception. These pieces are carved to resemble human figures and beasts, but they are carved mostly with crudeness and in the round. In searching for traces of foreign features on the sculptural forms of the Emperor's horse reliefs, one source of inspiration was possibly Sasanian rock reliefs.

Six Stone Horse Reliefs

Ordered by Emperor Tang Taizong, the images of the six famous horses were carved in stone during his reign from 636 to 649. The horses are represented on six separate limestone slabs, each measuring approximately 2.0 m long and 1.7 m high and 0.4 m thick. Each scene is carved in low relief with a raised border connected with a square cartouche, either on its top right or left corner. Three horses, shown in a flying gallop posture, have their legs extended to the utmost, before and behind. Each of the six horses is sporting a three–notched mane and a knotted tail; each is further adorned with full trappings of a complete bridle, a four–striped saddle blanket, a high–bridged saddle and a pair of stirrups. Four out of the six horses are depicted pierced with arrows ranging in number from one to nine. The two reliefs at the University of Pennsylvania Museum of Archaeology and Anthropology[3] were given the names of *Saluzi* 颯露紫 and *Quanmaogua* 拳毛騧 (Figures 1–2) by the Emperor. Saluzi is depicted standing, enduring pain from an arrow in the chest. He is accompanied by General Qiu Xinggong 丘行恭 (586–665), who is extracting the arrow. *Quanmaogua* is depicted treading steadily despite being shot with nine arrows, six in the front and three in the back. The six reliefs were placed three to a row, facing each other, on the spirit road closest to the Emperor's burial chamber at the Zhao Mausoleum (Figure 3).

Figure 1. Saluzi 颯露紫, Emperor Tang Taizong's favorite horse, is depicted with a three–notched mane. He is enduring pain while General Qiu Xinggong 丘行恭 is extracting the arrow from his chest. Tang dynasty, 636–649. Collection and courtesy of the University of the Pennsylvania Museum of Archaeology and Anthropology (C395).

Figure 2. Quanmaogua 拳毛騧, Emperor Tang Taizong's favorite horse, is depicted with a three–notched mane. Tang dynasty, 636–649. Collection and courtesy of the University of the Pennsylvania Museum of Archaeology and Anthropology (C396).

Sasanian Rock Reliefs

Monumental Sasanian reliefs are concentrated at two sites near Persepolis, nine reliefs at Naqsh–i Rustam, and four at Naqsh–i Rajab (Schmidt 1953–70, 13, 122).[4] Dated to the third and fourth centuries, they were carved on the cliff surface, high above the ground. As Herrmann points out, these Sasanian rock reliefs are considered "an official art form" and "equivalent of advertisement boarding." They

[2] Scholars think that it was not until the Han dynasty (205 BCE–220 CE) that stone sculpture became a recognized visual tradition in Chinese life, especially in tomb sculpture. The Han stone sculptures featured line carving. Based on this carving tradition, the Buddhists introduced pictorial representations on cliffs and steles that conveyed an impression of depth lacking in the linear Han treatment. Some Buddhists were trained foreign sculptors who might have brought new ways of visualizing figures inspired by a Greco–Roman background that was then transmitted to China via India (Mahler, 1959; Lerner, 2005; Paludan, 2006).

[3] The two horse reliefs came to the University of Pennsylvania Museum of Archaeology and Anthropology as a loan in 1918. Eldridge R. Johnson, the member of the Museum Board of Managers, contributed the funds for their purchases in 1921. Records of the payment, made in three installments, are kept in the Museum's Archives (Zhou 2001). The other four stone horse reliefs are currently housed at the Beilin Museum, Xi'an, China.

[4] According to Herrmann (2000, 36), 28 out of the 34–35 Sasanian reliefs are in Fars and most of them are at these two sites.

Figure 3. The North Gate of Zhaoling under excavation in 2002. The Emperor Tang Taizong's six stone horse reliefs were lined three in a row and placed on the east and west sides of the spirit road, as shown by the modern reproductions. Courtesy of Zhang Jianlin, the excavator of the North Gate of Zhaoling.

demonstrate "one of the most coherent and remarkable periods of rock relief art in Iran" (2000, 36, 40).

The Sasanian rock reliefs are carved out of the rough uncut cliff surface, which forms a border along the scene. Most of the rock reliefs depict a single scene that carries powerful political implications. They portray Sasanian kings wearing crowns with flying ribbons and riding on fully caparisoned horses. Three reliefs provide the best comparative material for this study with the following description drawn from Schmidt's introduction (1953–70, 122–135):

1). *The Investiture of Ardashir I* (r. 224–241), 6.3 to 6.65 m long and 4.2 m high, depicts Ardashir I, the founder of the Sasanian Empire (Figure 4). Mounted on a horse, he is wearing a crown and receiving a ring, the symbol of the right to rule. 2). *The Triumph of Shapur I over Valerian* (r. 241–272), approximately 7.95 m long and 6.10 m high, depicts Shapur I's victory over two Roman emperors, Valerian and Philip the Arab (Figure 5). The king wears a crown, carved beyond the frame of relief and filleted with two long, wavy ribbons. 3). *Equestrian Combat of Hormizd II* (r. 302–309), 8.4 m long and 4.10 m high, depicts Hormizd killing a foe (Figure 6). He wears a crown filleted with two long, wavy ribbons. Clad in armor scales below the waist with ring armor on his arms and legs, he is grasping a long lance and is accompanied by a quiver containing a bow and arrows. The horse is shown in a flying gallop wearing a globular forehead ornament. The saddle blanket shows waving tassels, and the horse's tail is doubled up and tied with a ribbon.

Figure 4. The Investiture of Ardashir I. Rock relief at Naqsh–i Rustam, Sasanian period, 224–241 (Schmidt 1953–70, Plate 81). Courtesy of the Oriental Institute of the University of Chicago.

Figure 5. The Triumph of Shapur I over Valerian. Rock relief at Naqsh–i Rustam, Sasanian period, 241–272 (Schmidt 1953–70, Plate 83). Courtesy of the Oriental Institute of the University of Chicago.

Comparisons

Emperor Tang Taizong's horse reliefs and the Sasanian rock reliefs are separated by several thousand miles in distance, and several hundred years in time. Although

Figure 6. Equestrian Combat of Hormizd II. Rock relief at Naqsh–i Rustam, Sasanian period, 302–309 (Schmidt 1953–70, Plate 91). Courtesy of the Oriental Institute of the University of Chicago.

Sasanian rock reliefs are almost three or four times larger than the Emperor's reliefs, but even so their presentations share obvious similarities.

They both utilize the same carving technique. Each of the Emperor's six reliefs, carved in low relief, is represented on a stone surface with a neatly cut border to frame the scene. The Sasanian reliefs, cut slightly deeper than the Taizong's horses, are also framed by borders using uncut surrounding space. The relief of *The Triumph of Shapur I* has a raised square cartouche at its top left corner; a similar cartouche also appears on either the top left or top right corner of each of Taizong's slabs.

Both sets of reliefs depict horses galloping, treading, or standing. The horse in *Equestrian Combat of Hormizd II* is depicted in a flying gallop,[5] as are three of the Emperor's horses. Without exception, horses are featured with complete and elaborate royal embellishments. The Emperor's horses are equipped with complete bridles, saddles, saddle blankets with four bands, and stirrups. Without stirrups, the Sasanian horses are adorned further with fancy tasseled saddle blankets and with ornamental disks, bud–shaped tassels, and flying ribbons.

Both sets of reliefs have inscriptions. For example, *The Investiture of Ardashir I* is inscribed with trilingual inscriptions on the chests of the horses. The text identifies the figure on the relief, to the viewer's left, as the king empowered by gods to rule. The Taizong's horse reliefs are engraved with the names and verses composed by the Emperor himself eulogizing the magic performance of the horses in battles. These writings are believed to have been carved in each of the square cartouches at the upper corner. Together with the horse images, these writings emphasize covertly that Taizong's ascendancy to the throne was owning to his military merits made on horse back.

Both the Sasanian and Taizong's reliefs portray the monarchs, either visibly or by implication. The Sasanian rock reliefs depict the Sasanian kings wearing crowns in the center of the scenes, and in many cases as riders; the horses on Taizong's reliefs are riderless, but one horse is accompanied by a general who rescued the Emperor from peril. The absence of the image of Emperor Taizong is by no means a denial of his ownership of these horses. On the contrary, the complete harnessing of the horses, the affixed saddles and the positioned stirrups, suggest that these horses are ready for or were ridden by their master. The sight of Taizong's galloping horses makes viewers imagine a skillful equestrian on these swift horses. The eulogistic verses, presumably carved on the upper corners, and the arrows pierced through the horses convey a sense of the life–and–death battles that Emperor Tang Taizong and his war chargers endured together. The subtle way of focusing on the famous war chargers by rendering them without riders differs from the direct depiction of the Sasanian images. Regardless of the presence or absence of the monarchs, the images and texts of both sets of reliefs are created to advocate for their monarchs.

Further, both the Sasanian and Taizong's reliefs took historical events as the subject. The relief *The Triumph of Shapur I* portrays the Sasanian king subduing two Roman emperors at his feet, a representation of well–known victories recorded in history. Emperor Taizong conquered a number of tough adversaries in several campaigns. Each campaign is recorded in Chinese history in detail, including the names of his horses. The portrayal of General Qiu Xinggong with *Saluzi* has captured the life–and–death battle scene in stone and frozen this historical event in perpetuity.

More importantly, the two sets of reliefs may also share similar political motives. Ardashir I, the founder of the Sasanian Empire, is "legitimizing his seizure of power [from Parthian] by illustrating his divine investiture with the right to rule" (Herrmann 2000, 38). They used sculpture located near Persepolis as propaganda. Tang Taizong was a usurper who seized the throne by killing his elder brother, the heir apparent, and forcing his father to abdicate. He used commemorative sculpture of the horses also as propaganda to create a visual justification of his usurpation and legitimization of his enthronement.[6]

The composition of these two groups of reliefs is surprisingly similar, particularly when one compares the Sasanian relief, *The Triumph of Shapur I* (see Figure 5), with Emperor Taizong's stone horse relief *Saluzi* (see Figure 1), both having a square cartouche at its upper left corner. The striking similarities strongly suggest some knowledge of the Sasanian rock reliefs in the production

[5] Bishop (1918, 264) thought that the flying gallop originated with the ancient art of Crete. "It spread, by independent routes, both to Sasania and to ancient China, apparently through the medium of the so–called Scythian culture."

[6] The author's article on examining the Emperor Tang Taizong's real motivation for commissioning the six stone horse reliefs and their construction dates is forthcoming.

of the Emperor horse reliefs, an issue that will be further discussed under the historical context.

Crenellated Manes on Chinese and Sasanian Relics

In the horse industry, it is common to treat manes by clipping, trimming, or notching. If manes are regularly notched, they are referred to as "crenellated manes" in this paper. One may notice that Emperor Tang Taizong's six horses all have crenellated manes, as do the horse images depicted on Sasanian silver plates.

As Maenchen–Helfen (1957–58, 85) pointed out, the practical function of the crenellated mane "[...] is nil. A horse does not run faster, a mounted bowman will not shoot better or farther whether the mane is crenellated or not." The fascinating practice of crenellation has served as an invaluable criterion for his study in establishing historical connections. Similarly, the study of crenellated manes on the Chinese relics will help trace the interaction between China and other cultural groups.

The practice of crenellation started with mobile pastoralists in the steppe, and the Chinese may have borrowed the practice from these neighbors at different time periods in history. There appear to have been two waves of crenellation in Chinese history. The first was represented by a dozen or so manes found depicted on the bronze and terracotta horses from the accompanying pits of the tomb of the First Emperor of Qin, who unified China and ruled from 221–210 BCE, and on a group of ceramic tiles from Henan, also dated to the third century BCE. From their styles, notched in square and curved triangles, it is plausible to associate them with the practice of crenellation by ancient cultural groups, such as the Scythians.[7] After the first wave, crenellation disappeared in China for almost a thousand years. It resurfaced, however, with abundant examples during the early Tang of the seventh century and formed the second wave of mane crenellation, the focus of the current study.

Crenellated Manes on Tang Relics

In the early Tang, crenellated manes appeared on Emperor Taizong's stone horses and other Tang relics. Each Emperor's horse has three tufts, also known as *sanhua* 三花 [three flowers]. They stick out 0.12 m from the clipped mane and are arranged in a row separated by 0.06–0.07 m in between. Each tuft is cut into a cluster of hair 0.04–0.05 m wide and projects 0.12 m from the base. The tufts, tapering at the top, are curved backward.

In addition to this distinctive group of images of crenellation, there are other examples. From Qianling, the mausoleum of Taizong's son, Emperor Gaozong 高宗 (r. 650–683), there is a free standing stone horse with a mane that is crenellated into three low, square notches (Figure 7). Three–notched crenellated manes are also found on the ceramic horses from the tombs of Prince Yide 懿德太子 and Princess Yongtai 永泰公主. Both tombs are dated to 706. Additionally, there are ceramic horses with one or three square notches in their manes from the tomb of a high official, Xianyu Tinghui 鮮于庭誨, dated to 723 (Figure 8) and horses with crenellations of irregular shapes

Figure 7. Stone horse showing a mane with three low notches. Qianling (the Mausoleum of Emperor Gaozong), Shaanxi; Tang dynasty, late seventh century (Chavannes 1909, Plate 457).

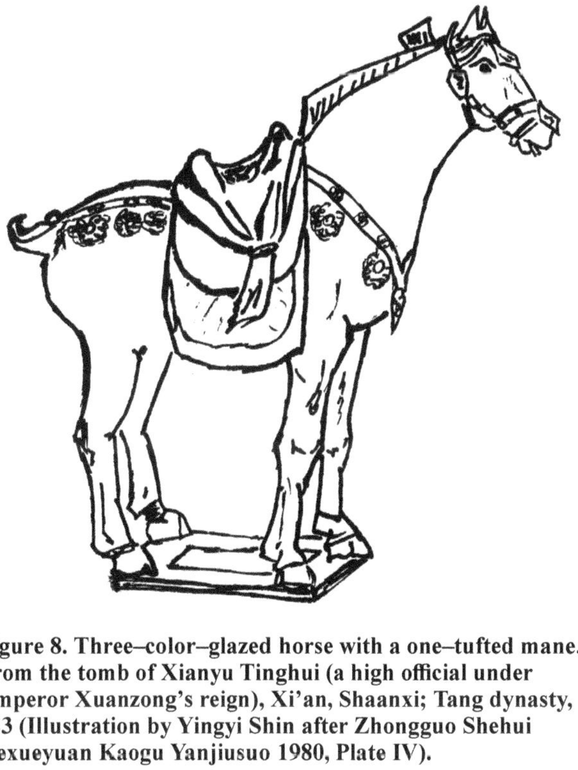

Figure 8. Three–color–glazed horse with a one–tufted mane. From the tomb of Xianyu Tinghui (a high official under Emperor Xuanzong's reign), Xi'an, Shaanxi; Tang dynasty, 723 (Illustration by Yingyi Shin after Zhongguo Shehui Kexueyuan Kaogu Yanjiusuo 1980, Plate IV).

[7] Details see Chapter Five in Zhou 2009.

Figure 9. A pottery horse showing a mane with three irregular notches. Tang dynasty, 618–907 (Illustration by Yingyi Shin after Maenchen–Helfen 1957–58, 117, figure 27).

on a copy of the eighth–century Tang imperial painting, "Lady Guoguo on a Spring Outing," and on Tang pottery horses (Figure 9).

Between the fifth to seventh centuries, crenellated manes also appeared on Turkic petroglyphs in the Altai Mountains.[8] Another significant group of crenellated manes is found on Sasanian silver plates that are directly related to this study.

Crenellated Manes on Sasanian Silver Plates

Nine out of 38 Sasanian silver plates carefully studied and published by Harper and Meyers (1981)[9] have been selected for this study. They show two styles of crenellation: Type A with three slightly curved tufts in a row; and Type B with single, multiple square or irregular tufts in a row.

Type A is represented on two plates, one depicting Shapur II (r. 303–309) hunting boar, now in the Freer Gallery of Art (Figure 10), and the other illustrating Peroz (r. 459–484) or Kavad I (r. 488–496; 499–531) hunting rams, now in the Metropolitan Museum of Art (Figure 11). Both are dated to approximately the fifth century. The mane of each horse is trimmed and three tufts stick out from an angular projection with long waving hair falling on the neck. Harper classifies them as "central Sasanian" (Harper and Meyers 1981, 5), believing they were made by the Sasanian court workshop. They are of extremely high quality and are the most exquisite examples of Sasanian silver plates surviving today.

Type B is found on seven silver plates with one, two, three, or four square–or angular–shaped tufts, and in one case, a round one (Figures 12–14). Although most of the figures wear royal crowns, they may actually depict crown princes or local leaders rather than the king himself. This group is categorized as "provincial works" (Harper and Meyers

Figure 10. Horse showing a mane with three curved tufts. Silver plate, Sasanian period, fifth century. Freer Gallery of Art, Smithsonian Institution, Washington, D.C.: Purchase, F1934.23.

Figure 11. Horse showing a mane with three curved tufts. Silver plate, Sasanian period, fifth century. Image © The Metropolitan Museum of Art.

1981, 8) and is dated to late or post Sasanian eras, probably around the seventh century.

Comparisons

The stylistic depiction of the Emperor Taizong's horses resembles the crenellation on the "central Sasanian" Type A plates, but with some modifications. The projection, placed between the crest and tufts on the Sasanian silver plates, has disappeared on Emperor's stone horses, although the manes still project. The tufts on the silver plates start to taper from the lower section, making them

8 Ibid.
9 Examples selected from this book include plates 10, 15, 17, 18, 21, 23, 25 & 26 and figure 46. Additional examples can be found in the same book and *Sasanian Silver* by Oleg Grabar, University of Michigan Museum of Art, 1967.

Figure 12. Horse showing a mane with three square tufts. Silver plate, Post–Sasanian era, ca. seventh century. Pushkin Museum (Illustration by Yingyi Shin after Harper and Meyers 1981, 222, Plate 21).

Figure 13. Horse showing a mane with two round notches. Silver plate, Post–Sasanian era, ca. seventh century. New York private collection (Illustration by Yingyi Shin after Harper and Meyers 1981, 227, Plate 26).

short or thin. Taizong's horses' manes, on the other hand, taper from the upper section, making them wide and thick and, without the projection, even longer. Despite these variants, the crenellated manes of Taizong's horses still are closest to that of Type A of the Sasanian silver plates. Both groups, limited in numbers and consistent in motifs, were commissioned by or for the sovereigns themselves, and their fine qualities suggest that they were produced under close court supervision.

Unlike the Emperor's horse reliefs, the stylistic depiction of crenellation on the Tang horses after Taizong's reign

Figure 14. Horse showing a mane with one notch. Silver plate, Post–Sasanian era, ca. seventh century. New York private collection (Illustration by Yingyi Shin after Harper and Meyers 1981, 140, Figure 46).

shows greater variation. Completely departing from the consistent long, thick, and projecting crenellation style, post–Taizong horses show low, square, round, half–circle, or even irregular tufts. Also, the number of tufts varies between one and three. No longer made exclusively for sovereigns, these horse effigies were produced for other royal family members or even high–ranking officials. This situation and the irregular styles match perfectly with that of the post–Sasanian era and of the Type B silver plates made during this time period. These similarities cannot simply be interpreted as coincidence. The interaction between China and Sasania during this period should also be examined in the broader historical context.

Historical Contact between China and Iran

This article focuses on China and Sasania, yet it is still beneficial to expand the historical review starting from the third century BCE, the Han Empire (206 BCE–220 CE) in China, and the Parthian Empire (248 BCE–224 CE) in ancient Iran. Although intermittent, since the third century BCE contact continued, and at times intensified, through later the Sasanian Empire (224–651) and well into the early Tang dynasty.

The discussion of contact between China and ancient Iran unavoidably leads to the "Silk Road," the ancient connection between China and many areas west of China, known as *xiyu* 西域 [Western Regions]. Scholars believe that unofficial contact via ancient routes can be "traced to the thirteenth or twelfth century BCE" (B. Wang 1993, 2). Zhang Qian 張騫 (d. 114 BCE), an envoy sent by the court, however, marks the first official endeavor of the Western Han (206 BCE–8 CE) to explore the Western Regions. He was sent on this mission twice, the first lasting thirteen years (139–126 BCE) and the second, five years

years (119–114 BCE). He made the first known Chinese report on the Parthian Empire.

In Zhang's accounts, Parthia is named Anxi 安息, a transliteration of 'Arsacid,' the name of the Parthian dynasty. His accounts are preserved in *Shiji* 史記 [Records of the Grand Historian]:

> 安息在大月支西可數千里. 其俗土著, 耕田, 田稻麥, 蒲陶酒. 城邑如大宛. 其數小大數百, 地方數千里, 最爲大國. 臨媯水, 有市民，商賈用車及船, 行旁国或數千里. 以銀為錢, 錢如其王面, 王死輒更錢, 效王面焉 (Q. Sima 1997, 123.3162).

Anxi is situated several thousand *li* west of the region of the Great Yuezhi (in Transoxonia). The people are settled on the land, cultivating the fields and growing rice and wheat. They also make wine out of grapes. They have walled cities like the people of Dayuan (Ferghana), the region containing several hundred cities of various sizes. The kingdom, which borders the Gui (Oxus) River, is very large, measuring several thousand *li* square. Some of the inhabitants are merchants who travel by carts or boats to neighboring countries, sometimes journeying several thousand *li*. The coins of the country are made of silver and bear the face of the king. When the king dies, the currency is immediately changed and new coins issued with the face of his successor (Q. Sima 1993, II, 235).

> 初, 漢使至安息, 安息王令將二萬騎迎於東界. 東界去王都數千里.... 漢使還, 而後發使隨漢使來觀廣大, 以大鳥卵及黎軒善眩人, 獻於漢 (Q. Sima 1997, 123.3172).

When the Han envoys first visited the kingdom of Anxi, the king of Anxi dispatched a party of 20,000 horsemen to meet them on the eastern border of his kingdom. The capital of the kingdom is several thousand *li* from the eastern border.... When the Han envoys set out again to return to China, the king of Anxi dispatched envoys of his own to accompany them, and after the latter had visited China and reported on its great breadth and might, the king sent some of the eggs of the great birds which live in the region, and skilled tricksters of Lixuan, to the Han court as gifts (Q. Sima 1993, II, 243).

Following Zhang Qian's embassy and report, contacts between China and Parthia and the Western Regions increased. During the first century BCE, quite a few Chinese missions were sent to the west and they were also recorded in *Shiji*:

> 漢始築令居以西, 初置酒泉郡以通西北国. 因益發使抵安息, 奄蔡, 黎軒, 條枝, 身毒国.... 諸使外國一輩大者數百, 少者百餘人....其後益習而衰少焉. 漢率一歲中使多者十餘, 少者五六輩, 遠者八九歲, 近者數歲而反 (Q. Sima 1997, 123.3170).

The Han first built fortifications west of the district of Lingju (northwest of Gansu province) and established the Jiuquan Prefecture in order to provide a safe route to the lands of the northwest. As a result more and more envoys were sent to Anxi [Parthia], Yancai [ancient name of Alains], Lixuan [Hyrcania], Tiaozhi [Mesopotamia], and Shendu [India].... The largest of these embassies to foreign states numbered several hundred persons, while even the smaller parties included over 100 members, though later, as the envoys became more accustomed to the route, the number was gradually reduced.... In the course of one year, anywhere from five or six to even ten parties would be sent out. Those traveling to distant lands required eight or nine years to complete their journey, while those visiting nearer regions would return after a few years (Q. Sima 1993, II, 240, 241).

During the Eastern Han (25–220), such contacts continued, probably at a reduced rate and size. It is recorded that in the thirteenth year of the Yongping 永平 reign (70) "the state of Anxi sent an envoy to offer lions and large birds from Tiaozhi 條支" (*Hou Hanshu*, 4.18). In the second year of the Zhanghe 章和 reign period (88), "the state of Anxi sent an envoy to present lions and *fuba* 符拔 [antelopes]" (Fan 1987, 4.4; Q. Wang 1960, 968.11378).

Chinese general Ban Chao 班超 (32–102), stationed in the Western Regions for 31 years, won numerous battles in defending the Chinese control of the Western Regions. His troops went as far west as the Caspian Sea and established direct military contacts with the Parthian Empire. Around 101, he dispatched his son, Ban Yong 班勇 (d. ca.128) "to escort the embassy from Anxi to present lions to the central plain [inland]" (Fan 1987, 77.14).

In addition to sending exotic gifts to China, Parthia also played a role in the Silk Road transmission of Buddhism from Central Asia to China. In the year 148, An Shigao 安世高 (fl. 148–170), a Parthian prince, gave up his inheritance and devoted himself to life as a Buddhist missionary in China. He came to the capital city, Luoyang 洛阳, where he established temples and became the first man to translate Buddhist scriptures into Chinese (Ciyi 1989, 2394).

After the fall of the Han dynasty, the contact was suspended due to frequent change of hands of political powers, such as the rise and fall of the Three Kingdoms 三國 (220–265) and the Sixteen States 十六国 (317–420). During the first half of the fifth century, contact was resumed with the Sasanian Empire and embassies were sent out with unprecedented frequency. In response to the envoys with tributes representing various states in the Western Regions, the Northern Wei 北魏 (386–581) sent 20 missions to the Western Regions in the first year of the Taiyan 太延 reign (435) and dispatched six more missions the following year (T. Yu 2004, 307). The mission sent to Bosi (Sasanian Persia) 波斯 (as recorded in Chinese historical records) was headed by Han Yangpi 韓羊皮 (S. Wei 1974, 102.2263) and Zhang Daoyi 張道義 (Linghu 1971, 36.642). Like the Parthians, the Sasanians pursued active foreign relations with the Northern Wei and frequently sent envoys to China. Historical records report at least twelve Sasanian embassies to China "to offer tribute."[10]

During the Shengui 神龜 reign (518–520), one of the tributes was accompanied by a memorial, which reads:

> "大國天子, 天之所生, 願日出處常為漢中天子. 波斯國王居和多千萬敬拜." 朝廷嘉納之. 自此每使朝獻 (S. Wei 1974, 102.2272).
>
> "The Son of Heaven of the great nation, whom Heaven begat, may Your Majesty always be the Son of Heaven in the Han land over which the sun rises! The king of Bosi, Juheduo [Kavad I, 488–496 (499–531)][11] salutes innumerable times in respect." The imperial court accepted this with praise. From then on Bosi often sent envoys to present tribute [T. Yu 2004, 329].

During the Northern Zhou 北周 (557–581), embassies were exchanged between the Northern Zhou court [Yuwen Zhou 宇文周] and the Sasanians with the possible plan of forming an alliance against the Turkics (Zhang 2003, 1053). Chinese envoys to Sasanian Persia must have had an audience directly with the Sasanian king, because they provided such vivid descriptions of the Sasanian king upon their return:

> 其王姓波氏, 名斯. 坐金羊床, 戴金花冠. 衣錦袍, 織成帔, 飾以真珠寶物 (S. Wei 1987, 102.2271).
>
> The king is surnamed Bo and named Si. He sits on a golden–sheep throne and wears a golden–flower crown. He is clad in a brocade robe and woven drapery [ribbon], both adorned with pearls and precious objects (translated by the author).

Contacts with Western Regions also extended to the Southern Dynasties 南朝 (420–589). The *Liangshu* 梁書 [History of the Liang] reports that the Persian envoys presented Sakyamuni's tooth to the court in the second year of the Zhongdatong reign 中大通 (530) (Yao 1973, 54.815), and came to offer tribute in the fifth year of Zhongdatong reign (533) and in the first year of Datong reign 大同 (535) (Q. Wang 1960, 968.11385).

Two paintings of the Southern Dynasties, probably later copies, *Fangguoshi tu* 方國使圖 [The portraits of envoys from various states] ordered by Pei Ziye 裴子野 (468–530) (Yao 1971, 30.445) and *Zhigong tu* 職貢圖 [Tribute–paying] are attributed to Xiao Yi 蕭繹 (508–555), later Emperor Yuan of the Liang 梁元帝 (r. 552–555). They have preserved images of tributes being brought to the Chinese court, including an envoy from (Sasanian) Persia (Jin 1960b, 14; T. Yu 2004, 341).[12]

The Sui dynasty 隋朝 (581–617), though short–lived, also exchanged embassies with the Sasanian Empire. During the Yangdi's reign (605–618), probably in 616 or 617 (Harper and Meyers 1981, 22, note 53), Li Yu 李昱 was dispatched to Sasania and in return the Sasanian Empire sent their own agents to offer tributes (Z. Wei 1973, 83.1857).

The early Tang witnessed another wave of frequent contacts with the Western Regions. In the twelfth year of Zhenguan 貞觀 reign (635), Emperor Taizong issued an edict to allow the (Sasanian) Persian monk Rabbon to preach Nestorianism and establish the first Nestorian temple, staffed with 21 priests, in Chang'an (P. Wang 1955, 49.864).

According to records, the exchange of embassies is portrayed also in Tang paintings, such as *Wanghui tu* 王會圖 [The gathering of kings] by an unknown artist (G. Sima 1956, 193.6068), *Waiguo tu* 外國圖 [The painting of foreign states] and *Bunian tu* 步輦圖 [On the sedan chair] both attributed to Yan Liben 閻立本 (600–673) (J. Yu 1985, 1439). Unfortunately, only *Bunian tu*, depicting Emperor Taizong in audience with Tibetan envoys, has survived. The mural excavated from the tomb of the crown prince Zhanghuai 章懷太子 (706) showing the "Tang officials greeting foreign envoys" cannot be identified with certainty as a Sasanian Persian envoy (Ji 2006, 150 f.).

The seventh century witnessed major political power changes, which greatly affected the contacts between the Tang and Sasanian Empires. The newly established Tang dynasty firmly grasped power and led China toward prosperity; the Sasanian Empire was in a state of emergency. When Yazdegerd III (r. 632–651) assumed the throne in 632, attacks from Muslim Arabs began (Litvinsky 2002, 385–88). He sent an envoy to the "Chinese court for offering tribute" in 639 (Q. Wang 1960, 970.11399), which seems a normal occurrence in the manner of Chinese historical records, but by that time the Sasanians had already lost several battles with the Muslims. In 642, Yazdegerd III amassed all his troops in Nihavand (modern Hamadan province in Iran) to launch a major campaign. Again he was defeated in this decisive battle. After that, the Sasanian king lacked the power to organize any meaningful resistance, and the counter attacks were handled locally. The king sent envoys two years in a row, in 647 and 648 (Q. Wang 1960, 970.11400 f.), to desperately "seek assistance from the Tang court with the hope to form a new army" (Chavannes 1900,

[10] S. Wei 1974, 5.115 (太安元年 [455]); 5.120 (和平二年 [461]); 6.126 (天安元年 [466]); 6.128 (皇興二年 [468]); 7a.142 (承明元年 [476]); 8.205 (正始四年 [507]); 9.225 (熙平二年 [517]); 9.228 (神龜元年 [518]); 9.232 (正光二年 [521]) and 9.233 (正光三年 [522]). Linghu 1971, 50.920 (廢帝二年 [553]) and 5.74 (天和二年 [567]).

[11] T. Yu assigns this visit to the first reign of Kavad I (488–496). The present author believes, since the tribute is recorded as being taking place during the Shengui reign (518–520), the visit must have taken place during the Kavad I's second reign (499–531).

[12] According to Jin, the painting was originally painted by Xiao Yi around 540, and not by Yan Lide or Yan Liben of the early Tang. The extant painting now in the collection of the Nanjing Museum is a Song copy.

172; Litvinsky 2002, 390; Rong 2002, 59). In 651, the king fled to Merv in Tokharistan and was murdered there.[13]

His successor, Peroz III (d. 677), in exile in Tokharistan, sent envoys to China to relay the situation and undoubtedly sought help in the fifth year of the Yonghui 永徽 reign (654) and in the first year of the Longshuo 龍朔 reign (661) (Rong 2002, 59). In 661, as a response to these requests, the Tang court established the city, Jilingcheng 疾陵城 [Zaranj], in Tokharistan, as the Bosi Dudufu 波斯都督府 [Sasanian Persian Military Commander Prefecture], and Peroz III was appointed as the *Dudu* 都督 [Military Commander]. Peroz sent envoys to Chang'an in 667 and 671. He came himself to the court in 673 and returned to Chang'an in 674. He was given the title *Youweiwei jiangjun* 右威衛將軍 [General of the Right Awesome Guard] and died in Chang'an in 677. His successor, Narsieh (d. 708), who remained in Chang'an, inherited the throne in 678. One year later, the Chinese army accompanied Narsieh in order to restore him to the Sasanian throne[14] and stopped in Tokharistan, where Narsieh fought against the Muslim Arabs for more than twenty years. In 708 he at last returned to Chang'an and was given the title *Zuoweiwei jiangjun* 左威衛將軍 [General of the Left Awesome Guard]. After his death in 708, his descendants and entourage remained in China (Liu 1975, 198.5312, 5313; Rong, 2002, 59 f.).

The official documentation in support of the contact between China and Parthia, and later the Sasanian Empire, is vast. These contacts can be described as frequent and official. This long–honored relationship and possible political alliance must have led the Sasanian kings and princes to turn to China to seek both military help and life–long exile in the middle of the seventh century. The history of contact lasting over one thousand years must also be a history of cultural interaction. It is under this historical context that we will discuss further the two already introduced topics, the influence on the sculptural form and the crenellated mane.

Discussion on the Sculptural Form

Although there is no literature directly linking the Emperor Taizong's six stone horses to the Sasanian rock reliefs, the subject of the mausoleum or stone horses could have been brought up during these frequent court–level contacts. Likewise visual models of such imagery could have passed into court treasuries at many points because of that contact. When Emperor Taizong ordered that the design for his mausoleum include real images of his favorite war chargers, it would have been natural that his ministers would contribute their input, or those in charge of the project would seek suggestions. Possibly they sought opinions from foreigners including Sasanian envoys. The curiosity of the Chinese people toward the people from the West was not limited to their rare or exotic products, but also extended to their culture, particularly their burial practices. The narrative of Zhang Qian's exploration, as recorded in *Shiji*, touches on multiple aspects of the social life of the Anxi people [Parthian], which included the issuance of a new coin at the death of a king. *Weishu* introduces the Sasanian memorial services in which they dedicated the "twentieth day of the first moon to the sacrifice to the ancestors" (S. Wei 1974, 102.2272). Xiao Yi or Emperor Yuan of Liang wrote in the preface to the painting, *Zhigong tu*, now at the Nanjing Museum, as follows:

> 臣以不佞，推轂上游，夷歌成章，胡人遙集，款開蹶角，沿泝荊門，瞻其容貌．訊其風，如有來朝京輦，不涉漢南，別加防采，以廣聞見．名為職貢圖云爾 (Jin 1960b, 15).

> Your vassal, who is a man of no ability, was appointed to guard the upper reaches [of the Yangtze River], where the songs of the Yi people flow as if from the pen of a master, and the *hu* people gather from distant places. [They] sincerely submit to [Your Majesty], coming and going at Jingmen along the river. [I] observe their appearances and have them tell their customs. If someone coming directly to the capital to present tribute does not arrive in Hannan, I will gather information separately to widen the knowledge, for that which is named *Zhigong tu* (T. Yu 2004, 341).

If the court officials who were sent as envoys explored the funeral customs of the people who lived in the West, there is no reason to doubt that other court officials would have had the same curiosity when they encountered foreign envoys in China. The Sasanian envoy was likely depicted in a mid-sixth century Chinese painting and the Sui delegation paid an official visit to the Sasanian court in 616 or 617. The subject of Sasanian rock reliefs could have been mentioned during such contact.

More than a few court officials served as envoys to the Western Regions during the Northern Dynasties. Some of them might have had opportunities to visit the famous sites of Naqsh–i Rustam and Naqsh–i Rajab and shared descriptions of them on their return to China. Official envoys from Sasania also could have spread information about Sasanian rock reliefs. It should be noted that the necklace found in the tomb of Li Jingxun 李靜訓 (599–608), believed to be of western Central Asian or possibly Sasanian origin (Harper 2006, 116), landed in the hands of the Sui royal family. Therefore, it is possible that various art forms, including knowledge of the Sasanian relief, could have been introduced to China by Sasanian envoys

[13] According to Litvinsky (2002, 390), there are several explanations to Yazdegerd III's death. Yazdegerd III fled to Merv, but was in conflict with the Merv governor (kanarang), who decided to leave Sasania and collaborate with the Turkics. Yazdegerd III demanded high taxation from the Merv governor and also planned to replace him. People in Merv generally did not welcome him. With no army and only few bodyguards, Yazdegerd III fled and hid in a Christian Mill where he was spotted and murdered by the Mill owner. Yazdegerd III's death marked the official end of the Sasanian Empire.

[14] Jiu Tangshu (Liu 1975, 84.2802) reveals that Chinese used the escorting of Narsieh to Iran as a pretext in order to conquer two western tribes on the way.

or by Chinese officials who had been there even before Emperor Taizong's reign.

There is also the possibility that the relief format could have been made known around the time when Taizong and his court were discussing burial plans for two members of the royal family, his father in 635 and his empress in 636. Taizong and his officials might have encountered Sasanian ideas and objects through contact with the Nestorians who built the Nestorian temple in 635, as well as during the Sasanian embassy in 639. People who traded Sasanian products to the Chinese were a diverse group. Besides official envoys, "many of them [were] Nestorian Christians" (Harper 2006, 116). Polo originated in Persia, was introduced into China during the Emperor Taizong's reign, and soon became popular among the Tang elite (Luo 1955, 136–66; Xiang 2001, 79–81).

In addition to the exchange of ideas on the court level, there might have been contacts on other levels, such as among artists. Chinese carvers must have also encountered Buddhist monks–some of whom might have been "trained foreign sculptors," strong in Greco–Roman backgrounds. As Paludan writes, at Dunhuang 敦煌 in Gansu 甘肅 and Yungang 雲岡 in Shanxi 山西, the Gandhāran style came along with "motifs from Sasanian Persia." At Longmen 龍門 in Henan 河南, foreign features including "Greek, Persian and Central Asian motifs" are evident (Harper 2006, 205 f., 211). The stone columns erected along the spirit road of the tombs of the Southern Dynasties show the Greek, Persian, and Indian art forms (Lin 1984, 55), such as concave fluting, possibly introduced originally by "foreign sculptors imported to work on the great northern Buddhist sites (Paludan 1991, 76)."

The artist Cao Zhongda 曹仲達, who was active during the Northern Qi 北齊 (550–577) and developed his own style featuring Western Regions' techniques, *chaoyi chushui* 曹衣出水 [Cao's treatment of garment and its folds like floating water], was originated from the Cao state 曹國 of the Western Regions (J. Yu 1985, 894). Several late sixth century stone sarcophagi, carved in low relief, were recently excavated and most likely commissioned by non–Chinese. The couch of An Qie 安伽((517–579) depicts the funerary ceremony of Zoroastrian, a national religion under Sasanian rule. The sarcophagus of Yu Hong 虞弘 (ca. 550–592) is decorated entirely with Sasanian Persian scenes. Yu Hong was from the Yu State, probably in Central Asia. He was Rouran's 柔然 envoy to Sasanian Persia, and later served in the Northern Qi, Northern Zhou, and the Sui courts (Shanxi et al 2001, 50).

There were also artists who had access to or were directly connected with the Western Regions and provided services to Emperor Taizong. Yan Lide 閻立德 (d. 656) was in charge of the construction of Zhaoling and certainly had a role in designing the form of the stone monuments and their placement at the mausoleum. His brother and court painter, Yan Liben, as mentioned above, is said to have made the drawings of the horses that were to be carved in the relief format. Both of them were the successors of their father, Yan Pi 閻毗 (563–613). Yan Pi, who was married to a Northern Zhou princess, possessed excellent painting and craft skills. Serving both the Northern Zhou and Sui courts, he was a contemporary of Yu Hong, the owner of the sarcophagus with Sasanian Persian imagery. Yan Pi could have encountered non–Chinese people, such as Yu Hong or others from the Western Regions, and shared what he learned with his two sons.

Yuchi Yiseng 尉遲乙僧 (fl. 639–710), whose fame is said to have equaled that of Yan Liben (Jin 1960a, 64; Sirén 1973, 72), and who was also active during the Taizong's reign, was a native of Yutian or Tokharistan. He was recommended by the king of his state to Emperor Taizong for his "*danqing qimiao*" 丹青奇妙 [marvelous red and blue] painting skills. It has been suggested that he was sent by the King of Yutian to serve the Tang court as a hostage (Q. Wang 1960, 968.11378; Ouyang and Song 1975, 221a.6235). He must have been an artist, however, serving either in Yutian or in Tokharistan before he came to serve in the court of Emperor Taizong during the first thirteen years of the Zhenguan reign (J. Yu 1985, 794–795; Jin 1960a, 64). Edward Schafer (1963, 32) unambiguously claims that he was a foreigner, a Khotanese, with the Saka name of Viśa Īrasangä and the Chinese name of Yü–ch'ih I–seng [Yuchi Yiseng]. He came to the Chinese court about the middle of the seventh century, recommended by his king, bringing with him a new painting style of Iranian origin, in which modeled and shaded polychrome figures seem to stand out in relief, or even to float free from their background.

Schafer's view is supported by *Zhongguo meishujia renming cidian* 中國美術家人名辭典 [Dictionary of names of Chinese artists]. There it is stated that his works on figures, flowers, or birds "resemble things foreign and are alien to Chinese appearance" (J. Yu 1985, 794). Yuchi was raised in the Western Regions where he developed his artistic style. He would have been familiar with artistic styles and skills that prevailed in the Western Regions, and the famous Sasanian rock reliefs may have been in his realm of study. He may have visited or had contact with "his elder brother, who also was good at painting but did not come to serve the Tang" (J. Yu 1985, 794), and remained in the Western Regions.

Since Yuchi was a contemporary of Yan Liben at the Tang court and the construction of Taizong's tomb was conducted under government supervision by involving the best people from all fields, Yuchi might have played a role in that project. Schafer's attribution of his roots as "Saka" and his painting style as of "Iranian origin" clearly provide a possible link between knowledge of the Sasanian rock reliefs and the Emperor Taizong's six stone reliefs.

These examples of interaction between China and ancient Iran are both numerous and continuous. During the early Tang, the capital Chang'an was truly cosmopolitan, filled with foreigners and merchants including many from the

Western Regions. With this historical context, it is quite plausible to think that famous sites, like Naqsh–i Rustam and Naqsh–i Rajab at Persepolis, where Sasanian rock reliefs are situated, would have been made familiar to the open and receptive Tang court and subsequently to sculptors. This could have provided inspiration for the sculptural form of the Emperor Taizong's six horse reliefs.

Discussion on Crenellated Manes

The historical interaction described above is not only valid for the examination of the sculptural form. It holds equal importance to the discussion of the presentation of crenellated manes on Chinese and Sasanian horses. As mentioned above, the styles and treatments of crenellated manes appearing on the Tang horses demonstrate similarities with those presented on the "central Sasanian" silver plates produced through central government–controlled workshops and the "provincial works" produced outside of government control. Specific deliberation on the circulation of Sasanian silver plates will shed light on this discussion.

"Central Sasanian" silver plates are of extremely high quality. These magnificent silver plates served varied functions, including "gifts to high personages" or those sent "abroad as official gifts" (Marshak 1998, 90; Harper and Meyers 1981, 127). Most of the Sasanian silver plates studied have been found outside Iran.

Although there is no way of determining how or when these plates left the Sasanian Empire, it is quite possible that they were sent abroad as diplomatic gifts. The plate depicting Peroz or Kavad I hunting rams (see Figure 11) "would have served as an appropriate gift from the Persian king to the Byzantine ruler Zero, who had provided the ransom demanded by the Hephthalite captors of Peroz" during the late fifth century (Harper and Meyers 1981, 127).

It is feasible that King Yazdegerd III (r. 632–651) sent "central Sasanian" silver plates as diplomatic gifts along with his envoys, who went at least on three occasions, 639, 647, and 648 to plead with Taizong for military assistance against the Arabs.[15] The most likely occasion on which to present the silver plates would have been in 639, since this could predate the stone horse reliefs. Coming empty–handed does not lead to successful diplomacy.[16]

It is also possible that the "central Sasanian" silver plates were brought to the Chinese court before the Tang, as Sasanians frequently sent envoys to the courts of the Northern Dynasties and the Sui dynasty. The Sasanian silver plate excavated from the tomb of Feng Hetu 封和突 (438–501) in Datong, Shanxi (Yuji Ma 1983), depicts a Sasanian king, probably Bahram I (r. 273–276), hunting boar on foot (Figure 15). It is suggested that the plate was presented to Feng after the Northern Wei court had received it as a tribute; or that the plate was in Qizil first and then was taken as a trophy by the Northern Wei when they sacked Qizil (Y. Ma 1983, 12). Harper believes the plate "was probably made in a provincial Sasanian center in the east Iranian world" and "may have traveled to China in the hands of one of the trade or diplomatic missions to the Northern Wei kingdom in the fifth and early sixth centuries" (2006, 123 f.).

Silver wares arrived in China by other means in the beginning of the sixth century. *Luoyang qielan ji* 洛陽伽藍記 [Record of Buddhist monasteries of Luoyang] records that Chen 琛 (fl. 497–524), the Duke of Hejian 河間王, while serving as the Governor of Qinzhou 秦州 (modern Tianshui 天水 area, Gansu 甘肅), sent people as far as Sasanian Persia for horses. He often showed off his possession of "more than 100 gold vases and silver vessels; many urns, plates, and bowls" "all coming from the West" to the imperial members (Yang 1987, 4.12). In all likelihood these included Sasanian silver plates.

One or more of these sources might be responsible for circulating the "central Sasanian" silver plates or spreading the iconography on silver plates to China, thus making the "central Sasanian" silver plates a reference for the design and decoration of the six stone horse reliefs.

The provincial silver plates, imitations of "central Sasanian" plates, were made by princes and local noblemen since the 630s when the Sasanian government became weak. These

Figure 15. Sasanian–style silver plate from the tomb of Feng Hetu, Datong, Shanxi. Northern Wei, 501. Collection and courtesy of the Shanxi Museum, Taiyuan, Shanxi, China.

[15] According to the Jiu Tangshu (Liu 1975: 198.5312), the Sasanian envoy presented an exotic bluish snake as a tribute to the Tang court in 647. Chavannes (1900: 172) attributes the tribute of the same bluish snake by the Sasanian envoy to the year of 638.

[16] J. A. Lerner (2001, 224) states "Only with the upswing in diplomacy between the Sasanian and Chinese courts does Sasanian art begin to impact on Chinese art, and then, not until the Tang period."

plates still carry traditional royal images and attributes, but with modifications and usually of lesser quality (Harper and Meyers 1981, 8, 138). They show manes crenellated into single or multiple tufts and in square or round shapes. Their variants also appear in Tang China on the stone and terracotta horses datable to the Gaozong's reign and later. This phenomenon coincided with the political situations of these two empires during the second half of the seventh century.

The Sasanian Empire officially ended in 651, but their princes kept the Sasanian court alive in other parts of the world, such as Chang'an, Tokharistan and Tabaristan, until the eighth century. Several Sasanian princes and their entourages were in exile in China and probably resided in a Persian community, one of the foreign communities established in Chang'an. A Zoroastrian or Nestorian temple was requested in 677 by Peroz III to be built for their religious activities (Rong 2002, 72). Noticeable figures originating from Iran were not few. Abraham or Wahrām 阿羅憾 (616–710) was the grand chief of the Sasanian Persian state in Tang China (Rong 2002, 65; X. Ma 2002, 101) and Li Yuanliang 李元諒 (723–784) was a Tang military leader recorded in the Tang official history (Ouyang and Song 1975, 156.4901; Zhang 2003, 1080–1084). Study of the epitaph of Li Su 李素 (744–817) reveals that he might be associated with, if not directly descended from Peroz and his queen, both of whom died in Chang'an (Rong 2001, 245). The epitaph of the wife of Su Liang 蘇諒妻 (d. 874) claims that the deceased and her husband, Su Liang, were both the descendants of Sasanian noblemen (Shaanxi 1964).

These princes and their entourages maintained communication between Chang'an and their provisionary government headquarter in Tokharistan and possibly had contact with other fleeing court members in the West. While in Tokharistan, as mentioned above, Peroz III sent envoys to Chang'an in 667 and 671. He came himself to the Tang court in 673 and returned to Chang'an the following year and eventually died there. His successor, Narsieh (d. 708), was in China before returning to Tokharistan for twenty years. He, too, eventually returned and died in Chang'an. Other Sasanian royal family members held court in Tabaristan, a region along the southern coasts of the Caspian in Iran, until 765. The fleeing courts sent envoys to the Tang court for as many as seventeen times between 719 and 771 (Rong 2002, 60 f.). Such communication and travel would have given them access to provincial Sasanian silver plates. They, in turn, may have facilitated the circulation of provincial silverwork and their iconography to China. This makes it reasonable to assume that the direct contact with Sasanian princes and their followers was responsible primarily for the diffusion of the provincial silver plates depicting crenellated manes to Tang China. There may be other possible agents, such as the Chinese who fought in the West, or who escorted the princes to the West, or other people traveling to the West who could have brought back provincial silver plates.

The spread of the crenellated mane as depicted on the silver plates and the actual practice of the crenellation, however, are two separate matters. Based on the interpretation of his name *Shifachi* 什伐赤, one of Emperor Taizong's six horses, was believed to have come from Sasanian Persia (Yi 1985, 119). This attribution is subject to further review, however, importing horses from the Sasanian region was not impossible as it was one of the 34 places in the West from which the Tang imported horses (Ma and Wang 1995, 70 f.). The ancient Iranian peoples of the Assyrian and Achaemenid periods trimmed their horses' manes, as revealed on many stone reliefs. The presence of crenellation on the Sasanian silver plates indicates that Iranians had started practicing crenellation during the Sasanian period, or simply copied patterns of crenellation, or used mane cover, purely for embellishment.

It is plausible that direct contact between the Sasanian Empire/post–Sasanian period and China might have made the artistic presentation of crenellation on the "central Sasanian" silver plates and "provincial works" known to the Chinese court and elite and consequently inspired the creation of various styles of crenellation in art works in China. It is doubtful, however, that such an interaction would have had much bearing on the practice of crenellation at the Tang state–run pasturelands. There is evidence that non–Chinese played a major role in horse management under the Tang, but such evidence lacks support for a visible role attributed to Sasanians. Rather, it reveals a significant role played by other ethnic people, such as Turkics, a topic to be addressed separately.

Summary

The study facilitates the reconstruction of the cultural interaction between ancient Iran and China. This is most clearly visible in Emperor Taizong's stone horse reliefs. It is evident that Sasanian art had exerted a strong influence, either in their sculptural composition or embellished horse motifs. As there are striking similarities in the sculptural forms and horse representations in Sasanian rock reliefs compared to the Emperor Taizong's stone horse reliefs. The documented continuous and frequent historical contacts between ancient Iran and China at different periods and levels might have made the famous Sasanian rock reliefs known to the open and receptive Tang China, and provided inspiration for the composition and representation of the Emperor Taizong's stone reliefs.

The similarities in treating the manes on the "central Sasanian" silver plates and the Emperor Taizong's six stone horses strongly suggest a close connection. Direct and indirect sources might be responsible for circulating the Sasanian silver plates or spreading the iconography of mane treatment on silver plates to China, which could then serve as an artistic reference to the design of the Emperor's stone horses. The presentation of the crenellated manes on the Tang relics after Emperor Taizong's era coincided with those styles depicted on the post–Sasanian "provincial works." These resemblances resulted from a typical political situation in China and Sasania and the special interaction and relationship forged during that time.

Artistic Sasanian influence was successfully absorbed by the Chinese based on their established stone carving tradition, featuring line and painting–like effect and prevailing since the Han dynasty. Those Chinese art works integrated with non–Chinese elements are exceptional and carry extraordinary cultural significance. Emperor Taizong's six stone horse reliefs are such excellent examples, showcasing the encounter of the two cultures, Sasanian and Chinese.

Interacting with non–Chinese cultures was the norm for several thousand of years of Chinese history. The foreign influence was not the exclusive privilege of Sasanian art, however, this study helps to support the statement that "Nowhere is the appeal of Sasanian art more remarkable than at the imperial court of Tang China, from the seventh to the ninth centuries" (Azarpay 2000, 71).

Acknowledgements

I would like to thank Dr. Sandra L. Olsen for reviewing my paper and inviting me to the symposium "Equids in the Ancient Near East, Egypt, and Arabia," held at the International Museum of the Horse, Lexington KY in June 2010, where the paper was presented. Special thanks go to Dr. Katheryn M. Linduff for her several reviews and comments. The paper is adapted from a section in Chapter Five of my dissertation (University of Pennsylvania, 2008).

Bibliography

Azarpay, G. 2000. Sasanian art beyond the Persian world. In *Mesopotamia and Iran in the Parthian and Sasanian Periods. Rejection and Revival c. 238 BC–AD 642*, edited by J. Curtis, 67–75. British Museum Press, London.

Bishop, W. C. 1918. The horses of T'ang T'ai–Tsung. In *The Museum Journal* 9 (September–December) 244–272.

Chavannes, É. 1900. *Documents sur les Tou–kiue (Turcs) occidentaux, recueillis et commentés, suivi de notes additionnelles*. Librairie d'Amérique et d'Orient and Adrien–Maisonneuve, Paris.

Chavannes, É. 1909. *Mission Archéologique dans la Chine Septentrionale*. Vol. 2 *Planches*. Leroux, Paris.

Ciyi 慈怡 1989. *Foguang da cidian* 佛光大辭典 [Buddha–light Great Dictionary] 4th ed., 7 vols. Foguang chubanshe, Gaoxiong (Taiwan).

Fan Ye 範曄. 1987 [445]. Hou Hanshu 後漢書 [History of the Later Han]. In *Siku quanshu* 四庫全書 [Complete library in Four Branches of Literature], nos. 252–253. Shanghai guji chubanshe, Shanghai.

Grabar, O. 1967. *Sasanian Silver: Late Antique and Early Mediaeval Arts of Luxury from Iran.* University of Michigan Museum of Art, Ann Arbor.

Harper, P. O. 2006. *In Search of a Cultural Identity: Monuments and Artifacts of the Sasanian Near East, 3rd to 7th century A.D.* Bibliotheca Persica, New York.

Harper, P. O. and P. Meyers. 1981. *Silver Vessels of the Sasanian Period.* Metropolitan Museum of Art, in assoc. with Princeton University Press, New York.

Herrmann, G. 2000. The Rock Reliefs of Sasanian Iran. In *Mesopotamia and Iran in the Parthian and Sasanian Periods: Rejection and Revival c. 238 BC–AD 642*, edited by J. Curtis, 35–45. British Museum Press, London.

Ji Dongshan 冀東山. 2006. *Shenyun yu huihuang: Shaanxi lishi bowuguan guobao jianshang* 神韻與輝煌: 陜西歷史博物館國寶鑑賞 [Appreciation of the National Treasures at the Shaanxi History Museum]. Tangmu bihua juan 唐墓壁畫卷 [The Volume of the Tang Tomb Murals]. Sanqin chubanshe, Xi'an.

Jin Weinuo 金維諾. 1960a. Yan Liben yu Yuchi Yiseng 閻立本與尉遲乙僧 [Yan Liben and Yuchi Yiseng] *Wenwu* 文物 4: 61–69.

Jin Weinuo 金維諾. 1960b. "*Zhigongtu*" de shidai yu zuozhe "職貢圖" 的時代與作者 [Dating and Artist of the Painting 'Making Tributes'] *Wenwu* 文物 7:14–17.

Lerner, J. A. 2005. *Aspects of Assimilation. The Funerary Practices and Furnishings of Central Asians in China.* Sino–Platonic Papers, 168. Department of East Asian Languages and Civilizations, University of Pennsylvania, Philadelphia.

Lin Shuzhong 林树中. 1984. *Nanchao lingmu diaoke* 南朝陵墓雕刻 [Tomb Sculptures of the Southern Dynasties]. Renmin meishu chubanshe, Beijing.

Linghu Defen 令狐德棻. 1971 [636]. *Zhoushu* 周書 [History of the Zhou]. 3 vols. Zhonghua shuju, Beijing.

Litvinsky, B. A. (ed.) 2002. *Zhongya wenmingshi* 中亞文明史–III [History of Civilizations of Central Asia III]. Translated by Ma Xiaohe. Zhongguo duiwai fanyi chuban gongsi, Beijing.

Liu Xu 劉昫. 1975 [945]. *Jiu Tangshu* 舊唐書 [Old History of the Tang]. 16 vols. Zhonghua shuju, Beijing.

Lu Xun 鲁迅. 1973. *Lu Xun quanji* 鲁迅全集 [Complete Collection of Works by Lu Xun]. 10 vols. Wenxue yanjiushe, Hong Kong.

Luo Xianglin 羅香林. 1955. *Tangdai wenhua shi yanjiu* 唐代文化史研究 [A Study of Tang Cultural History]. Taiwan Shangwu yinshuguan, Taibei.

Ma Junmin 馬俊民 and Wang Shiping 王世平. 1995. *Tangdai mazheng* 唐代馬政 [Tang Dynasty Horse Administration]. Northwestern University Press, Xi'an.

Ma Xiaohe 馬小鶴. 2002. Tangdai Bosiguo daqiuzhang Aluohan muzi kao 唐代波斯國大酋長阿羅憾墓志考 [A Study of the Tang Epitaph of the Persian Grand Chief, Abraham]. In *Zhongwai guanxishi: xinshiliao yu xinwenti* 中外關係史:新史料與新問題 [Chinese and External Relations: New Materials and New Questions], edited by Rong Xinjiang and Li Xiaocong, 99–127. Kexue chubanshe, Beijing.

Ma Yong 馬雍. 1983. Bei Wei Feng Hetu mu jiqi chutu de Bosi yinpan 北魏封和突墓及其出土的波斯銀盤 [The Persian Silver Plate Excavated from the Northern Wei Tomb of Feng Hetu]. *Wenwu* 文物 8: 8–12, 39.

Ma Yuji 馬玉基. 1983. Datongshi Xiaozhancun Huagetatai Beiwei mu qingli jianbao 大同市小站村花圪塔臺北魏墓清理簡報 [A Brief Report on the Cleaning of the Northern Wei Tomb from Huagetatai, Xiaozhancun, Datong]. *Wenwu* 文物 8: 1–4.

Maenchen–Helfen, O. 1957–58. Crenellated mane and scabbard slide. *Central Asiatic Journal* 3: 85–138.

Mahler, J. G. 1959. *The Westerners Among the Figurines of the Tang Dynasty of China*. Serie Orientale Roma 20. Istituto Italiano per il Medio ed Estremo Oriente, Roma.

Marshak, B. I. 1998. Decoration of some Late Sasanian silver vessels and its subject–matter. In *The Art and Archaeology of Ancient Persia. New Light on the Parthian and Sasanian Empires*, edited by V. S. Curtis, R. Hillenbrand and J. M. Rogers, 84–92. Tauris Publishers in association with the British Institute of Persian Studies, London and New York.

Ouyang Xiu 歐陽修 and Song Qi 宋祁. 1975 [1060]. *Xin Tangshu* 新唐書 [New History of the Tang]. Zhonghua shuju, Beijing.

Paludan, A. 1991. *The Chinese Spirit Road: The Classical Tradition of Stone Tomb Statuary.* Yale University Press, New Heaven and London.

Paludan, A. 2006. *Chinese Sculpture: A Great tradition.* Serindia Publications, Chicago.

Rong Xinjiang 榮新江. 2001. *Zhonggu Zhongguo yu wailai wenming* 中古中國與外來文明 [Medieval China and Foreign Civilizations]. Sanlian · Hafo, Yanjing xueshu congshu. Shenghuo, dushu, xinzhi sanlian shudian, Beijing.

Rong Xinjiang 榮新江. 2002. Zhongguo yu Bosi: Liangzhong wenjua zai Tangchao de jiaorong 波斯與中國：兩种文化在唐朝的交融 [Persia and China: Interaction of the Two Cultures During the Tang]. In *Zhongguo xueshu* 中國學術 [China Scholarship], edited by Liu Dong, vol. 3, no. 4, 56–76. Shangwu chubanshe, Beijing.

Schafer, E. H. 1963. *The Golden Peaches of Samarkand. A Study of Tang Exotics*. University of California Press, Berkeley.

Schmidt, E. F. 1970. Persepolis III. *The Royal Tombs and Other Monuments*. Oriental Institute Publications, 70. The University of Chicago Press, Chicago.

Shaanxi sheng wenwu guanli weiyuanhui 陝西省文物管理委員會. 1964. Xi'an faxian wan Tang xianjiaotu de Han, Poluobo wen hebi muzhi 西安發現晚唐祆教徒的漢, 婆羅鉢文合璧墓誌 [A Late Tang Epitaph of Zoroastrian Disciple Written in Chinese and Pahlavi Found in Xi'an]. *Kaogu* 考古 (9): 458–461.

Shanxi sheng kaogu yanjiusuo et al 山西省考古研究所等. 2001. Taiyuan Suidai Yu Hong mu qingli jianbao [Excavation Report of Yu Hong Tomb of the Sui Dynasty from Taiyuan, Shanxi] 太原隋代虞弘墓清理簡報. *Wenwu* 文物 (1): 27–52.

Sima Guang 司馬光. [late eleventh century] *Zizhi tongjian* 資治通鑒 [Comprehensive Mirror for Aid in Government]. 20 vols. Zhonghua shuju, Beijing.

Sima Qian 司馬遷. 1993 [91 BCE]. *The Records of the Grand Historian: Han Dynasty*. Rev. ed. translated by B. Watson. 2 vols. Columbia University Press, Hong Kong and New York.

Sima Qian 司馬遷. 1997 [91 BCE] Shiji 史記 [The Records of the Grand Historian]. In *Ershisi shi* 二十四史 [The Twenty–Four Histories]. No. 1, 20 vols. Zhonghua shuju, Beijing.

Sirén, O. 1973. *Chinese Painting: Leading Masters and Principles*. 7 vols. Hacker Art Books, New York.

Wang Binghua 王炳華. 1993. *Sichou zhilu kaogu yanjiu* 絲綢之路考古研究 [A Study on the Archaeology Along the Silk Road]. Xinjiang renmin chubanshe, Urumqi.

Wang Pu 王溥. 1955 [961]. *Tang huiyao* 唐會要 [Important Documents of the Tang]. 3 vols. Zhonghua shuju, Beijing.

Wang Qinruo 王欽若. 1960 [1013]. *Cefu yuangui* 冊府元龜 [Outstanding Models from the Storehouse of Literature]. 12 vols. Zhonghua shuju, Beijing.

Wei Shou 魏收. 1974 [554]. *Weishu* 魏書 [History of the Wei]. Zhonghua shuju, Beijing.

Wei Zheng 魏征. 1973 [636]. *Suishu* 隋書 [History of the Sui]. 6 vols. Zhonghua shuju, Beijing.

Xiang Da 向達. 2001. *Tangdai Chang'an yu xiyu wenming* 唐代長安與西域文明 [Tang Dynasty Chang'an and the Civilization of the Western Regions]. Hebei jiaoyu chubanshe, Shijiazhuang, China.

Yang Xuanzhi 楊衒之. 1987 [530]. Luoyang qielan ji 洛陽伽藍記 [Record of Buddhist Monasteries of Luoyang]. In *Siku quanshu* 四庫全書 [Complete Library in Four Branches of Literature], no. 587. Shanghai guji chubanshe, Shanghai.

Yao Silian 姚思廉. 1973 [636]. *Liangshu* 梁書 [History of the Liang]. 5 vols. Zhonghua shuju, Beijing.

Yi Mu 乙木. 1985. Zhaoling liujun shihua 昭陵六駿史話 [Historical Tales of the Six Horse Stone Reliefs from Zhaoling]. In *Tang Taizong yu Zhaoling* 唐太宗與昭陵 [Tang Taizong and Zhaoling]. Shaanxi sheng shekeyuan, Xi'an.

Yu Jianhua 俞劍華. 1985. *Zhongguo meishujia renming cidian* 中國美術家人名詞典 [Dictionary of Names of Chinese Artists]. Shanghai renmin meishu chubanshe, Shanghai.

Yu Taishan 余太山. 2004. *A History of the Relationships between the Western and Eastern Han, Wei, Jin, Northern and Southern Dynasties and the Western Regions*. Sino–Platonic Papers, 131. Department of East Asian Languages and Civilizations, University of Pennsylvania, Philadelphia.

Zhang Xinglang 張星烺. 2003 [1977]. *Zhongxi jiaotong shiliao huibian* 中西交通史料彙編 [Collection of Materials Regarding Transportation between China and the Western Regions]. Zhonghua shuju, Beijing.

Zhongguo shehui kexueyuan kaogu yanjiusuo 中國社會科學院考古研究所. 1980. *Tang Chang'an chengjiao Sui Tang mu* 唐長安城郊隋唐墓 [Sui and Tang Tombs at the Suburbs of Tang Chang'an]. Wenwu chubanshe, Beijing.

Zhou Xiuqin 周秀琴. 2001. Emperor Taizong and His Six Horses. In *Orientation*, February, 40–46.

Zhou Xiuqin 周秀琴. 2009. Zhaoling: The Mausoleum of Emperor Tang Taizong. In *Sino–Platonic Papers,* no. 187, April. Department of East Asian Languages and Civilizations, University of Pennsylvania, Philadelphia/PA.

Index

Index of Objects and Names

www.ingramcontent.com/pod-product-compliance
Ingram Content Group UK Ltd.
Pitfield, Milton Keynes, MK11 3LW, UK
UKHW061847041026
15591UKWH00015B/959

9 781407 316437